Penguin Books

Task Force: The Falklands War, 1982

On a visit to France and Belgium in 1967 Martin Middlebrook
was so impressed by the military cemeteries on the 1914–18
battlefields, that he decided to write a book describing just one
day in that war through the eyes of the ordinary men who took
part. His book, *The First Day on the Somme*, was published by
Allen Lane in 1971 and received international acclaim. Martin
Middlebrook has since written several other books which deal
with important turning-points in the two world wars; these are
The Kaiser's Battle, *Convoy*, *The Peenemünde Raid*, *The Battle
of Hamburg*, *Battleship* (with Patrick Mahoney). *The
Schweinfurt-Regensburg Mission*, *The Nuremburg Raid*, *The
Bomber Command War Diaries* (with Chris Everitt), *The Berlin
Raids* and *The Fight for the 'Malvinas'*. Many of his books have
been published in the United States and West Germany, and
three of them in Japan, Yugoslavia and Poland.

Martin Middlebrook is a Fellow of the Royal Historical Society.
Each summer he takes parties of visitors on conducted tours of
the First World War battlefields.

Task Force: The Falklands War, 1982

Revised Edition

Martin Middlebrook

Penguin Books

PENGUIN BOOKS

Published by the Penguin Group
Penguin Books Ltd, 27 Wrights Lane, London W8 5TZ, England
Penguin Books USA Inc., 375 Hudson Street, New York, New York 10014, USA
Penguin Books Australia Ltd, Ringwood, Victoria, Australia
Penguin Books Canada Ltd, 10 Alcorn Avenue, Toronto, Ontario, Canada M4V 3B2
Penguin Books (NZ) Ltd, 182–190 Wairau Road, Auckland 10, New Zealand

Penguin Books Ltd, Registered Offices: Harmondsworth, Middlesex, England

First published as *Operation Corporate: The Story of the Falklands War, 1982* by Viking 1985
This revised edition published in Penguin Books 1987
10 9 8 7 6 5

Copyright © Martin Middlebrook, 1985, 1987
Maps and diagram drawn by Reginald Piggott from preliminary drawings by Mary
Middlebrook
All rights reserved

Printed in England by Clays Ltd, St Ives plc
Typeset in 10/12 Linotron Times

Only the dead have seen the end of war

PLATO

Contents

8   Contents

Plates

Photo credits: All photographs are from the Ministry of Defence or have been supplied direct by the units which served in the Falklands, except Plates 2 (F/Lt D. S. Davenall), 26 and 27 (*Soldier* magazine), 30 (London Express News Service) and 31 (P & O Group).

Maps and Diagrams

Maps by Reginald Piggott from preliminary drawings
by Mary Middlebrook

Introduction

This book was born on Tuesday, 4 May 1982, the day that H.M.S. *Sheffield* was hit by an Argentinian Exocet. The news of that incident, coming hard on the heels of the sinking of the Argentinian cruiser *General Belgrano* two days earlier, marked the end of the diplomatic war over the Falklands and the start of the shooting war. It seemed unlikely that this would end until the battle was fought to its conclusion. I do not care to write about politics and diplomacy, but war – where decent young men of both sides die for patriotism, principle and the failure of politics – fascinates me. The fighting ended forty-one days later. It was not immediately possible for me to start my book; I still had to write one and a half books already under contract. That did not worry me. I much prefer the dust to settle and some degree of historical perspective to arrive. I did not fancy interviewing war-shocked servicemen just back from the fighting.

I arrived at the Ministry of Defence and asked permission to visit units and talk to the returned servicemen.

My request was granted, under certain conditions which I did not find unreasonable. I was able to carry out more than 200 interviews and met with much willingness to cooperate. I found that the greater the lapse of time, the more objective and more helpful the witnesses were. One naval officer said, 'I find that I say more each time I talk about it.' And the documents! This was the first British war of the photocopy age; I was showered with documents by officers anxious to see the efforts of their units properly covered. I thank the Ministry of Defence again for allowing me to fly down to the Falklands to visit some of the battlefields and to record the experiences of the islanders, who were in the war longer than anyone, from the first day to the last, and who had everything to lose if the Task Force had not sailed or if the war had been lost.

One ambition was not fulfilled. It is my custom to cover battles from both sides, to talk to the survivors of both sides. But my application for a visa to visit Argentina – made when the military *Junta* was still in power – was ignored; my book was half written by the time the new civilian government came to power. I wrote to several potential helpers and tried to obtain material by post. But my protests of an impartial approach were not accepted. 'Yours has been an unfair cause,' wrote one. 'The physical and emotional scars have not had time to heal,' said another. Useful new material is coming out of Argentina all the time but I would have liked to talk to Brigadier-General Menéndez and his soldiers.

The Revised Edition

Task Force is a revised version of the book which was published in 1985 as *Operation Corporate*. That edition has stood the test of time well and only a few minor corrections are needed for events or names, some because I was misinformed when carrying out my original research, some because of my own carelessness. If I say that one of the most serious of these factual corrections is that 45 Commando did not receive the benefit of a short helicopter lift to Sapper Hill in the closing stages of the war, as quoted in the original version, that will give some idea of how minor the corrections are.

The main purposes of the revision are three-fold:

1. To include a description of the three incidents where British troops were killed by the accidental action of other British forces – the 'blue-on-blues'. I knew of all of these when the original book was written but was asked by the Ministry of Defence not to include them 'for the sake of the families'. All of these incidents have since become public knowledge and are now included on pages 282, 322–3 and 299–300.

2. To include new material on two events – the possible work of Special Forces on the Argentinian mainland (on pages 192–3) and on the last Vulcan bombing raid to Stanley airfield (on pages 352–3).

3. To include my views on several aspects of the war, particularly the handling of the southern thrust of the advance across East Falkland just before the final series of battles. This and other matters have become much clearer in my mind with the passage of time and, I acknowledge, the considerable benefit of

hindsight not enjoyed by the commanders in the field. These new views are to be found following page 395.

Task Force has been chosen as the title of this revised edition because few people now remember that the original title, *Operation Corporate*, was the code-name for the operation to regain the Falklands. It remains basically a description of the British side of the conflict. Later this year, I hope to visit Argentina to prepare a new book covering the Argentinian side of the war.

Martin Middlebrook,
Boston, England.
April 1987

CHAPTER ONE

The Issues

It started at Mullet Creek, one of the innumerable inlets on the coast of the large island of East Falkland. It was about 4.30 a.m. Falkland Island Time, when one of the toughest and best-trained units of the Argentinian armed forces landed at the creek and deployed on the open, level ground around it. The Argentinian unit was the *Buzo Táctico*, an elite force of naval commandos, described in Argentina's press as 'underwater astronauts'. The exact manner in which the Argentinians came ashore is not known; it is probable that an advanced party landed by small boat but the noise of helicopters was heard later. The Argentinian press claimed that the first men were ashore four hours earlier. The commander of the unit was thirty-four years old *Capitán de fragata* Pedro Giachino (although that exact grade of captain's rank, also quoted in the Argentinian press, may have been a posthumous promotion, for Pedro Giachino would be one of the first to die in the war for the Falklands).

The sixty or so men (accounts vary) prepared to move. They wore neoprene overalls and dark woollen hats; their faces were smeared with black camouflage cream. They were liberally festooned with weapons and grenades but carried no large packs; their task was to be a brief one. Two assault teams were quickly formed. Lieutenant Bernardo Schweitzer led the larger party off to the north-west, towards the British Royal Marine barracks at Moody Brook; Captain Giachino's group went north-east, in the direction of Government House at Stanley. The night was dry and calm and there were no obstacles to the progress of the Argentinians over the tough grass and heather which covers most of the Falkland Islands. It was about three miles to each of the objectives. The two parties passed either side of a small rise, Sapper Hill, upon which a solitary Royal Marine had been posted as a look-out. But the Argentinians were moving with professional silence and Marine Michael Berry, who might have made a name for himself for firing the first shot of the Falklands war, saw nothing.

Five hours later, after fierce exchanges of fire at several places, and with Argentinian troops still being landed in large numbers at Stanley airport, His Excellency Rex Hunt, Governor of the Falkland Islands and Commander-in-Chief of the British forces, ordered those forces to cease fire and surrender. Sixty-seven Royal Marines and eleven men of the Royal Navy became prisoners of war, either on that fateful day of 2 April 1982, or when they were rounded up within the next few days. What the Argentinians called *Operativo Azul* – Operation *Blue* – was over. Blue had been chosen because that was the colour of the sky and of the sea from where the invading troops descended upon the Falklands and it was also the colour of the robe of the Virgin Mary, the protectress of the Argentinian armed forces. A British crown colony had been seized by force. There had been no ultimatum or declaration of war. Several Argentinians were dead but no Falkland Islanders or British servicemen had been killed, a fact which enabled the Argentinians to claim that they had 'liberated the Malvinas, peacefully and without bloodshed'. A disconsolate civilian population found their streets filled with armed troops and military vehicles, their governor and protective garrison expelled and a distinctly foreign and unwanted flag flying over them.

The issue, on the face of it, was that of sovereignty. Whose flag was to fly over Governor Hunt's flagpole? Whose government was going to rule these islands whose population was too small to rule itself? The regular population before the Argentinians came was only about 1,800 people. Comparisons with European statistics are so wide as to be almost ludicrous. The average population density of the United Kingdom is around 600 per square mile; Ireland has just over 100, the Highlands of Scotland less than fifty. The average number of people in the Falklands was *less than 0·4 per square mile*, and even that figure is misleading because more than half of the Falklanders lived in their capital, Stanley. That two great nations should have fought a war and lost their young men in the 1980s for the right to fly their flag over this land was incomprehensible to most of the rest of the world.

What sort of place was it for which two modern armies fought? (The early history of the islands will be covered later.) The community which lived here before the 1982 war was directly descended from a mainly British stock which was established in the 1830s. The colony became an important coaling station and refuge for the many ships using the Cape Horn route to the Pacific. The men operating the coaling stations took up sheep-farming as a sideline and, when the Falklands coal trade declined after the Panama Canal was opened in 1914, sheep-farming became the main activity. That is why the largest landowner and trader in the Falklands, the Falkland Islands Company, is now a subsidiary of a coal company based in Yorkshire, and why many of the other sheep properties are owned by Yorkshire-based companies or families. There are more than 300 sheep for every Falkland Islander but, even then, the pasture is so poor that the sheep population is calculated in acres per sheep rather than sheep per acre. A good property can carry one sheep for every two and a half acres, but the poorer ones need three acres or more per sheep.

The first thing that a visitor to the islands is asked to explain to 'the folks at home' is that the Falklands are not a group of storm-tossed rocks in the South Atlantic with a vile climate. It is understandable that anyone with the life-style of a London journalist who was forced to live outdoors for several weeks would paint a bleak picture – but the Falklands climate is none the less a temperate one. The weather rarely gets as cold as in a British winter nor does it get as hot in summer. There is a lot of cloud, but little rainfall (Stanley's annual average is 25 inches) although there is plenty of what the Irish would call 'soft weather'. But the wind blows strong and often (mean annual speed 17 knots, compared to 4 knots in Southern England). There is nothing to break the wind; the islands are quite treeless except for a few carefully nurtured specimens in gardens. When the weather clears, however, the scenery is beautiful and the air is crystal clear; there is no pollution.

But life for the Falklanders never was idyllic. All food except mutton and a few vegetables, all building materials except rock, all fuels except peat, had to be imported, some from Britain, some from the mainland of South America, and that usually meant from Argentina. Prices are high; the standard of living is low. The population was steadily falling. Even the sheep industry was in decline under pressure from rising costs, world recession and competition from synthetic fibres. The islands' second largest source of income before the war was the philatelic 'industry', with a team of women in Stanley's shabby post office ('Too much wind and not enough paint' could be Stanley's motto) forever posting new issues of stamps to customers all over the world. This trade received a huge boost with the war and the income from postage stamps then overtook that of the sheep industry – more than £500 per year for every Falkland islander!

There had been several attempts by the British to stimulate the economy. An airport was built at Stanley, but with a runway only long enough to take medium-haul jets from South America. This allowed a modest tourist trade, with small groups of visitors flying in from Argentina during the summer months to look at the penguin beaches, the peat bogs and generally to enjoy the coast and hills around Stanley. The British Government provided most of the finance for a £1½ million refrigerated mutton-packing plant in a remote place called Ajax Bay so that the Falkland Islands Company and other landowners could send their sheep in for slaughter instead of wasting most of the meat as in the past. The plant was ready for production in 1953 but never worked to full capacity and soon fell into disuse as the men of initiative sent to run it lost heart and drifted away. Many a wounded British or Argentinian soldier would later bless the protection of that abandoned meat plant. There is plenty of fish in the water around the islands but attempts by the British Government to send part of the hard-pressed British fishing fleet to Stanley were rebuffed by the local people. The only fishing was done by Polish and Russian deep-sea trawlers permanently stationed off the islands. There is undoubtedly oil under the waters around the Falklands but no international company would risk an investment in an area where there has been political tension over so many years.

There are two types of community in the Falklands – the village capital of Stanley and the sheep-farming settlements. The local people do not like the term 'Port Stanley', so often used in the press; the true name is simply Stanley, although both the British and the local post offices insist on using Port Stanley.

Stanley has been described many times as being like a Scottish fishing village, on its fine harbour, and that description cannot be improved. The officials who run the services are either local people or have been sent out from Britain under contract by the Ministry of Overseas Development. The locally born officials are steady, conscientious, decent people who form a stable factor in Falklands society, but they are handicapped by their lack of experience of the modern world. The Falklands, before the war, were really a 'toy town' community; the expenditure was little more than that of a parish council in Britain, although the problems were a great deal more varied. Most of the professional and technical expertise was provided by the contract personnel from Britain. The type of person drawn to a life 'down the other end of the world' varied. There were some dedicated people who fell in love with the islands, who renewed their contracts several times and who often settled permanently; there were also some duffers who could not get a decent job in Britain and were not much use in Stanley. Over all this ruled the Governor, appointed by the Foreign Office. When the Argentinian soldiers stormed Government House on 2 April 1982, the incumbent was Rex Masterman Hunt, C.M.G., a fifty-five-year-old former R.A.F. pilot (Spitfires, Tempests and Vampires), with thirty-four years in the Overseas Civil Service in Uganda, Borneo, Indonesia, Vietnam and Malaya behind him.

Perhaps the most interesting aspect of Falklands life is 'the camp', which is everything outside Stanley. The word 'camp' comes from the Spanish word *campaña*, meaning an open grassland prairie. The camp is composed entirely of sheep-farming properties. There are no villages; every acre of ground – the hills and coastland as well as the vast open grasslands – forms part of a sheep property, except for a small amount of common land near Stanley. Each property has boundaries of wire fences; these would prove to be useful navigational landmarks for the troops who fought in the war. The largest property – Goose Green – has 400,000 acres but a permanent population of fewer than a hundred people. The much criticized Falkland Islands Company owns nearly half of the Falklands. Most of the remaining properties are owned by the equally criticized absentee landlords abroad. The basis of the criticisms is that the Falkland Islands Company holds a monopoly over much of island trade and that the other owners are only interested in their investment, and return few of the profits to the local community. It is a delicate subject on which I am not qualified to comment but it is part of the background of the land for which Britain and Argentina fought a war. Ironically, two of the island properties in West Falkland – Weddell Island and Saunders Island – were partly owned by Argentinians.

The heart of each property is the settlement – and the names of some of the settlements are now part of Britain's military history. The manager's house, the store, the sheep buildings, the shepherd's houses, a bunkhouse for the visiting workers in the shearing season (mostly from overseas) in the larger settlements, a few tracks between the house, a jetty, a grass airstrip – that is all there is to a Falklands settlement. The manager's house always has spare bedrooms and most managers are given hospitality allowances for looking after visitors. Comparisons with 'outback' life in Australia or New Zealand are strong although the scenery and climate are Hebridean rather than Antipodean. By coincidence, the Falkland accent is very similar to that of New Zealand. There are no roads between the settlements, sometimes a Land-Rover track, but horses and, more recently, trials motorcycles are the best forms of overland transportation. There are plenty of horses ranging free over the properties and brought in for work as required. Radio was the everyday link with the outside world. Everybody heard everything; there are no secrets in the camp. The Falkland Islands Company coaster *Monsoonen* called every three months to restock the settlement store and take away the wool clip. The main lifeline for emergencies and for passenger travel was the government air service, which had two Beaver float-planes and an Islander for the short grass airstrips, but this form of transport was expensive. A visit to Stanley – for the races, for shopping, or to stay with friends – was a major outing.

There were many drawbacks to settlement life. Many of the communities are tiny, with perhaps no more than thirty people. Life, for the women in particular, was one of stultifying boredom, at least before the introduction of television videotapes, which are exchanged with great enthusiasm. The diet is boring in the extreme; the settlement shop stocks only basics. Meat is mutton *ad nauseam*, with each family being given one sheep per week; at the end of the week, when the best cuts have gone, the carcass is slung on a fence near the kennels for dog food. The average age of the settlement people is rising; few young men are prepared to accept the life-style. Bad workmen cannot be sacked; there are no replacements. There is no unemployment in the Falklands; most people in Stanley appear to have two or three jobs.

Education faces major problems. A few of the larger settlements have a resident teacher but teachers with good qualifications have been hard to find. Most children had to leave home and come into Stanley for the normal three school terms, living in a boarding house. Local education only goes as far as the O level stage. There is an official link with England whereby East Sussex County Council teach A-level students from the Falklands at a school in Rye, but few families manage to struggle over all the hurdles to achieve this and subject their children to the long absences from home, and fewer still go on to university. This means that the native Falklands population produces few qualified professional people, and many who do succeed are not willing to return home and become the leaders of such an isolated community.

So, that is the place in which the young men died. The village capital of Stanley, the scattered settlements. The wide open spaces with their mostly absent owners. A backward and declining economy, a falling population with few natural leaders and with little ambition to join the modern world. But there is a quality in the Falkland Islands that should not be overlooked. The people are quiet and slow of speech. Most of them are very deep and sincere people who have grown up in their own community where a slow pace of life is the norm, who have no desire to join the rat race, who are satisfied with a simple life and few possessions. They are fanatically pro-British. Their willingness to accept British aid and protection was sometimes derided, but there are as many names on Stanley's War Memorial – twenty-two for 1914–18 and eighteen for 1939–45 – as on comparable English ones. They have preserved a most English way of life. To walk through Stanley is to walk through an English village of thirty or more years ago; an English visitor can feel far more at home in Stanley than in most large English towns or cities.

*

The claim to sovereignty over the Falklands is held by the people of Argentina with passionate and unswerving intensity. A timetable of events may help to an understanding of that passion. In its compilation, I have tried to be as impartial as possible but I realize that some points will be contested by one side or the other.*

1540. Up to this date, the islands were uninhabited and undiscovered. There was a possible sighting in this year by a Spanish ship, now referred to as the *Incógnita* because neither the ship's nor the captain's name has survived. This ship was part of a small flotilla dispatched by the Spanish Bishop of Plasencia to trade in the Pacific. After losing her sister ship in the Strait of Magellan and running before a westerly gale, the *Incógnita* sheltered for ten months in a bay which its log called Bahía de las Zorras – Fox Bay – until the ship sailed back to Spain. It is assumed that a landing was made. The description in the log of the surrounding area is similar to that of the present-day Fox Bay area in the island of West Falkland.

1592. Possible sighting and landing by Captain John Davis in the British ship *Desire*. This is not accepted by the Argentinians, who say that the 'sighting' was invented to enhance Davis's reputation and that the description of the sighting – by one of Davis's crew – is full of inconsistencies.

1594. Sighting by the Dutch Captain Sebald de Weert of two outlying islands, the Jasons, to the north-west of the Falklands but forming part of the main island group. This claim is accepted by all parties.

1690. Captain John Strong, in the *Welfare* from Plymouth, landed in the islands and named the passage between the two main islands Falkland Sound after Viscount Falkland, Treasurer of the Navy (equivalent to the First Sea Lord) of the time. (The Falkland title came from a small village in Fifeshire.) The main group of islands were called Hawkins Land for a short period but then became known as the Falkland Islands.

1698 onwards. Regular visits made by French seal-hunters from Brit-

---

* Both sides rushed out publications during 1982, upholding their version of the dispute. The British Government published a pamphlet, *The Falkland Islands and Dependencies*, in March and two booklets, *The Falkland Islands, The Facts* in May and *The Disputed Islands* later in the year. As far as I am aware, only one British publication, *The Falklands War* by the *Sunday Times* Insight team, has attempted to look beyond the British version. The Argentinian version appeared as *The Malvinas, The South Georgias and The South Sandwich Islands – The Conflict with Britain*, by Rear-Admiral Laurio H. Destefani, published in August 1982.

tany, who named the islands 'les Îles Malouines' after their home port of St Malo.

1764. First French settlement under Louis Antoine de Bougainville, a diplomat and sailor who had obtained the permission of the French Government to colonize the islands. A fort and village were built at Port Louis on the main east island (still named Port Louis). A smaller settlement is believed to have been made on the Choiseul Sound in the south of the same island.

1765. Captain John Byron, sent by the British Government, made a landing at Port Egmont in a sheltered bay surrounded by islands just off the northern coast of the main west island and eighty miles away from the French settlement at Port Louis. Captain Byron landed on Saunders Island, raised the Union flag and proclaimed that the whole island group – 'The Falkland Islands' – were the property of King George III. He then departed.

1766. Captain Macbride established a settlement of a hundred people in Port Egmont, probably at the place now known as Saunders Settlement. It is believed that the British Government were not aware of the French settlement at Port Louis but a British ship did visit Port Louis soon afterwards and repeated the British claim.

1767. After learning of the French settlement at Port Louis, Spain objected to the French Government, presumably on the basis that the islands were what could now be called an offshore group of the Spanish colony on the mainland known as 'The Royalty of la Plata', which covers present-day Argentina, Bolivia, Paraguay and parts of Brazil and Chile. The islands were 300 miles from the mainland. The French agreed to hand over their settlements to the Spanish on payment of a sum of money to compensate de Bougainville for his expense. A Spanish governor was appointed and the French name of Malouines was changed to Malvinas. The main east island was named Isla Soledad and the west island Isla Gran Malvina. A Spanish governor took up residence at Port Louis, which was renamed Puerto Soledad.

1769. A Spanish ship met a British ship in 'the San Carlos Straits' (Falkland Sound) and then discovered the existence of the British base at Port Egmont.

1770. Six Spanish ships and 1,400 men sailed to Port Egmont and demanded that the British leave. After a week of parleying and then a brief exchange of fire, the only British ship present, the *La Favorita*, surrendered.

1771. After intense pressure by France on Britain's behalf and under the threat of the dispatch of a large British naval force, the Spanish agreed to allow the restoration of the British settlement at Port Egmont but also proclaimed that this act 'cannot or should not in any way affect the question of the prior right to sovereignty over the Malvinas Islands, also called the Falklands'.*

1774. The British withdrew their settlement at Port Egmont (on grounds of economy) but left a leaden plaque which claimed British sovereignty over 'the Falkland Islands, with this fort, the storehouses, wharfs, harbours, bays and creeks'. The plaque was later removed by a Spanish naval officer and taken to Buenos Aires.

1774-1811. Further Spanish governors, mostly junior naval officers, ruled at Puerto Soledad to make an unbroken line of thirty-two governors since 1767.

1810. Spanish rule in the La Plata Royalty was overthrown, leading to the establishment of independent states, including Argentina, in 1816. Argentina claimed all former Spanish possessions in the region.

1811. Spanish settlements in the Malvinas were withdrawn. A leaden plaque was left in a chapel at Puerto Soledad which stated, 'This island, together with its ports, buildings, outbuildings, and everything in them belongs to Fernando VII, King of Spain.'

1811-26. The islands were unoccupied.

1820. The islands were visited by an Argentinian ship and claimed for that country. Military commanders were appointed but there was no permanent settlement.

1826. An Argentinian settlement was established at Puerto Soledad (Port Louis) with Louis Vernet as governor. Britain protested.

1831. The American warship *Lexington*, commanded by Captain Silas Duncan, destroyed the fort at Puerto Soledad and evicted Vernet after the Argentinians had prevented American ships from hunting seals in the islands. Captain Duncan then declared the islands to be 'free of all government'.

1832. A new Argentinian governor, Major Mestivier, arrived in the islands but his men mutinied and Mestivier was murdered.

1833. A further governor, a naval officer, Commander Pinedo, was appointed and was attempting to establish order when a British warship, H.M.S. *Clio* (Captain John Onslow), landed her men, proclaimed British sovereignty again and forced the Argentinians to leave. The present British community commenced in that year, with the main settlement being at Stanley harbour.

The claim made by the Argentinians in 1982, indeed the same claim they had been making since 1833, was based on their claimed inheritance of the sovereignty for the whole of the former Spanish La Plata Royalty. They point out that other

* Declaration of the Prince of Masserano, Spanish Ambassador in London, 22 January 1771.

former Spanish colonies have been able to keep their offshore islands. Ecuador retains the Galapagos Islands and Brazil retains Trinidade Island, both more than 600 miles from the mainland; Chile has the Juan Fernández Islands – 400 miles off shore – and Easter Island, which is more than 2,000 miles out in the Pacific. The British seizure of the Malvinas and the eviction of Commander Pinedo in 1833 was, say the Argentinians, a straightforward occupation by force of a group of islands which, in the ordinary course of events, would have passed smoothly to Argentinian sovereignty, and the British community which inhabited the islands from 1833 to 1982 was implanted by force on Argentinian soil. The British claim in 1833 was based on a mere five years of settlement at one place in West Falkland nearly seventy years earlier, the leaving of a leaden plaque and then the eviction of the Argentinian governor by Captain Onslow. This compares with a continuous Spanish presence of forty years between 1767 and 1811 and an intermittent and unstable Argentinian one of seven years from 1826 to 1833.

Some merit must be granted to the original Argentinian claim. The British response in 1982 was that nearly a century and a half had passed since the last Spanish or Argentinian had lived in the islands and that the 1982 population was composed almost entirely of British stock, many of them in their fifth or more generation of residence. The Argentinians may have had a *de jure* claim to sovereignty in 1833 but the British had a 150-year-old *de facto* record of sovereignty. If all 150-year-old claims in the world were to be acknowledged and settled, the boundaries of half of the world's nations would need to be altered. Furthermore, a new method of deciding sovereignty was now deemed to take priority – the principle of self-determination.

During the 1982 war, an R.A.F. Vulcan bomber was forced to land at Rio de Janeiro with technical trouble. A Brazilian, talking to one of the R.A.F. men, said that he could not understand two major nations fighting over the tiny Falklands; it was, said the Brazilian, 'like two bald men fighting over a comb'. Many other people in the world felt the same way, but they were wrong; much more was at stake than the ownership of the Falkland Islands. The succession of British diplomats who have been appointed to the position of Governor at Stanley in recent times have borne this title: 'Governor of the Falkland Islands and the Dependencies of South Georgia, the South Sandwich Islands, the Shag Rocks and Clerke Rocks; High Commissioner for the British Antarctic Territory.' The Shag and Clerke Rocks are not of importance to the story, being so small as to be completely uninhabitable, but those other places – South Georgia, the South Sandwich Islands and the British Antarctic Territory – are all-important in any study of the origins of the war.

South Georgia is a large island (1,450 square miles), an island of mountains and glaciers and of fierce cold. The only animals are seals and wild reindeer and the only humans who would choose to live here were whalers – until so many whales were caught that the hunting of them became unprofitable – and scientists and soldiers. The island was sighted by various navigators between 1675 and 1756 and the British explorer, Captain Cook, made the·first claimed landing in 1775 and annexed the territory for Britain. There had been a continuous British presence in the form of a scientific research station backed by the Royal Navy since 1909. There were no formal objections to the British claim to sovereignty until 1927, when Argentina claimed South Georgia to be part of the South American continental shelf and therefore to be in the area under Argentinian sovereignty. South Georgia is 1,200 miles from the nearest point of the South American mainland but twice as far from Africa. The Argentinians also contested the accuracy of the Captain Cook landing claim in 1775 and countered by saying that the first reliable sighting was by a Spanish ship, the *León*, in 1756. The Argentinians also pointed out that the first settlement was by an Argentinian whaling company in 1904. The Argentinians were, however, one of several nations who were forced to apply for licences and to pay taxes to Britain for the right to operate whaling stations in South Georgia in the intervening years. Again the Argentinians have claimed a *de jure* sovereignty on the basis of Spanish sightings, a prior settlement and the fact of being the nearest mainland country. Britain, however, had exercised *de facto* sovereignty since 1907.

The South Sandwich Islands are a 150-mile-long chain of small volcanic islands 470 miles south-east of South Georgia, claimed by Britain because of Captain Cook's voyage in 1775.

The climate is Antarctic in nature with much packed ice and constant westerly storms. By all reasonable standards, they are uninhabitable and no British presence was ever established although a naval party did land briefly in 1964. The Argentinians had claimed sovereignty here in 1948, this time solely on the basis that Argentina was the nearest mainland country. In 1976, they established a small base on Cook Island, in the Thule Group at the southern end of the Sandwich chain. In appalling weather conditions, the Argentinian Navy built a permanent station. The Argentinians did not announce its presence and, when the British eventually discovered it, claimed that it was a purely scientific station and refused to leave. The result of this move was that the Argentinians were able to establish a permanent base in an area claimed by Britain and forming part of what Britain called 'the Falkland Islands Dependencies'. The Argentinians gained a *de facto* sovereignty which would last for five years.

To the south of all these places lies the frozen mass of Antarctica, a continent almost as big as the United States and Europe combined. British maps show a great segment of Antarctica to be 'the British Antarctic Territory'. The area claimed is bounded on the east and west by the 20° and 80° lines of longitude. Britain claims, therefore, approximately one sixth of the continent. The basis for this claim is early British exploration work from 1820 onwards and long-established scientific stations on the peninsula known as Palmer Land and Graham Land. The area is also claimed by Chile and Argentina, again on the basis of the fact that they are the nearest countries on the South American mainland. Argentinian maps call the area the 'Antártida Argentina'. In 1957–8, twelve nations took part in the International Geophysical Year, which particularly concerned itself with the Antarctic. Representatives of those twelve nations met in Washington in 1959 and signed the Antarctic Treaty, in which it was agreed that all territorial claims in the area south of 60° latitude – that is, on the Antarctic continent – should be held in abeyance. Argentina, Chile and Britain were all signatories of the treaty, which came into force in 1961.

There had been many attempts to resolve the sovereignty dispute since the Second World War. In 1947, Britain offered to

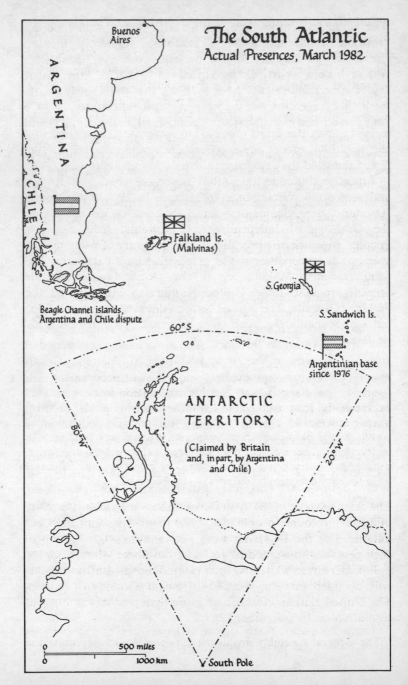

The South Atlantic
Actual Presences, March 1982

Buenos Aires

ARGENTINA

CHILE

Falkland Is.
(Malvinas)

S. Georgia

Beagle Channel islands,
Argentina and Chile dispute

60°S

S. Sandwich Is.

Argentinian base
since 1976

ANTARCTIC
TERRITORY

(Claimed by Britain
and, in part, by Argentina
and Chile)

80°W

20°W

500 miles
1000 km

South Pole

submit to the International Court of Justice at The Hague the dispute over the Falklands Dependencies – South Georgia, the South Sandwiches and the area now known as the British Antarctic Territory. Argentina did not accept this offer. Britain submitted the dispute over the dependencies unilaterally to the Court in 1955, applying for redress over Argentinian encroachments. The case was accepted and some preliminary work was carried out by the Court officials but Argentina announced that she would not accept any decision which might emerge and the Court removed the case from its lists in March 1956.

Attention turned to the United Nations of which both Britain and Argentina were founder members. In 1960, the General Assembly adopted a major policy document, 'A Declaration on the Granting of Independence to Colonial Countries and Peoples' in which was proclaimed 'the necessity of bringing to a speedy and unconditional end colonialism in all its forms and manifestations'. Making use of this declaration, Argentina formally requested the United Nations to intervene in the Falklands dispute. The United Nations debated the issue in 1965 but was immediately faced by a dilemma. Consider the 1960 declaration:

All peoples have the right to self-determination; by virtue of that right they freely determine their political status and freely pursue their economic, social and cultural development ... Immediate steps shall be taken, in trust and non-self-governing territories or all other territories which have not yet attained independence, to transfer all powers to the peoples of those territories, without any conditions or reservations, in accordance with their freely expressed will and desire, without any distinction as to race, creed or color, in order to enable them to enjoy complete independence and freedom.

The main purpose of the 1960 Declaration was to allow the many millions of coloured people in colonies to emerge as independent nations. But the Falklands were not inhabited by a group of such people who wanted to get out from under white rule; the Falklanders were white and their elected representatives repeatedly declared that they wished to remain a colony of Britain. The United Nations could do no more than pass this ambiguous resolution on 16 December 1965:

The General Assembly invites the Governments of Argentina and the United Kingdom of Great Britain and Northern Ireland to proceed without delay with the negotiations recommended by the Special Committee on the Situation with regard to the Implementation of the Declaration on the Granting of Independence to Colonial Countries and Peoples with a view to finding a peaceful solution to the problem, bearing in mind the provisions and objectives of the Charter of the United Nations and of General Assembly resolution 1514 (XV) and the interests of the population of the Falkland Islands (Malvinas).

The voting for this was 94 to 0, with Britain and the United States being among the fourteen abstentions. Argentina must have been bitterly disappointed that the United Nations, with its mass of ex-colonial members, could not have produced a firmer resolution in support of Argentina's case. Argentina's problem was – and remains – that the world now concerns itself less with the paper principle of sovereignty – in the case of the Falklands, a claim more than 150 years old – than with the principle of self-determination. The United Nations has never attempted to define sovereignty nor has it made any declarations on sovereignty, but it has pursued the principle of self-determination with much zeal. What the United Nations was not able to come to terms with was the problem of small colonies which would never be strong enough to govern themselves and which wanted to retain the protection of their colonial power.

Britain and Argentina settled down to talks in 1966. Nothing resulted. There are signs that the British Foreign Office would have liked to have been released from the Falklands problem by transferring sovereignty to Argentina; the dispute was endangering British relations and trade interests in a large part of the South American continent. But the British were bound by the clearly stated wishes of the Falkland Islanders, and the British Government had no great desire to withdraw from the South Atlantic completely. It might have been better to persuade the 1,800 Falklanders to leave their islands completely and settle elsewhere – there was already a Falkland 'settlement' near Auckland in New Zealand – but to acknowledge the Argentinian claim in that way would have set a precedent which would affect other places. It was a complex situation which defied facile solutions. A shared responsibility of some kind was probably the best answer but neither the Argentinians nor the Falkland Islanders were prepared to make many concessions. The Argentinians were unable to obtain much sympathy from the rest of the world because a transfer of the Falklands to their rule would result in an even worse form of colonialism, the rule of a long-established community by a clearly unwanted foreign power. Britain gave nothing away. Argentina gained nothing. The rest of the world did not care too much as long as the two sides did not escalate the situation and create a problem which might involve others. The latest round of talks between Argentinian and British representatives reached its usual inconclusive end in February 1982, only a few weeks before Captain Giachino's men captured Government House in Stanley.

Meanwhile, down in the South Atlantic, the British relied on a small party of Royal Marines – two officers and forty men – to form the garrison in the Falklands and the lightly armed patrol vessel *Endurance* to maintain a naval presence among the dependencies and off the Antarctic coast. These modest forces were intended to act as no more than a 'trip-wire', forcing the Argentinians to open fire and commit the first aggressive act if they ever attempted to recover any part of the area by force. There were no R.A.F. aircraft to carry out maritime reconnaissance. The nearest British military base was at Belize, nearly 5,000 miles away; the small garrison there was protecting a small ex-colony which had asked Britain to help it defend itself. The Argentinians were in the happy position of being able to make threatening moves at sea whenever they wished, to test the British reaction and keep the issue alive. There were several such tests and the Labour Government of 1977 certainly dispatched a submarine, two frigates and a tanker to counter such a threat; there was no fighting and in neither country was the public told of the incident at the time. British naval officers talk of regular moves by the Argentinians which stopped as soon as the Royal Navy appeared, whereupon hospitality was amicably exchanged between the two sides – but that version of events may be apocryphal.

Unfortunately, Mrs Thatcher's more recent administration had felt itself forced by financial pressure to cut back on Britain's military presence in the South Atlantic and there had been signs of a seeming weakening of the British will to remain in the area. In 1981, *Endurance*'s helicopter had flown over the Argentinian base in the South Sandwich Islands and reported that there appeared to be accommodation for more than forty men, more than required by the purely scientific party which the Argentinians claimed was all that was present. Captain N. J. (Nick) Barker, *Endurance*'s captain, requested permission to move the Argentinian 'intruders' but was refused. The British scientific headquarters on South Georgia moved from its base at Shackleton House to a smaller, more easily maintained building; unfortunately this was wrongly interpreted as a major cut-back. The building of proposed new barracks for the Royal Marines party at Stanley was postponed. Even more important, the 1981 defence review decided that the ageing *Endurance* should be retired after her current tour of duty and not replaced. The House of Lords rejected, by a one-vote margin, a bill granting full rights of residence in Britain to every Falkland Islander in the same way as had been granted to citizens of Gibraltar. Finally, the British side at the latest United Nations-sponsored talks in New York in February 1982 was judged to have failed to 'signal' that any Argentinian move over the Falklands would be firmly resisted. This catalogue of events led the Argentinians to conclude that Britain's resolution to hang on in the South Atlantic was weakening.

In contrast to the feuding at diplomatic level, civilian relations between Stanley and the mainland had been improving. An agreement signed in 1971 produced a vital twice-weekly air service with the mainland operated by LADE (Líneas Aéreas del Estado, a part of the Argentinian Air Force which operates commercial aircraft on non-economic routes), a freight service by ship, mainland help with serious medical problems and depots for Argentinian-supplied petrol, oil and bottled gas. Employees of LADE and the fuel companies lived in Stanley with seeming content. Argentinian teachers were provided to teach Spanish to the island children, it being thought by the Falklands Government that closer links with the mainland would be beneficial. Some Falklands families took advantage of an Argentinian offer of scholarships at a special bilingual school in Argentina. The Falkland Islanders realized that their future would depend more and more on this growing cooperation and this would have led, in time, to the islands falling naturally into an Argentinian sphere of influence. Harold Rowlands was the locally born Financial Secretary to the Falkland Islands Government.

They were establishing themselves so well in the Islands and I believed that we would be increasingly living as good neighbours, with benefits to us, the Argentinians and to Britain. I always thought there could be a way found by negotiation and agreement but I did not realize at that stage that we were dealing with an uncivilized country which was after sovereignty rather than good neighbourliness. I thought, in the end, both sides would give and come up with some sort of answer and, just before the invasion, I thought we may be able to come up with the idea of combining the Falkland Islands and the Dependencies and the other islands and possibly Tierra del Fuego into a new territory, which perhaps could be called the South Atlantic Territories, which could be developed by Britain, Argentina, Chile and possibly some other appropriate nations. It could have been a potentially rich area. We were all willing to carry on talking.

I could hardly believe that they would be mad enough to take military action and I also thought they were not so uncivilized as to do so. But I wouldn't ever want to talk to the Argies again – that all changed dramatically on the 2nd of April 1982.

So, the situation on the eve of the Argentinian invasion in April 1982 was that the mainland of Antarctica was enjoying a period of tranquillity under the 1961 treaty whereas, to the north, there were disputes over every island along a 1,600-mile spread of the South Atlantic. (It has not yet been mentioned that Argentina and Chile had a separate dispute over the Beagle Channel in Tierra del Fuego which was not settled until 1984.) The Argentinian flag flew over the eastern part of the chain of islands – at its station in the South Sandwich group – and in the west on Tierra del Fuego. The British flag flew over the Falklands and South Georgia.

When Argentina and Britain fought the war of 1982, both countries were undoubtedly fighting for the principle of sovereignty of the Falkland Islands and of South Georgia and the South Sandwiches and, in Britain's case, for the principle that the Falkland Islanders should be allowed to choose their own destiny without the armed occupation of an unwanted power. But the outcome of that war decided whether it would be British or Argentinian flags which would fly through the Falklands, South Georgia and the Sandwiches in the ensuing years. What may eventually be at stake in the South Atlantic is nothing less than who is going to be in the position to exploit the world's last reserves of oil and minerals and the riches of the sea if the

Antarctic Treaty – due for renewal or revision in 1991 – should ever collapse. There could be as much oil, natural gas, coal and other minerals in that area as in any part of the world. The sea is certainly teeming with krill and other fish. The coastal shelf around the Falklands alone could well be the site of a rich oilfield but the Falklands are only a minor part in the potential of the whole area. The Arctic has no comparable reserves; there is only floating ice over deep water at the North Pole. It is true that the cost of working these resources so far south and the political uncertainty about the area make such exploitation uneconomical at the present time. But as reserves, particularly of oil, diminish in the more temperate climates, so then will the cost of extraction in the Falklands and the rest of that area become progressively more favourable. The North Slope oilfield in Canada is being successfully worked at a latitude which is more than a thousand miles nearer to the North Pole than the Falklands are to the South Pole. The sea route from the Falklands to Europe is more than 4,000 miles shorter than the Cape route for oil tankers from the Arabian Gulf!

When the temperate oilfields expire and if international agreement upon the exploitation of the Antarctic should fail and a free-for-all develop in a world desperate for oil, then whoever controls the Falklands, South Georgia and the South Sandwiches will be in a commanding position over that huge sector of the Antarctic mainland described variously as the British Antarctic Territory and as Antártida Argentina, and over approximately one sixth of the world's last oil and mineral reserves. This may all appear fanciful in the immediate aftermath of the 1982 war but it is an area of the world worth watching as further years pass.

CHAPTER TWO

The Invasion

The war started because of the actions of three men, two with conscious decisions but one without any realization of what would ensue. These three were President Leopoldo Galtieri, Admiral Jorge Anaya and Señor Constantino Davidoff.

There had been seven military takeovers in Argentina between 1930 and 1976, the elected government being overthrown each time and dismissed. The year 1982 found the country in the sixth year of its latest military rule. The legendary President Perón had been ousted in 1976 for the usual reason, inability to manage the country's economy. A *Junta* of armed service officers had ruled since then, firstly with General Roberto Viola as president; when Viola became ill in December 1981, General Galtieri took over. Galtieri was the Commander-in-Chief of the army, a position he retained on becoming president. As most key army positions were already filled by his supporters, Galtieri's base of support was secure. The naval member of the *Junta* was Admiral Anaya and the air force leader was Brigadier-General Basilio Lami Dozo.

Galtieri and his colleagues were in difficulties. The Argentinian situation in early 1982 has been described many times: falling industrial output, falling wages, rising unemployment, a massive inflation of well above 100 per cent and still climbing. Those who had property, goods and overseas accounts were cushioned, but the mass of the population, depending upon work and cash wages, were becoming desperate. The military had been able to establish a political stability of sorts, but only by the use of savage repression against those who opposed the regime. Various figures between 7,000 and 15,000 have been quoted for those Argentinian citizens who were killed in the 'internal war' of the late 1970s, often after brutal torture in military rather than police establishments. So many died – often simply disappearing without public announcement of their deaths – that the relatives of the dead now constituted a separate vociferous group opposing the continuing military rule. General Galtieri and his colleagues needed a diversion and the one issue which united all Argentinians was that of the Malvinas. It is difficult for other people to appreciate the depth of fervour and conviction with which every Argentinian regards the Falkland Islands to be their own Malvinas, taken by force 150 years ago by the British when Argentina was weak and still emerging as a separate nation and occupied 'illegally' by British settlers ever since.

Soon after his succession to the presidency in December 1981, Galtieri announced that 1982 would be 'the year of the Malvinas'. It is believed that the fanatical Admiral Anaya had made the repossession of the Malvinas, either by diplomacy or by armed force if diplomacy failed, a condition of his support for Galtieri's presidency. Galtieri and Anaya are believed to have agreed that, if the negotiations with Britain due to resume in February 1982 failed, then the Argentinian navy would transport a military force to seize and occupy the Falklands some time between July and October 1982. By then, the British patrol ship *Endurance* would have returned at the end of her current tour of duty in the South Atlantic and, as Britain had already announced, would not be replaced. At the same time, three disputed islands in the Beagle Channel, in Tierra del Fuego, would be seized from the Chileans. The occupation of South Georgia and the reassertion of the Argentinian claim to the South Sandwich Islands would automatically follow from the capture of the Falklands. Lami Dozo, the junior member of the *Junta*, did not oppose this plan in which his air force would have little part to play.

The carrying out of these military moves should have presented no problem. With *Endurance* gone, there was only the small Royal Marine garrison at Stanley which might be persuaded to surrender without fighting in the face of overwhelming force; the garrison would be easily overwhelmed if it did put up a fight. The gamble was all in the following phase, the reaction of the United Nations, the United States and Britain. Galtieri and Anaya believed that Britain would not mount the major naval and military expeditionary force which would be required to repossess the Falklands in the worst of the southern winter, that the United States would remain neutral in the affair, and that the United Nations would huff and puff about the Argentinian use of force but would, in the end, be pleased to see the long-running Falklands/Malvinas dispute settled. Argentina's diplomats were to portray her as a long-injured party who had been driven in desperation to act in this way after all negotiations had failed. At home, Galtieri could expect a release from domestic pressure and perhaps gain a glorious place in his country's history as the man who recovered the Malvinas. January 1983 was the final date allocated for the operation; that would be exactly 150 years after H.M.S. *Clio* had sailed into Port Louis and ousted the Argentinian governor of the Malvinas.

In the early weeks of 1982, an Argentinian tourist in Stanley called at Government House, said he was an architect, asked for a tour of the building and then for photocopies of its plans. 'Like a fool', says the then Governor Hunt, 'I gave him the photocopies.'

It was now that Señor Davidoff came on to the scene. Davidoff was an Argentinian of Greek extraction, an ambitious metal merchant who had signed a contract in 1979 for the dismantling of four disused whaling stations in South Georgia which were the property of the British Government. On 23 February 1982, Davidoff visited the British Embassy in Buenos Aires to apologize for minor irregularities in his documentation on a previous visit to South Georgia and to make sure that his documents were in order for a further visit which he was about to make. He offered to carry supplies for the small British Antarctic Survey station at Grytviken – twenty-five miles round the coast from Leith – and offered the services of the doctor and nurse in his party to the British scientists. The embassy official could not immediately answer all of Davidoff's questions; they had to consult with Governor Hunt in Stanley because South Georgia was a dependency of the Falklands. The time came for Davidoff and his party of forty-one workmen to sail before a reply was received from Stanley, so Davidoff simply sent the embassy a letter declaring his intentions and left Argentina on the naval support vessel *Bahía Buen Suceso*, which he had chartered. He reached his destination, the old whaling station of Leith. This was the start of the notorious South Georgia scrap-metal incident which led directly to the Falklands war, indeed can properly be deemed the opening of the war. The main events were as follows:

19 March   Four British Antarctic Survey scientists out on a field trip reported that Davidoff's party was ashore, that shots had been fired and that an Argentinian flag was flying. Davidoff had broken two British regulations: he had not gone to Grytviken to obtain formal landing permission from the leader of the resident British survey party, who also doubled as immigration officer, and his men were carrying arms and firing shots at wild reindeer for meat, which was not permitted. The Argentinians refused the verbal request of the British field party to comply with the regulations.

20 March   The British survey leader reported the incident to Governor Hunt at Stanley. Hunt replied that the Argentinians should be formally instructed to comply with the regulations. The Argentinians did lower their flag but refused to report at Grytviken. Rex Hunt then signalled to London and suggested that Davidoff and his party should be evicted, both because of his earlier infringement of regulations and because of this latest violation. The Foreign and Commonwealth Office protested to Buenos Aires, stating that the *Bahía Buen Suceso* should remove the Argentinian party or 'the British Government would have to take whatever action seemed necessary'.

20 March   H.M.S. *Endurance*, with nine extra Royal Marines from the Stanley garrison, making twenty-two marines in all, was ordered to sail for South Georgia.

21 March   The *Bahía Buen Suceso* left Leith. The Argentinian Government informed Britain that it assumed all the Argentinian workmen had also departed; in fact, twelve workmen were left on shore. H.M.S. *Endurance* was ordered to resume her normal duties.

22 March   The British scientists reported the continued presence of Argentinians at Leith. Governor Hunt recommended that *Endurance*'s marines should remove the Argentinians. *Endurance* was ordered to resume her voyage to South Georgia.

24 March *Endurance* reached South Georgia but was ordered to anchor at Grytviken and not to proceed with the eviction of the Argentinians at Leith. The British Government advised Argentina that the *Bahía Buen Suceso*, which was still in the area, could call back at Leith and remove the workmen; if this did not happen, *Endurance* would do so. Twenty-two Royal Marines, under Lieutenant Keith Mills, were put ashore.

24/25 March A fresh Argentinian naval vessel, the *Bahía Paraíso*, appeared at Leith and landed more than a hundred armed marines and many stores. The new Argentinian party was commanded by Lieutenant-Commander Alfredo Ignacio Astiz. A Royal Marines observation party later saw the *Bahía Paraíso* discharging stores, but did not realize that more than a hundred armed Argentinian marines were ashore. British Intelligence – on the basis of radio intercepts – reported that two further Argentinian ships were sailing towards the area. These were the frigates *Drummond* and *Granville* and it was assumed that, if *Endurance* did pick up the Argentinian scrap-men and take them to the Falklands, *Endurance* would be intercepted by these two warships.

Endurance remained off South Georgia from 26 to 29 March. Captain Nick Barker describes this period:*

I had landed the Royal Marines and we stayed off the coast, playing 'cat and mouse', sometimes in the icebergs. We met the two Argentinian ships sometimes and we spoke on VHF; I knew the captains well but neither side went further than a 'Good morning'. A helicopter flew right over us one day and I clearly saw Captain Trombetta of their Antarctic Squadron with whom I had often cooperated. I signalled to Fleet Headquarters and offered to go over and visit Trombetta and try defuzing the situation over lunch or a drink, but I only received a delaying answer.
On the 31st of March, I realized things were coming to a head.

---

* Interview with author. All such personal quotations are the result of conversations or correspondence with participants, unless otherwise stated. Ranks used are those held in the period March to June 1982.

Endurance was hiding in the entrance to Cumberland Bay, near Grytviken, not showing a chink of light; we were prepared to fight. But Northwood ordered us back to Stanley; by now they knew that an invasion there was imminent. We slipped past the *Bahía Paraíso* in the dark. I didn't like leaving Keith Mills and his men ashore; I realized they were vulnerable. We sailed all through the 1st of April into a force 10 gale and were much delayed. I was very agitated; I realized that the Argentinians landing on the Falklands would find us half-way between Georgia and the Falklands, unable to help at either place. My last contact with Rex Hunt was on the night of the 1st/2nd; he wished us the best of British luck. As far as I know, the nearest British naval vessel was at Gibraltar. We knew there were at least four Argentinian warships at sea. The feeling of isolation was very strong.

Northwood told us to 'make noises' as though one of our submarines was in the area.

Señor Davidoff was a genuine businessman. His chartering of the *Bahía Buen Suceso* from the Argentinian Navy had no sinister purpose; the Argentinian Navy was responsible for any Argentinian voyage of this kind in the Antarctic area. But the link-up of scrap merchant and naval vessel allowed Admiral Anaya to control events throughout the crisis. Thereafter the affair ran its course. The British felt that they could not afford to tolerate a semi-naval Argentinian presence at Leith which was defying British regulations; it smacked too much of the South Sandwich affair, which had left a supposedly scientific station on another of the Falkland Dependencies for the past four years. Galtieri and Anaya had decided that they were not prepared to accept the humiliation of seeing Argentinian citizens seized by Royal Marines, marched up a gangplank of a British naval vessel and evicted from territory which Argentina had claimed as her own since 1927. Galtieri and Anaya agreed that their plans to occupy the Falklands should be brought forward and that, using the pretext of the South Georgia crisis, they should occupy both the Falklands and South Georgia as soon as possible. This decision was taken at a *Junta* meeting held in the Libertad Building in Buenos Aires on 23 March, the day *Endurance* arrived at South Georgia. The staffs reported back three days later and stated that the landings – Operation *Azul* – could be launched in the early hours of 1 April if all went well. It has been suggested that a recent increase of internal turmoil was a major contribution to the *Junta*'s decision. There is little validity to this; there was a huge anti-government demonstration in Buenos Aires on 30 March but the Argentinian troopships and warships were well on their way to their destinations by then.

The detailed planning for the capture of the Falklands was carried out by the *Armada Argentina* – the Argentinian Navy – and the execution of the plan would largely be in naval hands and the fighting, if any, done by naval troops. Three separate groups of ships had been at sea since 28 March, most of the ships sailing from the main naval base at Puerto Belgrano (Buenos Aires), more than a thousand miles to the north. Ostensibly they were taking part in joint manoeuvres with the Uruguayan Navy. Task Force 40 was the Falklands landing force, based around the ex-American tank-landing ship *Cabo San Antonio*; she was carrying nineteen large American-built LVTB amphibious landing vehicles. Escorts, and gunfire support if required, would be provided by the destroyers *Santísima Trinidad* and *Hércules* and the frigates *Drummond* and *Granville* which had been called back from their voyage to South Georgia. The submarine *Santa Fe*, the icebreaker *Almirante Irizar* and the transport *Isla de los Estados* completed what became known as Task Force 40. Rear-Admiral Jorge Allara was in command, using the *Santísima Trinidad* as his flagship. Further warships, Task Force 20, grouped around Argentina's only aircraft-carrier *Veinticinco de Mayo* (the 25th of May), would give distant cover from the north. *

The landing forces were provided by the specialist *Buzo Táctico* marine-commando unit composed of sixty or so men (accounts vary) and the 600–700 strong 2nd Marine Infantry Battalion. The landing force was commanded by Rear-Admiral Carlos Busser, a marine officer. (Argentinian marines carried naval ranks.) The *Buzo Táctico* would land first and attempt to capture Government House and the Royal Marine barracks at

---

* An article written in *Proceedings/Naval Review 1983* by Dr Robert L. Scheina has been most useful in this section. Dr Scheina is an American specialist in South American naval affairs and was provided with information by the Argentinian Navy immediately after the war. He declares his sympathy with the Argentinian cause but his article appears to be reliable and without bias.

Moody Brook. The 2nd Marines would then come ashore by landing craft at two places, with one party clearing the airfield area and the second landing directly in Stanley Harbour. It was hoped that the *Buzo Táctico* could persuade Government House and Moody Brook to surrender without a fight; if this did not happen, then the 2nd Marines, with their overwhelming numbers, their armoured amphibians and their heavier weapons, could be called in to bring the fighting to a brisk conclusion. All the Argentinians were given strict instructions that no Falkland civilians were to be endangered.

When the runway of Stanley airport had been cleared of obstacles and any British resistance in the area mopped up, Argentinian air force C-130 Hercules and other aircraft would bring in the army's 25th Infantry Regiment and the 9th Engineer Company. The 25th Regiment deserves special attention. An Argentinian regiment was approximately 1,000 men strong, roughly equivalent to a reinforced battalion or a battalion group in the British or American armies. All Argentinian regiments were manned by one-year conscripts, led by regular officers and non-commissioned officers. Most regiments drew their conscripts from a particular province but the green-bereted 25th Regiment had been raised as a special, symbolic unit to represent the nation. Small parties of the best men were provided by every other unit in the army so that all regions in Argentina could be represented in the repossession of the Malvinas. The 25th Regiment was intended to be the main force of occupation when the naval troops who seized Stanley were withdrawn.

The army units involved were told that their participation would be entirely symbolic, but Admiral Anaya's obsession with the Malvinas was to commit these and following army units to the agony of the war and to humiliating defeat. It would be the first exterior war fought by the army since the Paraguayan War of 1865–70, when Argentina, Brazil and Uruguay defeated Paraguay in a conflict over borders.

The landing force took five days to sail from Puerto Belgrano. The voyage did not go entirely to plan. The intended route – round to the south of the Falklands, with a final approach from the south-east – had to be abandoned because of fierce weather, which reduced speed at one time to 6 knots. A Puma helicopter on the *Almirante Irizar* was damaged; 900 men – mostly marines – on the *Cabo San Antonio*, a ship with proper accommodation for only 200 men, had a particularly miserable voyage. The landing timetable had to be set back by twenty-four hours. Information came in that the Royal Marine garrison at Stanley had doubled in size; the annual changeover of the garrison was taking place and both parties were present, except for a few of the marines who had been sent to South Georgia. Then, on 31 March or 1 April, Governor Hunt was heard broadcasting to the population of the Falklands that a landing was imminent. So, with the likelihood of increased opposition and with surprise gone, Admiral Busser decided to concentrate all his forces at Stanley, and he abandoned an earlier plan to put small parties ashore at Fox Bay and Goose Green.

The British were ready. The embassy in Buenos Aires had been reporting heavy movements of Argentinian forces which did not fit in with the naval manoeuvres which had been announced and signal intelligence – the world-wide gathering and deciphering of coded military and diplomatic signals which is such a major element in all modern intelligence – was reporting the movements of the Argentinian ships; the recall of the *Drummond* and *Granville* from their mission to South Georgia and the delays and rerouteings of the main forces caused by the bad weather had forced the Argentinians to use their radios and this provided plenty of raw material for the decoders. The most precise piece of information was a signal sent on 31 March to the submarine *Santa Fe*, ordering her to carry out a surface reconnaissance of landing beaches near Stanley.

Major Mike Norman of the Royal Marines had seventy-nine men under his command – two other officers and sixty-six marines and two officers and nine men of *Endurance*'s survey section who had come ashore some time earlier to lessen *Endurance*'s accommodation problems and who had been writing up their season's survey notes at Government House. Major Norman and half of the marines had arrived at Stanley only three days before the Argentinian invasion; he took over operational control from the outgoing commander, Major Gary Noott, less than twenty-four hours before the Argentinians landed. At 3.30 p.m. on the afternoon of Thursday, 1 April, further information came in from London which made everyone realize that the Argentinian force might be carrying out more than a demonstration. Governor Hunt ordered that the official 'period of tension' was to be upgraded to the more serious 'period of imminent danger'. He gave orders that, if the submarine known to be in the area put a demonstration party ashore, it was to be arrested; if the Argentinians landed tactically they were to be fired upon but no fighting was to take place near the houses in Stanley. Hunt still expected that the former event was more likely; he believed that a demonstration party would land, probably fly an Argentinian flag and then allow themselves to be captured. He intended to bring them to Government House, give them a sharp lecture and a glass of sherry, and send them back to Argentina.

Mike Norman prepared his men for action.

I assembled all the men and gave them an initial briefing. I gave them the good news that tomorrow we would start earning our pay. They took it remarkably well but the sailors from *Endurance* became very wide-eyed. Most of them were very keen to get on with it and, although we were all military men, they took it as a personal affront. There was a 'Who the hell do they think they are!' attitude, although we all knew that we couldn't really stop them.

The marines' entire stock of defence stores – a few coils of barbed wire – was then loaded on to a lorry and taken to one of the best beaches near the airfied; it was entirely inadequate. The airport authorities were asked to block the runway with vehicles and any other obstacles available. The marines completed their preparations by drawing as much ammunition as they could carry and having a good meal.

The Governor called a conference in the late afternoon. The civil officials were given instructions to destroy certain documents and Rex Hunt announced that he would be making a public broadcast. There was a discussion about what action should be taken over the forty or so Argentinians in Stanley – the LADE airline staff and an unusually large number of workers recently arrived for work at the gas depot. It has been suggested that the gas workers had been deliberately sent in as a 'fifth column' to assist the landing force, but the time-scale of events does not support this theory. Major Norman wanted to arrest all the Argentinians at once; the Governor and first secretary, Dick

Baker, were unhappy that the arrests could be used as an excuse for later Argentinian moves if no actual landings took place. So, the Argentinians were left for the time being. The meeting broke up and Rex Hunt went down to the local radio station to tell his people that an Argentinian invasion was expected during the night or early the next morning.

Mrs Eileen Vidal was the radio-telephone operator for the islands; hers was one of the best-known voices in the Falklands.

I had just got home from work. It had been a normal day until then – a beautifully clear and calm autumn evening.

I had been expecting something since January; I'd been telling people that one of these mornings we'd wake up and find them here. I felt so strongly that, when our peat supply came in in January, instead of making one stack of it, I made it into a dug-out. I'd pictured the Argentinians herding people into the streets and I felt that if we could hide out for the first few hours we would be OK. I had lived in Chile for seven years and, going around among the Argentinians who worked here and the tourists who came, you could sense the way the wind was blowing.

I saw a Royal Marines lorry going along Davis Street, very fast, with barbed wire in the back, then a cement mixer being towed towards the airport; they were going to use it to block the runway. It was obvious what was going to happen and, when the Governor came on the radio at eight o'clock, he only confirmed what I had already guessed.

During his broadcast, the Governor called upon both serving and past members of the Local Defence Force to report for duty at their Drill Hall. Twenty-three men turned up, mostly the younger members. But this was an even more symbolic force than the Royal Marine garrison. The members had received little training and the marines had already taken most of the weapons because of the unusual number of marines and the extra *Endurance* sailors needing to be armed. The civilian volunteers were dispatched to various points as look-outs and took no significant part in the later fighting. Major Norman received offers of help from two other men. These were Jim Fairfield, who had been a corporal in the marine garrison at Stanley two years earlier and Bill Curtis, a Canadian who had come to the Falklands with his family to escape from the rat race of the outside world and the danger of being caught up in a nuclear war. Fairfield was given a rifle and became part of the

Government House garrison. Curtis, a former air-traffic controller, was sent to deviate the airport's directional beacon in such a way that any aircraft using it might crash in the sea; he could not do this properly so he smashed it with a sledge-hammer. Curtis was later to be a most outspoken opponent of the Argentinian occupation. Seventeen-year-old Tony Hunt, the Governor's enthusiastic son, just failed to get himself a job as a dispatch rider with the Local Defence Force.

Major Norman assembled his men at 11 p.m. and gave his final orders to them in the bar at Moody Brook barracks.

It was the most difficult set of orders in my life. I made it clear that an invasion force was definitely coming; it was likely to be big and we were likely to be killed. I thought they would be depressed and I would have to gee them up a bit but that was a bad estimation because, at the end, they were raring to go – mainly, I think, because it was a slight to their professionalism. Their attitude was that they knew they were going to lose but the Argentinians would know all about it before they did.

The marines were ferried to their positions by lorry and, by 2 a.m., 'everything was as ready as it could be'.

And so the night passed. Patrick Watts, the young manager of the local radio station, kept his station open all night. Bob Rutterford was one of his listeners. Rutterford was an expatriate schoolteacher who had just made up his mind to extend his contract by a further two years because he and his wife loved the island so much. Val Rutterford had just given birth to twins, the first to an island resident for many years, although the babies had actually been born in an Argentinian hospital.

Most people stayed up. Patrick was playing the usual light music but the Governor kept phoning him with the latest reports and other people kept phoning in, just like a Jimmy Young show. Some of the calls were not too serious and many were thanking Patrick for doing a good job. Then, there was a report that Jack Sollis, the captain of the *Forrest*, had the Argentinian ships on his radar.

It was like a bad dream. It was so unbelievable that a lot of people never thought it would happen. I didn't. Maybe they would sail by and fire a few shots into the mountains. I thought that they could have shown their power by putting two warships off the island and that might have frightened people enough to consider a settlement.

The *Forrest* was a small Falkland Islands Government ship

leased to the marines for visits to the settlements. The naval survey party was ordered out to assist the police in rounding up and guarding the Argentinians in Stanley. The Governor received a further message. President Reagan's final appeal to Galtieri to call off the invasion had failed. Rex Hunt made his famous remark, 'It looks as though the silly buggers mean it.' At some stage during the night, the Stanley radio station received a broadcast from the Argentinian flagship appealing for a peaceful surrender but this was refused. *Forrest* went out again and reported further movements, but still well out to sea. Then the lighthouse keeper at Pembroke Point reported that he could see ships approaching. So far, all the attention had been directed out to sea to the east and there was no evidence that any Argentinians were yet ashore. But, at 6.05 a.m. small-arms fire and what appeared to be grenade explosions were heard from the direction of the empty barracks at Moody Brook, two miles to the west. Major Norman says, 'It came as a complete surprise; we were completely outflanked. It was a very good plan, well executed.'

The men of the *Buzo Táctico* had come across the ground south of Stanley undetected and then split into two groups, intending to attack any Royal Marines at Moody Brook and to capture Government House in simultaneous operations. But the Moody Brook attack went in first, after the Argentinians had cut the telephone wires running to Stanley. About forty men, probably under Lieutenant Bernardo Schweitzer, carried out the assault. Although the barracks were deserted and there were no casualties, this was an important little action because it was a test of the Argentinian claim that minimum force was used. It is possible that the Argentinians surrounded the long, single building at Moody Brook and called out a demand for surrender. Not receiving a reply, they may then have decided upon the full-scale assault from room to room through the building, which left the interior severely damaged by bullets and by phosphorus and fragmentation grenades. This party of Argentinians later made their way to Stanley.

Things did not go so well for the Argentinians at Government House. Major Norman radioed to some of his outlying sections to move back and reinforce Government House but these men did not arrive in time and Norman only had thirty-one Royal Marines, eleven sailors and Jim Fairfield, the ex-marine, to defend the seat of government of the Falkland Islands and the Dependencies. Governor Hunt, the Commander-in-Chief, was present as also were several members of his staff, including Don Bonner, the Governor's driver and a member of an old island family, who was standing guard at the window of the rod room with a shot-gun, having sworn to shoot 'the first bloody Argie' who attempted to lower the Union Jack flying from the flagpole a few yards away. Mrs Hunt and her family and the female servants were all in a house in Stanley.

The second group of Argentinian commandos managed to get into position on the high ground behind Government House undetected. They were led by the unit commander, Captain Giachino. Their first attack started at 6.15 a.m. and lasted for half an hour. There were no calls for surrender at this stage but most of the firing seemed to be high and many stun grenades were used. This action may have been intended as a cover for what appeared to be a 'snatch squad' of six black-clad Argentinians who attempted to enter the buildings at the rear, possibly with the intention of capturing the Governor. Three of these men were seen and quickly shot by the three Royal Marines guarding the back of the house. The Argentinians lay groaning in the yard, only a few feet from the house. The other three Argentinians took refuge in the maids' quarters in an outer building and were captured later by Major Noott.

A fierce fire-fight then developed, with the Argentinian fire now definitely being aimed to kill. The Royal Marines were at the windows of the building and around the garden wall. The sailors were posted at the front of the house but this was never attacked. Major Norman had taken up position behind a solid old cannon (made by F. Kinman in 1807) at a side door. The firing died down just before 7 a.m. The Argentinians were obviously waiting for further forces to arrive and they reverted to sniping. Their calls to 'Mr Hunt' to surrender went unheeded except for some rude Royal Marine replies.

Help for the Argentinians was already on the way. At least nineteen Amtracs (Amphibious Tracked Vehicles) from the *Cabo San Antonio* were landing men of the 2nd Marine Infantry Battalion at Yorke Bay on the airport peninsula. Ten frogmen had landed from the submerged submarine *Santa Fe* several hours earlier to ensure that this beach was clear of defences. The nearest Royal Marines were two men more than a mile away on the next beach. A long column of the menacing Amtracs set off towards Stanley, six miles from the landing beach. A Sea King helicopter from the *Almirante Irizar* had landed a party of men who were clearing the airport runway. The first C-130s would land and start bringing in the 25th Regiment two hours later.

Small parties of Royal Marines on the way fought little delaying actions against the Amtracs and then fell back to Stanley according to their orders. There was a more serious action on the edge of town where Lieutenant Bill Trollope, Major Norman's second-in-command, and a party of eight men were in position, armed with light anti-tank weapons which were ideal for use against the Amtracs. Marine S. Brown opened fire on the leading vehicles with his large Carl Gustav 84-mm launcher, but the sights were badly zeroed and the first rocket landed a hundred yards short. The Amtracs came on until they were within range of the smaller 66-mm launcher. Marine Mark Gibbs fired and scored a direct hit on the passenger compartment of the leading vehicle. The Amtrac stopped and manoeuvred itself off the road into a hull-down position; as it did so, the Carl Gustav fired again and its larger round scored a direct hit, square on the nose. The Royal Marines opened fire with machine-guns and rifles and an Argentinian later told a civilian that ninety-eight bullets had come through the holes in the side of the Amtrac and were found in the vehicle. The three Amtracs following stopped out of rocket range and the Argentinians were observed to deploy from each vehicle and take up positions for a fire-fight. No one, neither the Royal Marines nor the local civilians, ever saw any survivors emerging from the damaged Amtrac. Trollope and his men then withdrew in the direction of Government House as ordered.

Little fights were now taking place in several parts of Stanley. There was a lot of noise, both from weapons and from the huge Amtracs clattering through the streets; several houses were damaged, some by gunfire and some after being struck by manoeuvring Amtracs. There was one humorous moment. Anton Livermore was on duty at the Police Station.

This is where we get the funny piece. The radio had been broadcasting all morning, appealing to people to stay in their homes. Work normally starts here at half past seven and this old chap, Henry Halliday, comes walking up the road. The people had been warned that anyone seen walking on the road would be arrested. The fighting was still going on; we had marines around the Police Station at that time. We shouted at him but he didn't take any notice at all, so we dragged him into the station. We were all very tense but he wasn't concerned; he kept saying he was going to work. He was a street cleaner at the Public Works and he was just about ready for his pension; he didn't want to miss work. It was brilliant. That made us laugh a bit – kept us all in one piece. We made him a cup of coffee and he settled down.

Many civilians found that they were now on the enemy side of the fighting lines. Eileen Vidal and her ten-year-old daughter Leona had spent the night with a neighbour and her small children.

We heard firing on the road from the airport. We had this peat dug-out which we had built in January, but it was open-topped and no good against mortar fire so we hid behind the sofa; the small children were under the table. We were praying. Leona said 'Please God save us' over a hundred times; I think she got stuck in a groove but the message seemed to get through – we also got through.

The Royal Marines were in the alley between the house we were in and the next one. We could hear them on the other side of the tin walls. They were firing up the road and shouting warnings to each other. A mortar shell landed in the yard and it sprayed the house with so much shrapnel that we thought the chimney had collapsed and was falling down the roof. The roof had holes in it but not the walls; if they had hit the wall, they would have taken bits of us with them.

The next thing we heard was Argentinian voices shouting the same sort of thing as our boys had done, but in Spanish. One minute the marines had been there, the next the Argentinians were there and our boys were running back behind the houses.

Basil Biggs was one of the lighthouse keepers at Cape Pembroke. He had been performing a useful function radioing information back to Government House.

About seven o'clock – it was broad daylight by then – we saw a line of Argentinian soldiers right across the Cape, coming towards the lighthouse. They stopped about a hundred yards away; we could hear them rattling their guns and saw them getting down behind the grass tussocks. An officer was shouting and bellowing. We went outside with our hands up; we knew we had to give up, we had no guns. They searched the lighthouse and sheds, then put an Argentinian flag up on the rail on the top of the lighthouse and said, 'Don't touch that; you must leave it.' I said, 'O.K. I'll leave it; it's only temporary. The British will soon sort this lot out.' He could speak perfect English and he said, 'No, No. This is for ever.'

After the town of Stanley had been secured – though Government House was still holding out – a small Argentinian landing craft came through The Narrows into Stanley Harbour. Corporal Stefan York and five Royal Marines had been patiently manning their position on the western side of the 250-yard wide entrance. Marine Rick Overall let fly with his Carl Gustav and the round went straight through the side of the landing craft which soon sank.

The end was approaching at Government House. Only six men from the outlying Royal Marine positions had managed to return and strengthen the garrison; most of the remainder were surrendering at several places in Stanley. The Argentinians, by contrast, were being steadily reinforced. The other half of the *Buzo Táctico* arrived at Government House from Moody Brook, observed to be looking very weary. The Amtracs were deploying around this final British position. Aircraft could be heard landing at the airport. There had been further invitations to surrender and further exchanges of fire but none of the British defenders had yet been hit. An attempt was made during a lull to help the three wounded Argentinians in the open at the rear of the house but one, probably Captain Giachino, had appeared to be trying to unpin a grenade, so the Argentinians were left. Giachino then died from the painful wound in his stomach. Rex Hunt was reproached after the battle by a senior Argentinian officer, 'You killed my best officer!' Hunt replied 'He shouldn't have been here.' Two further Argentinians were killed at the other end of the house, their bodies were seen being taken away in body-bags after the battle. At 8.30 a.m., Major Norman formally advised Rex Hunt that Government House could not be held indefinitely against the heavier weapons which the Argentinians were now bringing up. Norman said the Royal Marines and the Governor could attempt to break out and set up a 'seat of Government' elsewhere or continue fighting until Government House was overrun or a truce could be negotiated. The position really was quite hopeless; the history of warfare shows that any surrounded position has no answer against heavy weapons brought up but kept out of range of the defenders' fire. There were negotiations. At 9.25 a.m., Governor Hunt ordered the Royal Marines to cease fire.

The absence of any British casualties allowed the Argentinians the important propaganda gift of being able to announce to the world that the Malvinas had been repossessed without the shedding of a single drop of British blood and the impression was created that the Royal Marines had given up after only token resistance. This could not have been further from the truth. Major Norman and his men had offered a full-blooded resistance and made the Argentinians pay dearly for their success.

The Argentinian press was allowed by the military censor to report the death of Captain Giachino; he was made a national hero. But the other Argentinian casualties were not announced. It is believed that five Argentinians were killed at Government House. No one was seen to emerge from the Amtrac hit on the road in from the airport; the normal crew was ten men and a driver. Argentinian helicopters were later seen searching the kelp along Stanley waterfront at the east end of the town for bodies of the men from the sunken landing craft. An estimate of twenty to thirty Argentinian deaths seems reasonable.

Most of the Royal Marines were rounded up, photographed in humiliating positions face-down on the road, forced to watch the Argentinian flag being raised at Government House – and see the flag fall down at the first attempt to a ribald British cheer – and were then flown out that night to Argentina; they were in Britain, with Governor Hunt and his family, three days later. Jim Fairfield, the ex-marine, managed to get away from Government House and reached his home safely. Corporal Stefan York and the five men who had sunk the Argentinian landing craft escaped into the interior and were given shelter at a remote shepherd's home. Corporal York's intention was to remain at large as long as possible, so that a British military presence could still be claimed, but he and his men were hampered by lack of local knowledge – they had only just arrived in the Falklands –

and they were worried for the safety of their host, a Mr Watson, who had three small children. They gave themselves up a few days later and were also flown back to Britain.

The citizens of Stanley heard of the surrender with relief; they are a peaceful people who did not want lives lost on their behalf. They started to face the realities of occupation by an alien force. Bill Etheridge, the Falklands-born Postmaster, spoke for most when he described his feelings that morning. 'I knew that things would never be the same again. Our way of life was gone; there was no way of turning back the clock. I felt destroyed, completely finished. Where should I go – Britain, New Zealand, Canada? My whole way of life was gone – all the work I had done here since 1947 was finished.' A clutch of senior Argentinian officers walked proudly along the waterfront. In overall charge now was General Osvaldo García, who had been given the title of Governor of the Malvinas, South Georgia and the South Sandwich Islands. With him were Admiral Jorge Allara, who commanded the naval task force now lying off the islands, Rear-Admiral Carlos Busser, who was in command of the marine-landing troops, Brigadier-General Luis Castellano, commander of IX Air Brigade at Comodoro Rivadavia and Brigadier-General Américo Daher, Commander of IX Army Brigade. Most of these were only temporary visitors and would rapidly hand over to more junior officers.

The Argentinian troops sorted themselves out. The big Amtracs were lined up on the waterfront; they and most of the marines they had carried into action would soon return to the mainland, leaving the green-bereted soldiers of the 25th Regiment to provide the permanent garrison. The Air Force carried out a ceremonial fly-past, a Hercules and three small planes variously reported as Mirages and Pucarás. Most of the naval vessels would soon return to their bases, only the destroyer *Hércules* would remain as guard-ship and to provide air-traffic-control facilities until the VOR beacon wrecked by Bill Curtis was repaired.

General García had brought copies of his regulations typed in Spanish and English; only the date needed to be stamped on before being released to the leading citizens of Stanley and issued to the radio station for immediate broadcast.

MALVINAS OPERATION THEATRE COMMAND
COMMUNIQUÉ NO 1

The Commander of the MALVINAS OPERATION THEATRE, performing his duties as ordered by the Argentina Government, materializes heretofore the historic continuity of Argentine sovereignty over the Islas Malvinas.

At this highly important moment for all of us, it is my pleasure to greet the people of Malvinas and exhort you to cooperate with the new authorities by complying with all of the instructions that will be given through oral and written communiqués, in order to facilitate the normal life of the entire population.

COMMUNIQUÉ NO 2   RELIEF OF AUTHORITIES

As of now, the colonial and military authorities of the British Government are effectively relieved of their charges and shall be sent back to their country today, with their families and personal effects.

Furthermore, it is hereby made known that General of Division OSVALDO JORGE GARCÍA, on behalf of the Argentine Government, is taking power of the Government of the Islas MALVINAS, GEORGIAS del SUR, and SANDWICH del SUR.

COMMUNIQUÉ NO 3
INSTRUCTIONS FOR THE POPULATION

As a consequence of all the necessary actions taken, and in order to ensure the safety of the population, all people are to remain at their homes until further notice. New instructions will be issued.

The population must bear in mind that, in order to ensure the fulfilment of these instructions, military troops shall arrest all people found outside their homes.

To avoid inconveniences and personal misfortunes, people are to abide by the following:

1. Should some serious problem arise and people wish to make it known to the military authorities, a white piece of cloth is to be placed outside the door. Military patrols will visit the house so as to be informed and provide a solution.
2. All Schools, Shops, Stores, Banks, Pubs and Clubs are to remain closed until further notice.
3. All infringements shall be treated according to what is stated in EDICT NO 1.
4. All further instructions shall be released through the local broadcasting station, which shall remain in permanent operation.

COMMUNIQUÉ NO 4   GUARANTEES
The Governor notifies the population that:
Faithfully upholding the principles stated in the National Constitution, and in accordance with the customs and traditions of the Argentine people, he guarantees:
1. The continuity of the way of life of the people of the Islands.
2. Freedom of worship.
3. Respect for private property.
4. Freedom of labour.
5. Freedom to enter, leave or remain on the Islands.
6. Improvement of the population's standard of living.
7. Normal supply situation.
8. Health assistance.
9. Normal functioning of essential public services.

Furthermore, the population is exhorted to continue normally with their activities, as of the moment in which this will be stated, with the support of the Argentina Government, in an atmosphere of peace, order, and harmony.

ISLAS MALVINAS, 02 ABR 1982
OSVALDO JORGE GARCÍA
General de División
Gobernador de las Islas
MALVINAS, GEORGIAS del SUR y
SANDWICH del SUR

Patrick Watts repeated the instructions over the radio and a rash of white sheets appeared at windows or were draped over garden gates as nearly everyone in Stanley found a reason for going out on that first day. Soon, there were many civilians about, walking down the street holding a white handkerchief, being determinedly cool to the Argentinians. Bob Rutterford says, 'You can walk past a person without being aggressive, but letting them know they were not wanted. I believe many of the lower ranks of Argentinians were disillusioned to find no crowds in the streets waving Argentinian flags.'

Civilians required to keep essential services going started to receive their instructions. Bob Gilbert and Les Harris were taken from their homes to the power station. This is Bob Gilbert's account.

In the early afternoon a marine lieutenant and a small party of men escorted Les and I in. They manoeuvred us into a corner of the power station. The officer started out, in very good English, by saying not to worry, there's going to be no harm done to anyone. He told us that, although we had heard of all the terrible atrocities in Argentina, he was just like me, working as an engineer, and his country was being ruined by the Communists – and the gist of his speech was that it was the Communists who had started all the killing and the Government had been forced to take the severe measures. That was the reason he'd joined the forces, because of his patriotism, and we were not to believe all we had heard on the radio because it was really a war in Argentina. He finished by saying that now the Malvinas had been returned to their rightful owners and that it was a just war and he was pleased there had been no bloodshed. He stopped then, as though it was the end of a tape recording. It had all come out without hesitation, as though he had memorized it all.

Then he changed completely, as though he'd got the human rights thing off his chest. He apologized for all the tea and sugar his men had used that morning, before they brought us in. He said he would replace it but we never got it. Later in the day, he went to the trouble of seeing that my home phone was repaired; his corporal had noted that it was out of order when he came to my home that morning. My wife was expecting a baby at that time and he made sure it was working properly.

Many children from the settlements were at Stanley's boarding-school house and John Fowler, the Superintendent of Education, negotiated with the Argentinians for the radio-telephone service to be reopened so that the settlement parents could be reassured about conditions in Stanley. Eileen Vidal, the operator, was sent for and opened up in her little tin radio shack.

There were no Argentinians with us, just John and about thirty children – some crying, some excited. I started the process of getting in touch with the parents; they were frantic. Almost straightaway, *Endurance* came on. 'Stanley. Stanley. Stanley. This is *Endurance*. Can you tell me the situation in Stanley, please?' I think it was an ordinary operator who had heard me operating. I was really relieved; I thought they'd sunk her by then. I told them what I could, that there was about 5,000 troops here and there were these big troop carriers and the amphibious carriers as well. I told them about the Pucaras – my seventeen-year-old son had already identified them – and the Hercules and the jets and the helicopters and about the ships. He told me to hang on a minute and he came back with someone else – probably an officer, someone who sounded as though he'd got a hot potato in his mouth – and I, expecting the door to open and the Argentinians to come in and shoot me, had to repeat it all. Once was asking for trouble; twice was plain bloody stupid. When I'd finished, I said, 'For God's sake stay out of it.' I could see the *Endurance* sailing in and taking on the whole Argie fleet.

I went back to putting the children in touch with their parents; that took about an hour.

At the Upland Goose Hotel, the King family also had to face the new conditions. This is a joint contribution by Mrs Nanette King and her daughters, Alison and Barbara.

The nine young Argentinian gasworkers who had been rounded up the previous night were soon released and came back to their old rooms. They urged us not to buck the military, 'They can be dangerous. The best way to deal with them is to give them what they want.' That was the difference between us; whereas we consider our military there to protect us, they consider their military as a threat.

Then a naval officer, Captain Gaffoglio, arrived; he was a regular pre-war visitor. He wanted two rooms. We refused but he knew we had the rooms and he said, 'We know you are English,' and there was the indirect hint of deporting Daddy and leaving the rest of us here. They were very nice, the way they put it, but the threat was there. We put him and two army officers into the coldest room. Alison looked into Gaffoglio's room and was surprised to see a very thick book; she opened it and found the pages cut away and a gun inside.

We had four British reporters staying but they were put out like a shot. The Argentinians wanted all foreign press out, in preparation for a big splash job by a hundred Argie reporters who came in on a day trip, the next day, Saturday. One of them pinched our phone directory as a souvenir. They all came down the road in a big swarm, hoping to see happy Malvinas residents liberated, but no one wanted to be interviewed. Some of the more naïve really believed they would find a Spanish-speaking, repressed Malvinas population who would welcome the Argentinians. There was not much press coverage of this visit because of the disappointing response.

On the Sunday, Barbara King and her father attended a service at the Anglican cathedral.

The Reverend Bagnall preached a special sermon. He told us to keep our heads down and stand fast till liberation. Then we all sang *Auld Lang Syne*. We were all in tears.

The recent bad weather in the South Atlantic, together with refuelling problems for the Argentinian ships involved and uncertainty over the exact position of *Endurance*, prevented the Argentinians from making the Falklands and South Georgia invasions simultaneous, but Britain's last foothold in the area fell one day later. The grandly named Task Force 60 carried out the operation. The small *Bahía Paraíso* had landed a party of approximately one hundred Argentinian marines at Leith whaling station on the night of 24/25 March. These men had remained quietly ashore since then, dressed in civilian clothes. The Royal Marines had also remained passive and their presence was unknown to the Argentinians. A second Argentinian naval vessel arrived at Leith, presumably on 2 April. This was the *Guerrico*, a modern Exocet-armed missile frigate. The only British ship in the area, *Endurance*, was steaming hard towards South Georgia. The Argentinian ships each had at least one helicopter; British accounts suggest that three helicopters were used, two Alouettes and a Puma. The plan of Lieutenant-Commander Astiz, the local Argentinian commander, was to reconnoitre Grytviken by helicopter and then, if all was found to be clear, some of his marines would be landed by helicopter. If any opposition was suspected, the *Guerrico* would bombard the British positions.

Lieutenant Keith Mills, the twenty-two years old Royal Marines officer had an unenviable task. The total strength of his party was only twenty-two men, whose heaviest weapon was a hand-held Carl Gustav anti-tank missile launcher. There were thirteen unarmed civilians at the scientific station. Grytviken's hinterland was an uninhabited, barren, desolate range of mountains and glaciers. Mills had received a series of signals from London. The earliest had ordered him to 'fight it out', but successive signals had tempered this and he was now ordered to defend his positions in such a way as to ensure that the Argentinians were seen to have captured Grytviken by force but not to incur any loss of life if possible.

Lieutenant Mills had 'mined' the best landing beaches with drums of explosives, paint and petrol and had also prepared the wooden jetty for demolition the moment any ship came alongside. He disposed of his men in a series of half-completed trenches covering the best landing beach and the small bay on which Grytviken stood. The Royal Marines had been working hard all through the sleet, hail and strong wind of the previous day. A reconnaissance party sent out on a launch to an island off Leith had brought the bad news that a frigate had arrived. Now, on the morning of 3 April, the marines were in their positions when the first Argentinian helicopter made its reconnaissance. Most of the scientists had been sent to take shelter in the church; Steve Martin, their leader, had begged Mills not to risk his men uselessly in a battle now that the Falklands had fallen. There were some communications between the base and the Argentinians; Martin had been speaking for the British side to conceal the fact that troops were present. The Argentinians demanded that all the British line up in the open on the beach but Martin declined and finally warned the Argentinians that Grytviken was going to be defended.

Marine Steve Chubb was sent to man an observation post at the seaward end of the British positions.

It was a fine day, unbelievably so – calm, cold and dry. Suddenly the helicopters appeared, without any warning. There were two of them, Alouettes. They flew over the beach along the other side of the harbour and then along our side. We all kept out of sight and they seemed to disappear. We grabbed all of our equipment and as much ammunition as we could carry. We thought we were going to need quite a lot. Our morale was still good; we were laughing and joking as we ran to the trench.

We were nearly there when one of the helicopters came right over our heads. We went flat on the grass. Then the frigate appeared but we weren't in position to fire at it. It came on very slowly. She looked surprisingly big and the gun at the front was swinging round, pointing at our positions. Once or twice it looked as though it was pointing right at me. You feel slightly vulnerable when that happens; you are sure that someone must have spotted you.

We continued on to our trench and then I heard the first burst of fire from down at the survey buildings. We turned round and looked down. There was a helicopter about thirty to forty feet up and about twenty-five to thirty metres from our main positions. In other words, it was a fairly easy target for a machine gun. I saw a stream of tracer hitting it. The engine seemed to explode and smoke started belching out of it. It turned away and the pilot managed to get across the harbour and I watched him flop down – a crash landing – behind a small ridge on the other side. He must have been a good pilot to get it across. That was a Puma, a bigger troop-carrying chopper – British made.

That was the beginning of it really.

The action which followed was short and furious. One helicopter had already landed a small party of men near the survey station before this Puma was forced down by the British fire; an

Alouette which went to its assistance was also claimed by the British to have been hit but this is unlikely. Argentinian marines – probably the survivors of the crashed Puma, then opened fire from the other side of the bay. In the meantime, there had been another dramatic incident when the *Guerrico* appeared in the narrow opening of the bay and was immediately hit by the Carl Gustav. The 84-mm round scored a hit on the ship's side, underneath the Exocet launcher at which it was aimed, though there was probably no missile in the launcher at the time. An earlier round had fallen short but travelled through the water and exploded against the ship's side, holing a fresh-water tank. Two smaller missiles – 66-mm – also scored hits, as did a mass of machine-gun and rifle fire. The *Guerrico* came into the bay, turned round quickly and steamed off into the outer bay. She turned again at a safe distance and started bombarding the British positions with her 100-mm gun. Although much of the firing was inaccurate, because of damage caused to the gun's mechanism by the British fire, there was no way the British troops could answer this fire and, with Argentinian marines now working their way round to the British trenches, Lieutenant Mills told his men by radio that he had carried out his orders and, 'if no one had any violent objections', he intended to surrender.

The Royal Marines suffered one serious casualty; Corporal Nigel Peters was hit in the arm by two bullets while standing up to fire his 66-mm launcher. The captured British were searched and forced to have their photographs taken while the Argentinian flag was raised; they pointed out and then made safe the various demolitions, prepared so that no Argentinians would be hurt accidentally. The twenty-two Royal Marines and thirteen scientists were then put aboard the *Bahía Paraíso* and eventually reached England by air on 20 April. Many of the marines volunteered to join the Task Force, which had by then set off for the South Atlantic, and, after only four days' leave, formed part of 42 Commando, which fought in the later Falklands campaign. Marine Steve Chubb, who earlier described the Grytviken fighting, was to be badly wounded in the arm by shell fire on Mount Harriet. Thirteen further scientists, together with Cindy Buxton and Annie Price who were making a wildlife television programme, remained undetected and undisturbed at outlying field camps on South Georgia. The Argentinians suffered a Puma helicopter destroyed and the frigate damaged. Three Argentinians were killed and seven men were injured.

H.M.S. *Endurance*, earlier sent in the direction of the Falklands, had just failed to intervene in the action. Her Wasp helicopter had been preparing for a missile attack on the Argentinian ships when news came that Lieutenant Mills had surrendered. Fleet Headquarters then refused Captain Barker permission to carry out any offensive action and ordered him to conceal *Endurance*'s whereabouts. That afternoon the frustrated and deeply disappointed Captain Barker added this to his situation report to London: 'This is the most humiliating day of my life.'

CHAPTER THREE

The Lion Stirs

General Galtieri and his *Junta* colleagues had proceeded on the assumption that their occupation of the Falklands and South Georgia could not fail; this had proved correct, though the Argentinian casualty list was unexpectedly high. Now they had to see if the diplomatic part of their plan would succeed as smoothly. They had committed themselves on three assumptions: that Britain would not take any military action to repossess the islands; that the United States – which was always concerned at anything that happened in any part of the Americas – would remain neutral; and that the United Nations, representing world opinion generally, would fail to make any positive move and would be happy to acquiesce to the new situation and be rid of a long-standing problem. Galtieri and his colleagues were wrong on every count.

Sir Anthony Parsons, British representative at the United Nations, had been working fast, hard and to good effect. On the day the Falklands fell, he called for a meeting of the Security Council, that part of the United Nations machinery which remains at permanent readiness to cope with the world's crises. Britain was one of the fifteen members of the Security Council. Sir Anthony gave notice of a motion calling for condemnation of the Argentinian action but was persuaded to wait until the following day. The vital meeting thus took place while the Argentinians were capturing Grytviken, though that news did not arrive in time to affect the outcome of the meeting. Señor Costa Mendes, the Argentinian Foreign Minister, presented his country's case. The voting exceeded the wildest British hopes; the resolution was supported by 10 votes in favour, 1 against and 4 abstentions. Panama was the only country to give a positive vote against Britain. The United States supported Britain. Spain, with her heritage links with Argentina, disappointed the Argentinians by not voting against Britain, but Spain was negotiating to join the Common Market and possibly Nato at that time and felt it best not to annoy her future partners in those organizations; Spain abstained together with Russia and two other Communist countries.

The United States had been in a dilemma. She could hardly desert her old ally, Britain, but American foreign policy was based on support for any right-wing government in Central and South America which could stop Communist expansion. The Argentinian military government certainly came into this category, although the ferocity of its internal war and the lack of civil rights had caused a rift between the two countries. This was in the process of being healed when the Argentinians invaded the Falklands. The United States Government would be courted vigorously by both sides in the war which followed, as well as having to cope with the vociferous sentiments of its own people, who were often tired of their government's automatic support for right-wing leaders in South America who were so often corrupt or evil. The American position would be an important one. No one had forgotten the manner in which American pressure forced the collapse of the Anglo-French expedition to repossess the Suez Canal in 1956. It is said that 'the spectre of Suez' hung over the British War Cabinet throughout the Falklands war. The Americans were soon to throw themselves into a prolonged negotiating campaign to solve the Falklands problem without further fighting and would resume a neutral stance during that period. It would be some time before they were again forced to declare for one side or the other.

So, Britain scored a vital diplomatic victory in obtaining clear support from the Security Council of the United Nations. Resolution 502 read as follows:

Deeply disturbed at reports of an invasion on 2 April by armed forces of Argentina,

Determining that there exists a breach of the peace in the region of the Falkland Islands (Islas Malvinas):

1. Demands an immediate cessation of hostilities;

2. Demands an immediate withdrawal of all Argentina forces from the Falkland Islands (Islas Malvinas);

3. Calls on the Governments of Argentina and the United Kingdom to seek a diplomatic solution to their differences and to respect fully the purposes and principles of the Charter of the United Nations.

The Argentinians decided to ignore the call for the immediate withdrawal from the Falklands, thus following in a long line of United Nations signatory nations that have ignored such calls.

Galtieri did not know it, but his first assumption – that Britain would take no military action to repossess the islands – was being proved wrong even before the first Argentinian marine set foot in the Falklands. The British Government had been alerted on Saturday, 27 March, to the probability of an Argentinian invasion. News had reached London of the extraordinary activity at Argentinian ports, of movements of troops by air inside Argentina and of detachments of ships towards South Georgia from the supposed naval manoeuvres off Uruguay. Some preliminary British moves had already been made. The Ministry of Defence had announced on 24 March that *Endurance* would remain in the South Atlantic and the naval supply ship *Fort Austin* was ordered to prepare to sail south and resupply *Endurance*. *Fort Austin* sailed from Gibraltar at 10.00* on Monday, 29 March – the very first British unit to start moving south. There had been even earlier British military moves, though only tentative ones. The Ministry of Defence is properly responsive to political events and these often trigger off preparations which are usually stood down, when a crisis passes, without any public announcement ever having been made. The Commando Forces Headquarters in Plymouth was being 'unofficially stood up and down all that week', as ideas were being produced to reinforce the Royal Marines in the South Atlantic, ideas which ranged from sending the whole of 40 Commando by air to Ascension Island, half-way to the Falklands, and on from there in the *Fort Austin*, to a plan to send a few men of the Air Defence Troop in civilian clothes to Montevideo by civil airliner, this last being quickly abandoned when the problems of how their Blowpipe missiles were to be concealed on the final stage of the journey by Argentinian air liner to the Falklands were appreciated.

But the warnings of Saturday the 27th, that Argentinian

* Many of the times quoted in this book will be in the 24-hour clock system used by the armed services and will be the Greenwich Mean Time being used in London, the so-called Zulu Time of the services.

warships were moving south in strength, brought firmer and more realistic decisions. It was decided that submarines were to be sailed for the South Atlantic; the actual decision was taken by Mrs Thatcher and Lord Carrington, her Foreign Minister, on Monday the 29th during an air flight to a Common Market meeting in Brussels. These nuclear-powered submarines were the fastest ships in the Royal Navy, able to cruise indefinitely at around 25 knots, and their prolonged underwater capability enabled them to be used as a bluff against any navy, like Argentina's, which was without the advanced submarine tracking devices. It should be stressed that 'nuclear-powered' refers only to the submarines' engines; they had no nuclear weapons on board. John Nott, the Minister of Defence, was ordered to dispatch three of these submarines, a request the Navy considered unreasonable; a single submarine was a real deterrent, a second was a reasonable back-up but a third was a serious drain on other commitments.

The Press assumed that the submarine *Superb* was in the South Atlantic and the Ministry of Defence were happy to let that assumption pass uncorrected; *Superb* was actually in the North Atlantic at that time and sailed quietly into Faslane on 17 April. But, again, a naval unit was available off Gibraltar and was able to start its voyage south from an advantageous position. Vice-Admiral P. G. M. Herbert, Flag Officer Submarine, ordered the submarine *Spartan* to sail south as soon as possible. He wrote in his war diary: 'With twelve scrap-metal merchants creating a stir in South Georgia, it is difficult to believe that it is necessary to disrupt *Spartan*'s exercises.' *Spartan* sailed on 1 April; she would reach the South Atlantic eleven days later. To provide the other two submarines, *Splendid* was recalled from the North Atlantic, replenished with stores in a great hurry at Faslane and sailed the same day as *Spartan*, and *Conqueror*, who took on board men from the Special Boat Service, followed on the 4th. All three submarines had a most efficient anti-submarine and anti-surface-ship capability, could sail around the world without surfacing or refuelling, but their food would run out in fifteen weeks.

Meanwhile, however, the British had been forced to face the worsening news that an Argentinian invasion was imminent. On Wednesday, 31 March, signal intelligence (intelligence gained from the deciphering of coded signals) reported that a major Argentinian force of ships was approaching the Falklands. A fateful meeting was held in Mrs Thatcher's room at the House of Commons that evening. Present were Mrs Thatcher, John Nott, junior Foreign Office Minister Richard Luce and Humphrey Atkins – spokesman in the Commons for Lord Carrington, the Foreign Minister, who could not speak in the Commons. Absent were Lord Carrington, in Israel, and Admiral of the Fleet Sir Terence Lewin, Chief of the Defence Staff and the Government's principal service adviser, who was on a round-the-world tour taking in meetings of the Nato Nuclear Planning Group at Colorado Springs in the United States and of the Five Power Defence Agreement in New Zealand. Lewin was in touch with London each day by telephone but his Minister, Mr Nott, was anxious not to heighten the crisis by calling Lewin home prematurely.

The meeting was a hastily convened one. Air Chief Marshal Sir Michael Beetham, Chief of the Air Staff and Admiral Lewin's nominal deputy as spokesman for the services, had not been invited. The preliminary talk was gloomy, overshadowed by the fact that the Falklands were 8,000 miles away to the south, roughly the distance to Hawaii to the west or Borneo to the east; Gibraltar was less than one eighth of the way to the Falklands and there were no airfields south of Ascension Island which could be made available to those British aircraft which were able to land on the short runway at Stanley's airfield. But now there came on to the scene Admiral Sir Henry Leach who, as First Sea Lord, was the head of the Royal Navy but who ranked after Lewin and Beetham in the joint-services leadership. Admiral Leach's normal place of work was at the Ministry of Defence, a few hundred yards away from the House of Commons, but Wednesday was the 'free' day he normally reserved for visiting one or other of his naval shore establishments. He had spent that day at the Admiralty Surface Weapons Establishment at Portsdown Hill, near Portsmouth. He had returned by helicopter to Battersea and then to his office at the Ministry of Defence by car, to find a copy of the 'brief' on the South Atlantic situation submitted that afternoon to Mr Nott by his mixed staff of naval officers and civil servants. This was Admiral Leach's reaction.

I got two 'feels', one that tension was mounting more and more rapidly than in the recent past and, secondly, that our attitude to it was largely unchanged – mainly because of our other commitments and the distance to the Falklands. Knowing that there was no conceivable means of any U.K. agency doing anything about it unless they were got there by the Navy and protected by the Navy when they got there, it seemed to me that it was the wrong naval attitude to tell ministers what we couldn't do; we should at least say what was possible. That was not the tenor of the brief which had been prepared for Mr Nott that day. I thought it had taken the wrong slant, although it had been taken consistently on previous occasions. But the hardening intelligence gave me the clear feel that this time there was rather more to it.

I knew Mr Nott was being briefed. I grabbed the papers and went down the corridor to his office but found that the briefing, unusually so, was at the House of Commons. I grabbed a House of Commons pass and went over there. I was in my ordinary Number 5 working dress, not 'full regalia', as someone has said, no medals and sword.

Admiral Leach had only gone to the House of Commons to see his own Minister but, after a delay, he found himself in the Prime Minister's office and was asked to give his views on the situation. The general tenor of his advice was that nothing could be done to deter any invasion, but that the Navy, acting without the other services if necessary, could mount what Leach called a 'retrieval force' of a brigade of Royal Marine commandos with all necessary supporting arms, amphibious ships to carry out a landing, two aircraft-carriers, and the necessary warships to protect the aircraft-carriers and the landing ships. This force, Leach claimed, could be dispatched 'within days'; the time taken to get the aircraft-carrier *Hermes* out of dock at Portsmouth was a major factor in the date of dispatch. A minimum of three weeks would then elapse before any action could possibly take place in the South Atlantic. This would allow the United Nations and world diplomacy the opportunity to persuade the Argentinians to withdraw from the Falklands peacefully.

Admiral Leach had earlier had some well-publicized arguments with Mr Nott over proposed cuts in the number of ships in the Royal Navy and he has been portrayed as glorying in the opportunity to put to the Prime Minister a plan of action which would show that there could still be such an important role for the Royal Navy. It is my opinion that this aspect of the affair has been overdone and that Sir Henry Leach acted with utter conviction and his prime concern was for the national interest. It was a happy coincidence that his intervention would also help to secure the future of his beloved service. As one naval officer who fought in the Falklands said later, 'The Prime Minister thus got direct advice from this forceful officer, untrammelled by the can't-do merchants.' Mrs Thatcher accepted Sir Henry Leach's advice and the following day the firm decision was taken to make the necessary preparations for the dispatch of a force to regain the Falklands – one day before the Falklands were invaded! Sir Henry himself was delighted by the decision. 'Half would say "couldn't" and half would say "wouldn't". I had expected the U.K. to shout loud but do nothing but I was wrong and I was very satisfied with the outcome of that meeting.'

It might have seemed sensible to send, through secret diplomatic channels, a message to the Argentinian *Junta* to the effect that the approach of their forces to the Falklands had been detected and that a powerful British force was being prepared to sail south and reclaim the islands if the Argentinians did not turn back. But this was not done, probably on Foreign Office advice. Sir Henry Leach says: 'The F.O. were arguing hard that we should not be provocative; it was an absolute obsession with them.'

So was mounted Operation *Corporate*, the name given to 'the complete operation to move a Task Force to the area of the Falkland Islands, covering naval, ground and air battles around the Falklands and including the subsequent return to the U.K.'* Britain was going to war 'at the end of a seven and a half thousand miles long logistic pipeline, outside the Nato area, with virtually none of the shore-based air we normally count on, against an enemy of which we knew little, in a part of the world for which we had no specific plan or concept of operations'.†

The above statements encapsulate neatly the operation about to commence and the difficulties facing the British forces. Three comments must be made at this stage. Firstly, the force was being dispatched for possible use only in case diplomacy failed to remove the Argentinians in the interval. Secondly, further political decisions would be required before armed force was actually used. Thirdly, the term 'Task Force' referred to a great

---

* From a note provided by 29 Commando Regiment, Royal Artillery.

† From Major-General Sir Jeremy Moore's service presentation notes after his return from the Falklands.

variety of forces dispatched separately, often operating away from the main force and many of whose elements were not known to the public or the enemy.

The *Fort Austin* and the submarines had sailed south in secrecy. The next elements to move in that direction also operated out of the public limelight. The U.S.-built Lockheed C-130 Hercules is the standard transport aircraft of many air forces. At 00.01 on Thursday, 2 April, Hercules XV 196 took off from Lyneham airfield with the first load of stores for Operation *Corporate*. The pilot of the Hercules was Flight Lieutenant Roger Case; he and his crew were all from 24 Squadron. This first load was destined for Gibraltar, for the ships on manoeuvres in that area which would shortly be sailing south, and for Ascension Island, which was to become the Task Force's forward base. Roger Case's flight was the first of a massive effort by the Hercules force – four squadrons based at Lyneham – an effort which continued long after the war was over. Six more Hercules left England that day. The Hercules could not reach Ascension in one bound and would have to use Gibraltar as a refuelling stop for a further week. The much better placed airport at Dakar, in Senegal, was then made available to British military aircraft. The Hercules crews were 'surprised at the speed with which diplomatic clearance was obtained; it was very gratifying to have some friends along the way to give us staging facilities. Dakar saved us two hours of flying time.' The eighty crews of the R.A.F.'s Hercules force settled down to a prolonged period of flying; it required three crews for one thirty-hour round trip from Lyneham to Ascension and back again. The aircraft stood up to the work well, though their airframe life was being rapidly consumed, particularly with the take-offs from hot and humid tropical airports with maximum loads. The aircrews were worked hard and instructors from the Hercules Operational Conversion Unit soon had to be brought in as reinforcements. There were, of course, other commitments still to be maintained in the meanwhile, a condition which applied to many of the units involved in Operation *Corporate*.

Sixteen British warships of the 1st Flotilla were out at sea in two groups off Casablanca carrying out high-speed Sea Dart firing at unmanned drone aircraft in the annual *Springtrain* fleet exercises. The flotilla's commander, Rear-Admiral J. F. Woodward, was in the *Antrim*. His staff had for the past two days been preparing a SNORC – a Short Notice Operational Readiness Check – a supposedly paper exercise for a hurried deployment of ships to the Far East. The real destination was the South Atlantic. At 03.00 on 2 April, Woodward received instructions from England that Operation *Corporate* was now in progress and he should prepare seven of the *Springtrain* ships for a covert voyage south, together with three other ships which were in that area and would soon join him. These orders were received at almost exactly the same time that the first Argentinian troops were landing at Mullet Creek in the Falklands.

Fleet Headquarters had already indicated which ships were to sail. Immediate selections from *Springtrain* were the County class destroyers *Antrim* and *Glamorgan*, whose main armament was their French-built Exocet anti-surface-ship missile, the more modern Type 42 destroyers *Coventry*, *Glasgow* and *Sheffield*, with their Sea Dart 'open ocean' anti-aircraft missiles, the Type 22 *Brilliant* with its Sea Wolf 'point' air-defence missiles, well suited for the close protection of major units such as aircraft-carriers or troop-ships, and one general-purpose frigate, the *Arrow*. *Arrow* was only narrowly preferred to her sister ship *Active*, but *Active* turned up in the Falklands a month later. The remaining ships at sea with *Springtrain* – the older ships *Ariadne*, *Aurora*, *Dido* and *Euryalus*, and the modern *Battleaxe* which needed repairs – would all miss the war.

At first light on 2 April, the ships selected for *Corporate* were all paired with one of those due to return to England and a hurried transfer of stores took place by jackstay and helicopter; the home-going ships 'threw everything they had at the rest', items as large as a Lynx helicopter being exchanged or transferred without any authorizing paperwork. Fortunately the day was reasonably calm and, one by one, the transfers were completed. As the ships parted, the southward-bound ships were cheered off by the crews returning to England. Exactly fifty British seamen and two Hong Kong laundrymen would die in those departing ships. A further ship, the frigate *Plymouth*, had left *Springtrain* earlier, fully stored for a long tour of duty in the West Indies. She was ordered back to Gibraltar, to collect a stock of charts of the South Atlantic, and to join the Falklands-bound force that evening.

The nine warships, together with the tanker *Tidespring*, set sail at 02.30 on 3 April, just as the citizens of Stanley were going to bed on the first night of Argentinian occupation.

Meanwhile, a vast amount of work was being carried out in Britain preparing the further ships nominated for the task force. At 10.30 a.m. on 3 April, the Ministry of Defence ordered the Commander-in-Chief Fleet to get these ships to sea as quickly as possible. The first sailings would take place within forty-eight hours – a period which saw some major problems faced and solved. The smaller ships required were provided without difficulty. The major problems were the complete absence of any form of airborne early-warning aircraft, a shortage of fighter aircraft for distant defence and a chronic shortage of shipping space for the transportation of troops and for all types of aircraft, fixed-wing and helicopter. Many expedients were employed to solve the problem of shipping space but the lack of airborne early warning would never be rectified and the Royal Navy would pay the price for this.

There were two aircraft-carriers, the twenty-three years old *Hermes* and the almost brand-new *Invincible*, but these were 'anti-submarine-warfare' carriers, each with a squadron of Sea King anti-submarine helicopters but each with only a weak squadron of five Sea Harrier fighter-bombers. The ' V/STOL' (vertical or short take-off and landing) capability of the Harrier had resulted in both ships being fitted out in such a way that the high-performance fighter aircraft required for distant defence could not be operated. The last of Britain's real aircraft-carriers, the *Ark Royal*, with her Buccaneers and Phantoms, had been withdrawn from service at the end of 1978. There was a further dilemma in the case of *Hermes*. She had a secondary role as a commando assault carrier, capable of taking into action the Royal Marine commando brigade, which was available, and the necessary assault helicopters to put the brigade ashore on the Falklands. That landing role had been practised earlier in the year, but it was immediately decided that *Hermes* would carry Sea Harriers and anti-submarine helicopters and that the Royal Marines would have to be found other means of reaching the Falklands.

The work of loading stores on *Invincible* and *Hermes* and of getting the aircraft aboard proceeded apace all through that weekend. *Hermes* required the most work; she was two weeks into a routine six-week maintenance period. Chief Petty Officer Alan Taylor was on the *Hermes*.

Initially, the main activity was storing ship. We took in as much food as we could cram in, from first thing in the morning until late in the evening and sometimes into the night. Convoys of civvy and Naafi lorries were queuing up for delivery. For other stores, the departments on the ship were 'looking inwards', deciding what they wanted and asking for it. Someone else outside was sending stuff they thought we should have; if we forgot anything, they had already thought of it. I have never seen the supply system work so well, so quickly, and most of it was the correct stuff. I cannot think of anyone who didn't do a super job. The dockyard, who were often 'sticky', suddenly appeared 'lubricated'; the civvy lorry drivers were great and even the Ministry of Defence police had smiles on their faces for some reason.

The Fleet Air Arm squadrons came aboard, the Sea Harriers and some green assault Sea Kings from their base at Yeovilton and the blue 'pinger' anti-submarine Sea Kings from Culdrose. The hovering capabilities of the Harriers enabled them to come aboard while the work of preparation continued at Portsmouth; gone were days when an aircraft-carrier had to go out into the Channel and steam hard into the wind to land on her fixed-wing planes. The arrival of the Harriers and the constant movement of helicopters was observed with great interest by the people of the Portsmouth area. Every possible extra Sea Harrier was put aboard the carriers, mostly by taking over the aircraft of 899 Squadron, the training unit at Yeovilton. 800 Squadron on *Hermes* was thus reinforced to a strength of twelve Harriers, and 801 on *Invincible* to eight. An Australian and an American pilot who had been serving tours of attachments to the Royal Navy were ordered, like all overseas officers on attachment, to leave their units and did so with obvious reluctance; the American, a marine pilot, was particularly vociferous, thinking it 'most unfair to have a war and not invite the U.S. Marines'. The extra pilots required to man the reinforced squadrons came from a great gathering up of instructors, tour-expired Sea Harrier pilots who had moved on to other duties and one or two hurriedly qualified pilots from the training squadron. The average age was high for fighter pilots. Seven of the Harrier pilots were R.A.F. officers on current or recent attachment from the R.A.F.'s own Harrier force.

These twenty Sea Harriers and their pilots – fewer than thirty pilots in number – would be outnumbered ten to one by the Argentinian air force. Air Chief Marshal Beetham's view, that the navy was taking a great risk in committing the task force to action with such slender air cover, would later receive wide publicity but Air Marshal Curtiss, commander of 18 Group who would later be responsible for all R.A.F. operations in the South Atlantic, says 'that was a blinding glimpse of the obvious; we were all concerned'.

There were few problems with the helicopter squadrons. 820 Squadron and 826 Squadron each came aboard their regular carriers, *Invincible* and *Hermes*. Prince Andrew was a co-pilot in 820 Squadron in *Invincible*. The Fleet Air Arm's best assault squadron, 846, flew nine of its Sea King 4s in from Yeovilton and landed on *Hermes*. Seven more of this squadron's Sea Kings and the sixteen Wessex 5s of 845 Squadron would find their way south in penny packets, many aboard the Royal Fleet Auxiliary supply ships, although five were taken down to Ascension Island in heavy-lift Belfast air freighters hired from a private firm.

Accommodation was found in *Hermes* for a company of Royal Marines and some Special Air Service and Special Boat Service men, and *Hermes* and *Invincible* then sailed from Portsmouth to an emotional send-off by a huge crowd. These two ships would always be considered by the public to be the heart of the Task Force. Their sailing was certainly carried out in a blaze of publicity, as part of the campaign to impress the Argentinians with the strength of the force which was setting off to reclaim the Falklands. But the Ministry of Defence was keeping quiet about other ships, so as to keep the Argentinians guessing at the overall strength of the force. The frigates *Alacrity* and *Antelope* slipped quietly out of a foggy Plymouth the same day as *Hermes* and *Invincible* left Portsmouth and later met the two aircraft-carriers off Land's End. Two more Fleet tankers, the *Pearleaf* and the *Resource*, and the supply ship *Olmeda* also sailed on this day. The public knew nothing of the eight *Springtrain* ships steaming south from Gibraltar, nor of two more warships, the *Broadsword* and *Yarmouth*, which were in the Mediterranean, steaming towards duties in the Arabian Gulf, before being recalled and sent south. The old *Yarmouth* was ordered to reverse course first, leaving *Broadsword* to steam on 'in total disbelief that *Yarmouth* could be included and not us', but *Broadsword* was then ordered back also and both ships later joined the carrier group coming down from England. All of these sailings and additions to the task force took place on 5 April. So, only three days after the Argentinian invasion of the Falklands, there were two aircraft-carriers, seven destroyers, five frigates and four tankers or supply ships steaming towards the South Atlantic. These ships formed the original Falklands naval task force. The Ministry of Defence deception operation was working well. A well-known national newspaper published a list of ships believed to be with the task force; this contained all the ships actually in the force but also named seven others which were not!

The show of strength was not limited to the sailing of warships. The landing force which would be needed if diplomacy and naval action failed to dislodge the Argentinians was being prepared just as vigorously and the first troops would leave England only twenty-four hours after the two aircraft-carriers. The military preparations were centred on the 3rd Commando Brigade. This formation was mainly Royal Marines in content, was administered by the Royal Navy and commanded by a Royal Marine, Brigadier Julian Thompson. The green-bereted marines were among the fittest and toughest troops Britain possessed and events were to show that they were ideally suited to the operation, to which they were formally allocated at 3 a.m. on Friday, 2 April, when a telephone call roused Julian Thompson from his bed, telling him that the Argentinians were about to invade the Falklands and instructing him to prepare his brigade for a move in three days' time.

3 Commando Brigade was really an independent 'brigade group', designed to move independently by sea, with its own supporting arms and stores and ready to fight a limited action without lines of communication. There were three battalion-sized units of infantry, 40 and 42 Commandos based in the Plymouth area and 45 Commando at Arbroath in Scotland. 42 and 45 Commandos were trained in arctic and mountain warfare

techniques and carried out regular winter exercises in Norway. 40 Commando did not have this speciality but there was no unit in the brigade which was not well trained and all had been regularly rotated for duty in Northern Ireland, which was as near to active service as any Nato troops had ever come in recent years. The brigade's artillery was provided by 29 Commando Regiment, an army unit but all of whose members were trained to commando standards and wore the privileged green berets. The regiment was equipped with helicopter portable 105-mm guns firing a 33-lb shell and most of the unit were also arctic-trained. The other main elements of 3 Brigade were its own air squadron at Yeovilton, which would take nine Gazelle and six Scout helicopters to the Falklands for liaison and battlefield work, the 1st Raiding Squadron for light-boat coastal work (this was not the Special Boat Service), 59 Commando Squadron Royal Engineers, the Commando Logistic Regiment, an Air Defence Troop and a small electronic-countermeasures unit.

After having been told on the evening of 1 April that there was now 'no requirement' after earlier tentative plans to move minor units to the South Atlantic, Commando Forces Headquarters at Plymouth had been told to prepare 3 Commando Brigade for action, the order coming through in the early hours of 2 April. The earlier 'no requirement' was the result of the pessimistic briefing submitted to Mr Nott before Admiral Leach intervened. It was Leach's more positive view, expressed to Mrs Thatcher at the House of Commons, which had brought the new order to Major-General Jeremy Moore, the commander of the Commando Forces, who would have overall control of the frantic preparations which now commenced.

That week remains in my mind as one of the most exciting of my service ... It was a magic week and was my introduction to the astonishing and spontaneous support which all endeavours received from, it seemed, the entire British public – a potent factor for the maintenance of our morale.*

Individual units were facing and solving their problems. Most of 42 Commando's men were dispersed throughout the country, on leave after returning from an exercise in Norway; and twenty-

* From Major-General Moore's presentation to service units after the war.

five men were abroad, as far apart as one officer on his way to the Far East and another who was fetched back from his wedding reception in the United States. 40 Commando had just finished their annual weapons firing at Altcar range near Liverpool. Y Company of 45 Commando was in Brunei, practising jungle warfare, and Z Company was split up on adventure training in Scotland, Wales and the South of France. Men from several units were still returning from clearing up after the recent exercise in Norway. Most of these men were quickly brought back to their units and the whole brigade would miss its Easter leave. Only a few changes needed to be made to the basic brigade organization. Part of the Raiding Squadron had to be left in Northern Ireland. The small Mountain and Arctic Warfare training cadre had fortunately just completed a seven-month course for unit instructors, so the thirty-six instructors and recently qualified pupils became a special reconnaissance unit for the brigade. Pressure on shipping accommodation resulted in a priority for 'bayonets' rather than administrative personnel, so the Commando Logistic Regiment was ordered to leave half of its men and a third of its equipment – mostly vehicles – behind. Some of the men would catch up later, under the guise of being a makeshift company for beachhead defence, but there would at first be severe problems when the reduced regiment had to handle supplies for a force which grew to include battalion-sized units of infantry with their supporting arms.

There was a dramatic development on the second day of the preparation period, with the news that army units were going to be added to the marine brigade. News had now reached London of the size of the Argentinian force which had landed in the Falklands, this information almost certainly being based on the conversation between *Endurance* and Eileen Vidal in her radio shack in Stanley. It was now decided that another battalion of infantry, more air defence and some light armour should be added to 3 Commando Brigade. This decision was typical of the swift and positive handling – both political and military – of the whole Falklands repossession operation. The fresh battalion of infantry was the 3rd Battalion Parachute Regiment – 3 Para, as it is more popularly called. This battalion was at Aldershot and formed part of the newly formed 5th Infantry Brigade, which was designated as Britain's 'out of Nato theatre' reserve unit. 3 Para was at the top of the 'Spearhead Battalion' roster and thus most readily available for use. Lieutenant-Colonel Hew Pike, commander of 3 Para, came down to Plymouth on the Friday morning, 4 April, and was greeted warmly by Brigadier Thompson. Most officers at this level in Britain's small regular forces were on close terms; Hew Pike had been a student at the Staff College when Julian Thompson was a member of the directing staff and 3 Para had acted as 'the enemy' to 3 Commando Brigade in exercises on Salisbury Plain the previous year. The long-standing rivalry between the red and green berets in the lower ranks would spur the men of both sides on during the campaign.

The army's air-defence contribution was provided by T Battery of the 12th Air Defence Regiment, Royal Artillery, at Kirton-in-Lindsey. T Battery traced its ancestry back to the Indian Mutiny, when it was 'Shah Sujah's Troop' and 'Shah Sujah' was still incorporated in the battery's title. T Battery would take twelve Rapier anti-aircraft missile launch units to the Falklands and its men would find they were 'the most popular guys around' when the beachhead came under Argentinian air attack. The battery had been alerted as early as 2 April and was made ready in two days. Two non-commissioned officers – groom and best man – having been fetched out of a wedding reception at Doncaster, the battery drove down to Plymouth on 4 April – the day after the Argentinian capture of South Georgia, another indication of the speed at which the task force was forming. The battery commander, Major Graham Smith, later wrote:

> The sixteen-hour journey to Plymouth was tedious but made light in the closing stages by the spontaneous encouragement of motorists, motorway patrons and late-night revellers in Plymouth, one of whom appeared to have been trained to drop his trousers at the mention of the word Argentina.

The armoured support for 3 Commando Brigade was provided by another historic unit, the Blues and Royals (two famous cavalry regiments now merged), stationed at Windsor. Argentinian television pictures showing the huge Amtracs in Stanley prompted the decision to send four Scorpions, four Scimitars and

a Samson recovery vehicle from B Squadron. This small detachment was under a young subaltern, Lieutenant Mark Coreth,
and was to encounter much difficulty because it had no officer
more senior to represent it at 3 Commando Brigade Headquarters. The tanks – always called 'cars' by this old armoured-
car regiment – were moved to Southampton in container lorries.
Some marine officers were slightly scornful when they saw the
tanks and forecast that they could never be operated in the
Falklands away from tracks, but the Scorpions and Scimitars
were really very light vehicles and they operated in the Falklands
without great difficulty.

The strength of the reinforced brigade now stood at 5,500
men. At 4 p.m. on Tuesday, 6 April, Brigadier Thompson's staff
was flown out in three Sea Kings from the Long Room helicopter
pad at Stonehouse Barracks to the command ship *Fearless* in the
Channel. Brigadier Thompson followed two hours later.

The efficient assembly of a landing force was not so easily
matched by the provision of ships to carry the troops on their
long journey south. Britain's war plans – under the constant
pressure of annual budgets – no longer envisaged the likelihood
of a major amphibious operation being carried out, certainly not
one so distant from the United Kingdom. Once the decision had
been taken to use *Hermes* as an aircraft-carrier, there remained
only two multi-purpose assault ships, *Fearless* and *Intrepid*, and
six small landing ships of the *Sir Lancelot* class. When 3 Brigade
was alerted for sailing on 2 April, only the *Fearless* and four of
the 'Sirs' were immediately available and only *Fearless* could
take a reasonable number of troops; the 'Sirs' were primarily
vehicle and stores carriers. *Fearless*, due to be the command ship
for the Falklands landings, sailed from Portsmouth on 6 April,
collected her landing craft at Spithead, flew on various helicopters as she steamed down the Channel and, as has been
described, took on 3 Brigade Headquarters off Plymouth. Four
of the 'Sirs' – *Galahad*, *Geraint*, *Lancelot* and *Percivale* – sailed
from the military port of Marchwood at Southampton. These
were Royal Fleet Auxiliary ships, crewed by Merchant Navy
officers and Hong Kong Chinese ratings. These ships carried 29
Commando Regiment's guns, T Battery's Rapier units, the
Brigade Air Squadron helicopters, a few vehicles, and much

heavy stores. The second assault ship, *Intrepid*, was rapidly refitted and the two remaining 'Sirs' – *Bedivere* and *Tristram* – were fetched back from other duties and all followed as quickly as they could be loaded.

But few of the 3,000 infantry allocated to the operation had so far been provided with a ship. Britain had no regular troop-ships in an age when all long-distance trooping was by service or chartered aircraft, but plans were in existence to cover the gap, though the planners had never envisaged the extent – or the cost – of the operation now being mounted. This was STUFT, the name commonly applied to the Ships Taken Up From Trade process. The Falklands task force could never have been mounted without the ships of the Merchant Navy taken over in this way. Ships could be chartered on a freely negotiated basis, or requisitioned if the owners proved unwilling. Some of the ferry operators, with ships on regular runs, were most reluctant to release their ships for an indefinite voyage to the South Atlantic, but P & O, Stena, Townsend Thoresen and United Towing all earned themselves a reputation for being cooperative. The owner of the Stena oilfield repair ships, the Swedish Mr Stena Olsen, came to London and told the Admiralty Board that his company was 'one hundred per cent behind the task force' and declared his fervent belief that the Falklands operation should be carried through. No formal honour was given to Mr Olsen, as was given to some of the British shipowners, but he was 'wined and dined' by Admiral Sir Henry Leach. All foreign crews on the ships taken up from trade had to be replaced by English seamen, of whom there were not many left unemployed at the end of the campaign. A Swedish crew put off one of the Stena ships was very annoyed; they had been very keen to sail 'for a fight with them Argies'. Every ship had a naval party put on board, depending on the ship's need; these varied in size between the 160 naval personnel put aboard *Canberra* to two men – a signaller and a sick-berth attendant – put on to the salvage tugs which were requisitioned. Cooperation generally was superb but some of the naval parties experienced difficulty in persuading merchant masters and crews to face up to the problems of working in a potential war zone; one of the naval officers concerned says that 'defence measures was the hardest story to get across'. Only three of the masters – Captain Jackson of the *Queen Elizabeth 2*, Captain North of the *Atlantic Conveyor* and Captain Twomey of the *Atlantic Causeway* – had Second World War experience.

Again, in an age of mass air travel, there were no regular passenger liners to call in as troop-ships, only large cruise liners and North Sea and Channel ferries. Attention immediately focused on the cruise liners *Queen Elizabeth 2* and *Canberra* and the large North Sea ferry *Norland*. By most fortunate chance, both the *QE2* and *Canberra* were shortly due back at their home ports after long voyages. The 45,000-ton *Canberra*, being available a few days before *QE2*, was preferred. She sailed into Southampton on 7 April, was fitted with a helicopter pad and suffered other modifications to her luxury fittings, was loaded with 2,400 troops and sailed in the evening of the 9th. A thousand passengers booked on her next world cruise received telegrams to say that their holidays were postponed indefinitely. The *QE2* was not required at this time and carried on with her normal cruise programme. The *Norland* was selected for service and was initially earmarked to take 3 Para but this unit was squeezed into *Canberra* with most of 3 Commando Brigade and *Norland* was held in reserve. Only three companies of marines, some Special Air Service men and a few other small parties of troops were flown out to Ascension Island to join ships of the task force there. The fortunate availability of *Canberra* and the swift decision to requisition her enabled the bulk of the infantry of the amphibious force to be sailed comfortably, swiftly and with the dramatic publicity effect which was part of the plan to put pressure on the Argentinians to leave the Falklands quietly.

The second STUFT ship to sail was the 5,436-ton P & O ship *Elk*, which normally worked the North Sea run to Norway. She was loaded with 2,000 tons of ammunition and stores and sailed from Southampton with *Canberra*. These two ships were the first of nearly fifty merchant ships taken over in this way. (Appendix 1 lists all the merchant ships taken up from trade.) One aspect of these early sailings – both of the naval amphibious ships and of the merchant ships – should be stressed. Because the troop-ships and the supply ships were loaded and sailed as fast as possible in order to create the maximum momentum and publicity effect, none of the ships was loaded 'tactically', that is, ready for immediate sailing into a war zone. A large amount of the war stores had later to be restowed and many stores and some of the men needed to be transhipped between different parts of the task force. Nearly every one of the ships sailed in that first week of Operation *Corporate* could have been properly loaded in English ports, *sailed three weeks later* and still have reached the Falklands in time for the landings.

There was a lull after *Canberra* and *Elk* left Southampton on 9 April. The Royal Navy found two more frigates, *Argonaut* and *Ardent*, and these sailed from Plymouth ten days later. *Argonaut* was the first ship of the large but ageing Leander Class to be sent south; *Ardent* was a modern Type 21 frigate. The significance of the dispatch of these two ships is that, as the days passed without a diplomatic solution, the ships in the task force were starting to be sorted out into what might loosely be called a 'Battle Group', based on the aircraft-carriers, and an 'Amphibious Group', consisting of the ships that would land troops on the Falklands and their naval escort. *Argonaut* and *Ardent* were destined to be escorts in the Amphibious Group.

The next departure was another ship taken over from the Merchant Navy and was a sobering reminder that the chances of a peaceful solution were receding. The 17,000-ton P & O schools-cruise ship *Uganda* was in harbour at Alexandria when she was requisitioned as a hospital ship on 10 April. The 944 children and their teachers from the sixty or so preparatory schools which had booked *Uganda* for this voyage were mostly away on a bus trip to Cairo and the Pyramids when the message reached *Uganda*'s surprised captain. His passengers were not told until the ship left Alexandria the next morning, some astute schoolboys realizing that the ship was steering west towards Italy instead of north, as planned, towards Rhodes. The children lost five days of their cruise, were landed at Naples and flown home to England. A helicopter platform, a refuelling point to allow *Uganda* to replenish while at sea, and much other conversion work was completed in a hectic three day period at Gibraltar dockyard. Eighty-three naval medical staff – including nurses – and the twenty-strong Royal Marines band to act as stretcher-bearers were flown in from England and the new hospital ship, with large red crosses painted on its sides, received a big send-off when she sailed on 19 April. It was the first time Britain had sent a hospital ship to sea since the Korean War thirty years earlier.

There was some criticism that the Royal Yacht *Britannia* was not sent instead of *Uganda*; one justification for *Britannia*'s expensive retention was that she could be converted to serve as a hospital ship in time of war. But *Britannia* used heavy furnace oil, instead of the lighter diesel used by most modern ships, and would have required her own tanker while operating away from the task force as the hospital ship would be required to do. The small naval survey ships *Hecla*, *Herald* and *Hydra*, already fitted with helicopter platforms and capable of staying at sea for long periods, followed *Uganda* and would become 'ambulance ships', ferrying convalescent wounded between *Uganda* and whichever South American port would accept the British wounded.

The final part of the main task force did not sail from England until 26 April. There had been persistent anxieties over the number of Argentinian troops who could oppose a British landing; latest reports indicated the presence of nine units of battalion strength – against the mere four British battalions so far dispatched. The conventional superiority required for an attacking force was three-to-one over the defence; this the task force never would achieve but a fifth battalion was now added. 2 Para at Aldershot had been on tenterhooks ever since their comrades in 3 Para had sailed on *Canberra*. Lieutenant-Colonel H. Jones had dashed home from a skiing holiday at Meribel in France to badger the authorities, and his battalion was now allocated to the Task Force. Also added at this time were 29 Battery of 4 Field Regiment, 9 Parachute Squadron Royal Engineers, the Parachute Clearing Troop (a field surgical unit) and A Flight of 656 Squadron Army Air Corps with three Scout helicopters; these units were all from the 5th Infantry Brigade, except for the helicopter flight. 5 Brigade was now left with only one of its three original battalions, the 7th Gurkhas. These army units were all placed under the command of Brigadier Thompson and 3 Commando Brigade; in this way, the Royal Navy remained in full control of the whole Task Force.

2 Para and most of the men in the other units sailed on 26 April from Hull in the North Sea passenger ferry *Norland*, with

the guns, helicopters and other heavy equipment sailing in the Townsend Thoresen ship *Europic Ferry*.

So, the young men of Britain went off by sea to war, following in the footsteps of so many earlier expeditionary forces – three times to France in two world wars and to such distant places as Gallipoli, Korea and Suez. The actual send-offs of the troopships had a strong flavour of the Boer War, with 'Goodbye, Dolly Grey' being replaced by such tunes as the haunting 'I am Sailing' and 'Don't Cry for Me, Argentina'. There were other historical comparisons – to the Battle of the Falklands in December 1914 and of the River Plate in December 1940, each of which had brought the Royal Navy her first victory in a world war. Four of the ships now at sea, or soon to set out, bore the same names as warships in those earlier battles – *Invincible*, *Glasgow* and *Bristol* in the Battle of the Falklands and *Exeter* at the River Plate. There were less happy comparisons with the Suez expedition of 1956, when exactly the same Royal Marine Commando units and Parachute battalions now sailing south had landed successfully at Port Said with French forces to repossess the Suez Canal after its nationalization by President Nasser, but were forced to withdraw by hostile world opinion led by the United States. Counting *Endurance*, still on duty in the South Atlantic, approximately sixty-five ships were at sea by the end of April – twenty warships, eight amphibious landing ships and nearly forty other ships of the Royal Fleet Auxiliary and the Merchant Navy. There were some 15,000 men in the Task Force, of whom about 7,000 were the marines and soldiers of the potential landing force. The stores ships were carrying supplies for three months of operations.

The units had left with mixed emotions. Former members had begged to be taken on the strength again; regular members made sure they were not omitted. Most of the men had sailed with a good deal of pride and much moved by the public enthusiasm and emotion displayed towards every unit allocated to the Task Force, although few thought that it would come to fighting. Surely, everyone thought, the Argentinians would not choose to stand, to fight and face inevitable defeat. Marine Steve Oyitch, sailing in R.F.A. *Resource*, says, 'We never thought it would come to fighting – never. Nobody in 1982 takes on the British armed forces, except perhaps the Americans or the Russians.' Captain Philip Roberts had taken the *Sir Galahad* out of Plymouth.

The marines on board all lined the ship's side, procedure Alpha they call it. I had always wanted to do that since being an R.F.A. master. We felt all this publicity of ships sailing would frighten the Argies and we would all be on our way home in a week. That feeling stayed with us for a long time.

Marine Neil Young was in M Company, 42 Commando.

The whole Commando was paraded at Bickleigh Barracks. General Moore told us he was sorry he would not be going and gave us a pep talk. Then Colonel Vaux ordered, 'To the South Atlantic, quick march.' Then most of the unit marched past by companies to buses for Southampton but we in M Company were discreetly hidden in the gymnasium until the press and families were all gone. Then Colonel Vaux came in and told us that there would be an attack on South Georgia before the task force got to the Falklands and we were going to be the advanced body.

We were quite chuffed at the prospect that we might get a chance of getting ashore and into action before the rest of the war was called off. We were told we would definitely land on South Georgia whatever else happened, even if there was a cease-fire.

We were confined to barracks that night and we all got drunk in the Naafi. That wasn't unusual but it was quite unique that we were off to war in the morning. There was a lot of singing, the usual marine songs plus a couple we made up for the occasion. One was 'We're on the March with Maggie's Army', to the tune of the Scottish World Cup song of a few years ago when they played in Argentina. We had a good night.

Two days later we went in coaches to Brize Norton. There we were, off on an 8,000-mile journey and the bus driver lost his way on the first stretch!

CHAPTER FOUR

Task Force South

The Argentinians watched every move in the British demonstration of force but were not persuaded to give up easily what they had waited 150 years for with such emotion. Nor did diplomacy work. The Argentinians continued to ignore the United Nations resolution of 3 April, which had demanded their immediate withdrawal from the islands. Four days later, there started the famous 'Haig shuttle' in which President Reagan sent his Secretary of State, General Alexander Haig, to fly backwards and forwards between London and Buenos Aires in a sustained effort to negotiate a peaceful solution in the interval between the dispatch of the British forces and their arrival in the South Atlantic. The British indicated that they might be willing to halt the task force and talk about the future if the Argentinians observed the terms of the United Nations resolution and withdrew from the Falklands. But the Argentinians were not willing to leave unless their rights to full sovereignty were acknowledged. Haig could make no real progress and world tension increased at the prospect of full-scale military action over the islands. The United Nations, which had held back during the Haig negotiations, then took over the role of intermediary, but with no better success.

The attitudes of both sides were supported by the majority of the populations in their respective countries. General Galtieri, in an interview with the Argentinian magazine *Clarín* a year later, stated that he had proposed to his *Junta* colleagues that the Argentinian forces be withdrawn from the Falklands over a sixty-day period, in exchange for the halting of the British forces and the promise of further negotiations, instead of the firm guarantee on sovereignty which he publicly demanded. Galtieri stated that he and his *Junta* partners had 'no domestic political margin' bearing in mind 'the state of euphoria that the people were enjoying'. The *Junta* decided to hold their ground and send reinforcements to the Falklands, hoping that the United States or Russia would decide that they were not prepared to see world stability threatened by this argument or that the United Nations, representing general world opinion, would withdraw its condemnation of the Argentinian invasion or that Mrs Thatcher's public support in Britain would evaporate. None of these things happened. The euphoria of the Argentinian people dampened only slightly when their young men were sent off to reinforce the Falklands garrison and as the British ships drew nearer. Even the Anglo-Argentinian community, said to number 30,000 people, remained mainly in favour of the 'Malvinas for Argentina' solution.

The Argentinian garrison on the Falklands immediately after the invasion of 2 April numbered no more than 2,000 men, despite all the reports of higher numbers. It was the sailing of the British marines and paras in the *Canberra* on 9 April which pushed the Argentinian military machine into top gear. A general recall of recently released reservists was announced that evening; this would bring all major units up to full strength. Two brigade headquarters, eight further regiments of infantry and some support units were earmarked for duty in the Falklands.* A total of approximately 13,000 troops would eventually face the British. The Argentinian ships which had supported the invasion returned to their mainland bases and were prepared for further action. The Air Force could do little except deploy their main attack units at bases in the south of Argentina because the runway at Stanley was too short for use by their high-performance jets. A twenty-five-strong unit of Pucará propellor-driven ground-attack aircraft, together with a small number of other aircraft, was sent to the Falklands.

It was at this stage that the emotional story of the Argentinian *chicos* commenced. These were the direct equivalent of the eighteen-year-old British National Service men of the 1940s and 1950s and of the conscripts who form the basis of many armies – but not of the British forces now bound for the Falklands, which were entirely professional in content. Roughly three-quarters of the Argentinian forces were composed of conscripts, who were called up during the early months of each year for a

* Appendix 2 lists the Argentinian forces. A regiment was slightly stronger than a British battalion.

nominal twelve months' service (although some could be released before their year was completed, as a reward for good service). The 'Class of '63' – young men born in 1963 – had just commenced their year's service when the Falklands crisis erupted. Their officers and non-commissioned officers were long-service regular soldiers.

The Class of '62 was recalled to bring the Falklands-bound units up to strength. Argentinian regiments are recruited on a regional basis and the reservists reported to their local barracks. This was a unique event for Argentina but it was a scene which had so often been enacted when Europe's great conscript armies went to war. A book, *Los Chicos de la Guerra*,* based on interviews with young men of the Class of '62, gives an excellent description of the recall process and the mixed emotions with which these young men joined their even younger comrades of the Class of '63. Some press articles described hastily formed volunteer units, made up of even younger men hastily assembled and dumped on a foreign battlefield, but these stories were imaginary. The recall process went smoothly and the units were efficiently transported by civil airliner to bases in the south of Argentina and then on to Stanley, again by air, mainly in the ten-day period commencing on 11 April. Morale was good at this stage and most of the soldiers felt they were supporting a just cause, although they did not like the coldness either of the Falklands climate nor of the local people's attitude, and many were apprehensive over the prospect of having to fight for this strange land which their parents and leaders insisted was part of Argentina. All units were at full strength, mostly well equipped, and had a clear superiority in strength over the British force sailing down the Atlantic. But these soldiers were mainly of Spanish, Italian or native South American Indian blood, not temperamentally suited to prolonged military operations involving hard living and facing modern firepower. They were also members of an army which had no experience of modern war and were not trained to the high standard of Nato armies. The Argentinian units which would have been best suited to conditions in the Falklands were the mountain regiments, but these were all retained in Argentina because of a fear that Chile might

* By Daniel Kon, Editorial Galerna, 1982, English edition New English Library, 1983.

attack the long western border while attention was focused on the Falklands.

The Argentinian command in the Falklands had been changed immediately after the initial invasion. Brigadier-General Mario Benjamin Menéndez arrived on 3 April to become what the Argentinians described as the eighth Military Governor of the Malvinas, the first seven having ruled between 1821 and 1833. The day he arrived in Stanley was Menéndez's fifty-second birthday. He came from an old military family and is credited with a sound record of staff work. Poor Menéndez; news of the dispatch of British forces came within a few days of his appointment and he was forced to become Commander-in-Chief as well as Military Governor. The command hierarchy in Stanley was completely unsuited to the direction of war operations. In theory, there were subordinate commanders for all three services – Brigadier-General Américo Daher of the original army landing force, Vice-Admiral Edgardo Otero, and Brigadier-General Luis Castellano of the air force – but inter-service rivalry in Argentina continued at Stanley and the services operated on severely separate lines; there was little concept of a joint command.

The army chain of command in the Falklands ran from Menéndez, through Daher, who became roughly the Chief of Staff (Operations), on to the two brigade commanders. These were Brigadier-General Oscar Joffre of X Brigade, based mainly around Stanley with its headquarters initially at Moody Brook, and Brigadier-General Omar Parada whose much smaller III Brigade was responsible for the Goose Green area and the whole of West Falkland. Menéndez was the youngest of the four brigadier-generals. It would have been better if the Argentinians had moved a *general de división* into Stanley to take a tighter grip on the situation as soon as the prospect of a British landing developed. Instead, four brigadier-generals were left to sort out a command situation never clarified by Army Headquarters in Buenos Aires. Brigadier-General Joffre was more favoured by Buenos Aires and, although never formally appointed field commander, steadily rose in influence. The civilian government which took over in Argentina from the *Junta* in 1983 appointed the Rattenbach Commission, composed of senior officers, to investigate the conduct of the Falklands war and this criticized Joffre for being too rigid, for not allowing subordinate commanders to use initiative and for being responsible for poor morale in the fighting units.

The initial voyages of the scattered elements of the British task force were completed. All of the ships were making for Ascension Island, just south of the Equator, 3,700 nautical miles from Britain but still 3,300 miles to the Falklands. The warships which had been in the *Springtrain* exercise needed eight days to reach Ascension; the ships from England took three or four days longer. The random sailing of the various elements of the main Task Force meant that the arrivals at Ascension were spread over a twenty-five-day period, from the first arrivals of the *Springtrain* ships on 10 April to that of the landing ship *Intrepid* on 5 May. They were days of storms in the Bay of Biscay and of glorious sun in the more southern latitudes; there were crossing-the-line ceremonies, Easter Bonnet competitions, much intensive training and some boredom, anxiety about mechanical defects and personal affairs and concentration on the continuing negotiations with Argentina, but there were no major incidents.

Long-range Russian 'Bear' reconnaissance aircraft, operating out of Luanda in Angola, were met off the African coast and Russian merchant ships proved to be persistent shadowers, but there was never any evidence that the Russians were other than curious for their own purposes. General Galtieri told Alexander Haig, during Haig's negotiation attempts, that the Russians had offered to torpedo one of the British aircraft-carriers and allow Galtieri to claim the success for an Argentinian submarine. This must be considered to be another of Galtieri's inventions; it is most unlikely that the Russians would have risked becoming involved in this way.

There was one casualty. The hospital ship *Uganda* had the misfortune to lose Captain Biddick, who became his ship's first patient when he was operated upon for an old complaint which flared up; he had to be flown home from Freetown but he died soon after reaching England. Brian Biddick had been promoted to captain at an early age and had commanded the school-cruise ships *Nevasa* and *Uganda* for nearly five years, during which approximately 100,000 schoolchildren had sailed with him. Captain Jeffrey Clark was flown out to replace him.

The British could hardly have fought and won the Falklands war without Ascension Island. Ascension – only thirty-four square miles in extent – is situated 500 miles south of the Equator and 1,000 miles out in the Atlantic from the coast of Africa. It is a British colony administered as a dependency of St Helena, which is 700 miles further south. The terrain is entirely volcanic in nature; there are forty extinct volcanoes. It is like a range of old slag heaps; only one peak, Green Mountain, is high enough to catch sufficient rainfall for vegetation to grow. Before the Falklands crisis, Ascension was a quiet but secret and important place. There was a good airfield, built by the Americans during the Second World War for various purposes; it was called Wideawake Airfield after a species of local birdlife. Britain had the American space-exploration programme to thank for the good condition of the 10,000-ft-long runway in Operation *Corporate*, because the airfield was leased to Pan American airways for the supply of personnel and stores to the important satellite tracking station on the island. The few inhabitants of the island were American and British, with a handful of St Helenians; there was no indigenous population.

Ascension was brought into use as a base as soon as Operation *Corporate* started. St Helena, although having better beaches and being nearer to the Falklands, was ruled out because of its lack of a good airfield. No one in South America would have provided facilities and the fine naval base which the British had helped build at Simonstown in South Africa was now ruled out because of *apartheid*. So Ascension became the forward base for the task force, although the 'forward' was only relative; it was still a huge distance to the Falklands.

Captain R. McQueen, R.N., established the first British base organization on 6 April, with Group Captain J. S. B. Price, R.A.F., later becoming his deputy. A stream of urgent stores and personnel was soon arriving by air for the ships which had been sailed from Gibraltar or hurriedly from Britain. The pre-war number of air movements had been three per week. Pan American's manager, Don Coffey, described in a television interview what happened after he was asked if he could handle six arrivals in the first five days of the crisis.

We were pleased, of course; it would give us something to do. Well, the six aircraft ended up in the first few hours something like three hundred people and twenty-six aircraft and, believe it or not, Ascension Island went from there to become the busiest airport in the world. On Easter Sunday, we even surpassed Chicago O'Hare which had held the title for many years.

That vast number of aircraft movements – which peaked at 400 per day – included a large number of helicopter flights between the airfield and the task force ships.

The Ministry of Defence refused to give any information to the media about the activity at Ascension, partly to keep the Argentinians in ignorance of the island's true value until some defences were installed against an Entebbe-type commando raid and partly to conceal the American involvement during the Haig peace shuttle and to protect Pan American's interests in South America. But the climate of opinion has now cooled and the extent of the American help has been revealed. The American base commander, Lieutenant-Colonel W. Bryden, told Rear-Admiral Woodward over lunch that he had been ordered to give every help, 'but not to get caught doing it'. The Americans provided most of the food required at Ascension, flying the supplies in by their large C-140 transports to relieve pressure on British transport capacity. The British on Ascension thus fed well, even though their living conditions were very cramped. Aviation fuel was consumed in vast quantities and the R.A.F. aircraft at Ascension were sometimes on the verge of a shortage. American tankers kept arriving and pumping fuel ashore but, even so, all British transport aircraft flying in from Britain tried to land at Dakar in Senegal, or Banjul in Gambia, and top up with fuel to avoid drawing too heavily on the supply at Ascension. The other major shortage at Wideawake Airfield was that of aircraft parking space, particularly for the four-engined R.A.F. jets which would soon start operating there.

Most of the task force sailed south with little idea of what operations were intended if and when the South Atlantic was reached, but a meeting of senior officers on board *Hermes* off Ascension on the morning of 17 April brought an end to part of this uncertainty. Because some ships had sailed from the Gibraltar area and most of the potential landing units had sailed hurriedly from Britain, there had so far been no opportunity for all the commanders involved to meet and resolve some important matters. The general public at home believed that there was one task force – composed of warships, troop-ships and supply ships – steaming steadily south to land on the Falklands and repossess the islands. Because there were reporters on *Hermes*, many people believed that Rear-Admiral Woodward, who had brought the ships of his 1st Flotilla from Gibraltar and had then moved to *Hermes*, was the overall commander of the whole task force, but it was not as simple as that.

The chain of command for Operation *Corporate* was not taken from any existing model but it was based on recognized principles. The two guiding principles for the operation about to commence in the South Atlantic were that the ultimate authority and responsibility rested with the government of the day and that the detailed conduct of operations was placed in the hands of the service most involved, in this case the Royal Navy. Military endeavours sometimes fail because of divided or uncertain leadership. Operation *Corporate* was not about to suffer such a failure.

Governments normally direct the overall strategy of a war through a small group of senior ministers and such a group was quickly formed in London. Though it was never officially known as the War Cabinet, it will not be misleading to refer to it as such in this book. The regular members were Prime Minister Mrs Thatcher, Deputy Prime Minister William Whitelaw, Defence Minister John Nott, Foreign Secretary Francis Pym (appointed after Lord Carrington resigned over criticism of the Foreign Office's handling of the pre-invasion period) and Chairman of the Conservative Party Cecil Parkinson. The War Cabinet normally met at 10 a.m. each day throughout the war, more often in emergencies. Its role was to coordinate and direct the three separate campaigns – diplomatic, economic and military – which Britain was waging to force the Argentinians to leave the Falklands. This book is only concerned with the military campaign; the other methods were important and much effort was devoted to them, but it was the military solution that had to be fought out in the end.

A regular and vital attender at the War Cabinet was Admiral of the Fleet Sir Terence Lewin, Chief of the Defence Staff, who provided professional advice on behalf of all the services and passed on the decisions of the War Cabinet to the services. His post was one which was held in two-year terms by rotation by Royal Navy, Army and R.A.F. officers. It was only chance that an admiral held the position during this unique operation in which the Royal Navy played the dominant role and there have been no complaints that Lewin ever gave anything but impartial, tri-service advice. There was little press coverage of his position but he was an important figure. His greatest contributions were his clarity of thought, which in turn helped to force politicians to produce clear aims, and his realism. He was the only officer directly involved who had fought in the Second World War, his active service mostly being in destroyers on convoy duty. He had won a Distinguished Service Cross when H.M.S. *Ashanti*, in which he was the gunnery officer, fought off a long series of U-boat and aircraft attacks on a convoy returning from North Russia and then towed her crippled sister ship *Somali* for four days and nights until *Somali* sank in an Arctic gale. Lewin's wartime service would make him a steadying influence whenever bad news from the South Atlantic rattled the less-experienced members of the War Cabinet and of the services. The War Cabinet would be able to control every step in the escalation of operations. In the days before satellite communications, military forces were usually sent into action with 'no holds barred', but now it is possible to give instant decisions to commanders on what degree of force should be used. Task force officers state that they rarely had to wait for these decisions; their needs had usually been foreseen in London and the necessary 'Rules of Engagement' were always ready.

The next link in the chain of command bypassed the Ministry of Defence, which is an administrative rather than an operational organization, and went directly to Fleet Headquarters at Northwood, in the northern suburbs of London. The Commander-in-Chief Fleet was Admiral Sir John Fieldhouse, a fifty-four-year-old submarine specialist, a Yorkshireman. He became the operational commander of all land, air, surface vessel and submarine forces in the South Atlantic. He did this without leaving his headquarters; excellent satellite communications enabled him to exercise direct control at all times. There was some press coverage of Admiral Fieldhouse and his headquarters but much of the general public did not appreciate the extent of his responsibility or realize that he was the overall commander of the Task Force and not Rear-Admiral Woodward in *Hermes*. Fieldhouse's air commander for all R.A.F. operations south of Ascension Island was Air Marshal Sir John Curtiss, commander of 18 Group, the R.A.F. organization which controlled all maritime air operations. 18 Group Headquarters was also located at Northwood, indeed Northwood had been the old Coastal Command's headquarters before the Navy moved in with Fleet Headquarters. The land-forces side of Admiral Fieldhouse's Operation *Corporate* team had a less clearly defined role. Major-General Jeremy Moore of the Royal Marines moved from Plymouth to Northwood after 3 Commando Brigade sailed and became 'Land Force Deputy', but his role initially was more advisory than operational and he was not, at that stage, the direct commander of the land forces as Air Marshal Curtiss was of the R.A.F. units involved.

Operation *Corporate* was run from the underground Fleet Operations Room at Northwood, a long, narrow chamber, not at all impressive in appearance, with old tables, chairs and filing cabinets, and no obvious gadgetry in sight. A new wall map of the South Atlantic was installed, all surplus activities were banished to other places and Operation *Corporate* was directed with little extra cost to the taxpayer. The modern and sophisti-cated Nato Eastern Atlantic Command Operations Centre a few yards away was not used. Admiral Fieldhouse held briefings in the Operations Room each morning and evening; these could be attended by well over a hundred officers at times of peak action, 'crammed in just like Piccadilly Circus in the rush hour'. Fieldhouse's own office was alongside.

The commander of the main group of surface warships, centred around the two aircraft-carriers, was Rear-Admiral John Woodward, Cornish-born and due to reach his fiftieth birthday during the war. (The formal title of Woodward's command was Task Group 318.1 but it will be more convenient in this book to refer to it as 'the battle group', 'the carrier group' or 'the task force'.*) Sandy Woodward is the first to admit that he was lucky to be occupying this interesting position at that time, having arrived at it by a series of chance happenings in the naval-appointments system. Woodward himself should have been Flag Officer Submarines at that time but the posting had been delayed by one year, leaving him as Flag Officer 1st Flotilla. Even when Woodward took the ships of his flotilla down to Ascension, it would have been no surprise if an 'air admiral' had been sent to take over when *Hermes* and *Invincible* arrived from Britain. The Flag Officer 3rd Flotilla would have been the obvious choice but Rear-Admiral John Cox had just handed over to Rear-Admiral Derek Reffell. Another possibility was Rear-Admiral Edward Anson, Flag Officer Naval Air Command, but Admiral Fieldhouse was happy to let Woodward continue. So a submarine and small-warship specialist was put in command of a force of ships which was based on two aircraft-carriers and which would eventually cover an amphibious landing. Some officers say that Woodward's lack of aviation background was a benefit, ensuring that he viewed his command in overall terms and was not 'blinded by the aviators'. He was provided with an air staff officer sent from England and he consulted the captains of his two aircraft-carriers at least once every day.

The commander of the land forces at this time was Brigadier Julian Thompson, a forty-seven-year-old black-haired Cornishman who, as commander of the 3rd Commando Brigade with two parachute battalions and other units attached, found himself with the most interesting Royal Marine field command since the Second World War. Alongside Thompson was one of the war's 'secret' men, Commodore M. C. Clapp, the Commodore Amphibious Warfare. Mike Clapp, a former naval airman who once commanded a Buccaneer squadron, had a small staff which was responsible for planning and executing all ship-to-shore movements of troops and stores. The Royal Marines or the Army loaded the amphibious ships but it was Commodore Clapp who commanded those ships until the troops landed, the responsibility only passing back to the land commander on the 'shore

* The term 'Task Force' (capitalized) has been used to refer to the overall operations of all services, under the command of Admiral Fieldhouse. By contrast, 'task force' (lower-case) refers to the naval group under the command of Admiral Woodward, *en route* to or standing off the Falklands.

wet, shoes dry' principle after the landing. Reporters were not allowed to interview Clapp nor to mention his role; the public did not know of his existence. This secrecy was to ensure that the Argentinians received no hint of when the landing might take place. A smooth-running relationship between Woodward, Thompson and Clapp and a good deal of mutual trust and understanding were essential if the landings were to be successful.

The command ship *Fearless* arrived at Ascension early on Saturday, 17 April, only one day after *Hermes*. Brigadier Thompson and Commodore Clapp and members of their staffs were helicoptered across from *Fearless* to *Hermes*. Admiral Fieldhouse, Air Marshal Curtiss and Major-General Moore had flown out from England. The scene was set for the first conference between the British commanders. For several of them it was the first time they had ever met; Admiral Fieldhouse was the only one known to everybody. The importance of this conference, which lasted all day, should not be overestimated. Admiral Fieldhouse had already taken the most important decisions and parts of his plan were now in operation, with two groups of ships already sailing on south from Ascension. But it was important that all the leaders were personally acquainted with their commander's overall thinking and that he received the views of his subordinates.

The basic military plan was that the ships under Fieldhouse's command were split into five separate groups in order to achieve four main purposes:

The establishment of a sea blockade around the Falklands.

The repossession of South Georgia.

The gaining of sea and air supremacy around the Falklands.

The eventual repossession of the Falklands.

While these operations were developing, the diplomatic and economic campaigns being waged against Argentina would continue but it must be stressed that few military moves were ever delayed to allow more time for diplomacy. There were two reasons for this. Firstly, diplomacy had to be backed up with the knowledge that the British forces were moving fast into a position from which they could open 'the shooting war' if diplomacy failed and, secondly, once the British forces sailed onwards from Ascension, the time they could remain at sea would be limited. It would be a disaster if time ran out and the British forces had to return from the South Atlantic without fulfilling their mission.

These were the main elements in Admiral Fieldhouse's plan.

The Sea Blockade

The British Government had already announced that, from 04.00 on 12 April, a Maritime Exclusion Zone would exist in the area 200 nautical miles from the centre of the Falklands. The Argentinians claimed the move was 'an act of aggression'. There had been no declaration of war – there never was – but Britain claimed that the blockade was justified by Article 51 of the United Nations, 'nothing shall impair the inherent right . . . of self-defence if an armed attack occurs' and by the Argentinian refusal to withdraw from the islands as demanded by the United Nations resolution of 3 April. The exclusion zone applied only to Argentinian ships and was intended as a weakening move against the Argentinian supply and reinforcement position in the Falklands.

The Maritime Exclusion Zone had been in effect for five days when the *Hermes* conference took place. Its enforcement was not the business of the ships at Ascension but the application of the blockade can usefully be described here. The submarines *Spartan*, *Splendid* and *Conqueror* arrived in the South Atlantic on 11 April. *Conqueror* went to South Georgia and *Spartan* and *Splendid* took up station around the Falklands, *Spartan* watching the approaches to Stanley and *Splendid* patrolling between the Argentinian coast and the islands. Commander James Taylor of *Spartan* was the first to set eyes on the occupied Falklands when he made landfall on 12 April. It was a grey and cloudy day but the peaks of Mount Kent and Two Sisters appeared over the horizon and then the coastline around Stanley. There were no British surface vessels in the area around the Falklands; *Endurance*, hardly a blockading ship, was still hiding around South Georgia.

The Maritime Exclusion Zone which operated from 12 to 30 April was a partial success. It is probable that some Argentinian supply ships were deterred from sailing to the Falklands but at least one ship ran the blockade undetected and an 'air bridge' continued at full blast all through this period. The Argentinian naval-landing ship *Cabo San Antonio* was spotted off Stanley on four consecutive days, apparently laying mines, but *Spartan* was refused permission to attack, partly to conceal the presence of the submarine for attacks on larger targets but mainly to avoid opening the shooting war too soon and compromising the diplomatic efforts still being pursued.

The Repossession of South Georgia

The repossession of South Georgia had also been in hand since 7 April, when the destroyers *Antrim* and *Plymouth* were detached from the force of ships sailing south from Gibraltar and ordered to make speed for Ascension. They were accompanied by the tanker *Tidespring* and became Task Group 319.9. At Ascension, the ships took on arctic clothing and embarked the marines of M Company of 42 Commando, who had been flown to Ascension; D Squadron of the Special Air Service was later transferred from *Fort Austin*. This little force had sailed from Ascension on Easter Monday, 12 April, and the venture was designated Operation *Paraquet* (which was sometimes corrupted and appeared in signals as *Parakeet* or *Paraquat*) and was under the command of Captain B. C. Young in *Antrim*.

The Establishment of Sea and Air Control around the Falklands

Sea and air control was a subject of more relevance to the officers gathered in *Hermes*, because the warships at Ascension would have to carry out the plan. Before giving details, it is necessary to mention a prior event. While Woodward's ships had been approaching Ascension, a 'flash' signal from Northwood had, on 14 April, ordered five warships to top up with fuel and then speed on at 25 knots until their fuel supplies were down to 30 per cent. The purpose of this move was to get a naval force as far south as possible in case a standstill should be imposed on all military forces as part of diplomatic negotiations. Accordingly, the destroyers *Brilliant*, *Coventry*, *Glasgow* and *Sheffield*, and the frigate *Arrow*, were detached. It was estimated that, by hard steaming, they could reach a point about half-way between Ascension and the Falklands if the weather remained good. The ships were then to 'loiter' to allow the tanker *Pearleaf* to catch up and to await further orders. This group of ships was designated Task Unit 317.8.2 and was commanded by Captain J. F. Coward in *Brilliant*.

The main battle group under Rear-Admiral Woodward was to sail from Ascension as quickly as possible and attempt to establish superiority around the Falklands by drawing out the Argentinian warships and aircraft from the mainland and destroying them in battle – a prospect which both thrilled and chilled those British officers in the know. This phase was an important preliminary to any amphibious landings, all plans for which were based on the assumption that sea and air superiority would first be gained. Secondary roles for the battle group were the softening up of Argentinian positions in the Falklands by bombardment and air attack and the reconnaissance of possible landing sites. For this last purpose, part of G Squadron of the Special Air Service and most of the Royal Marines' Special Boat Service were placed under Woodward's command and five Sea King 4 helicopters of 846 Squadron were retained in *Hermes* when the rest of the squadron was transferred to the amphibious force at Ascension.

The battle group, which would sail from Ascension as soon as possible, would initially consist of the aircraft-carriers *Hermes* and *Invincible* with their twenty Sea Harriers and their anti-submarine helicopters, the destroyers *Broadsword*, *Glamorgan* and *Yarmouth*, the frigate *Alacrity* and the supply ships *Olmeda* and *Resource*. But Woodward would quickly be joined by the ships sent ahead in the diplomatic 'dash south' and later by those returning from the South Georgia operation. The battle group was expected to reach the Falklands area by 2 May and then to require two weeks to wear down the Argentinian naval and air forces and to carry out the reconnaissance work.

The Repossession of the Falklands

Much thought had been given to this last and most important part of the campaign, the repossession of the Falklands – if it should be required. No decisions had yet been taken and the *Hermes* conference spent much time discussing possible plans. There was no prospect of an immediate sailing for the amphibious ships. 2 Para in *Norland* and the second assault ship, *Intrepid*, with her landing craft, had not yet even sailed from England and would not do so for a further nine days. Commodore Clapp's ships would have to wait at Ascension for these laggards to catch up and for the battle group to carry out its preliminary operations around the Falklands. A tentative sailing date for the amphibious ships from Ascension was agreed for the end of April and the first possible date for a landing was calculated as being 16 May. There was much discussion over landing sites. These issues are described in more detail later (see Chapter 13).

Woodward's battle group sailed from Ascension on Sunday, 18 April, the day after the *Hermes* conference. The sailing was preceded by a submarine scare which lasted all morning. The supply ship *Olmeda* reported that she had seen a periscope 'feather' and a full-scale hunt followed in which *Alacrity* and *Broadsword* and at least three anti-submarine Sea King helicopters and a Nimrod aircraft took part. An underwater contact was followed to the south for some time but was later assumed to have been a whale. The main group sailed at noon and *Invincible* followed that evening after recovering the last Sea Kings from the submarine hunt.

There were now thirteen warships and four supply ships sailing south from Ascension in three separate groups. Radio silence was maintained; Nimrods of 42 Squadron provided air cover from Ascension during the early stages of the voyages. The ships heading for South Georgia reported no serious incidents and it is believed that their presence remained secret. Captain John Coward of the *Brilliant* describes what happened to the five ships sent ahead to get as far south on the way to the Falklands as possible in case of a diplomatically induced standstill.

The weather was very, very stiff. We decided to push on at an average speed of 25 knots and leave the tanker behind – he could only do 18 knots – and we were actually prepared to run out of fuel if necessary. We had not worked together before and there were some difficulties. *Coventry*'s steering motors both burnt out; she had to refuel in a rising gale, being steered manually by hand pumps. She did very well.

After about four days of hard steaming, we were almost level with Argentina. We had to be prepared for and actually expected the Argentinian Navy coming out to meet us. We were very disappointed not to meet anything; we really thought we were leading the fleet into action.

But nothing happened. *Brilliant* was ordered to join in the South Georgia operation on 22 April and the remaining ships met the carrier group three days later.

The main interest remained with Sandy Woodward's force of ships, which covered 3,000 nautical miles in the next twelve days, steaming steadily south-west, keeping about 600 miles out from the coast of South America. In theory, there should have been three time changes during this voyage, each one setting clocks back by one hour, but Admiral Fieldhouse decided that the whole of the widespread South Atlantic operation should be conducted on Greenwich Mean Time to avoid any error. This was an unpopular move with the men in the task force and some men took to wearing two watches, one showing the operational time being used and the other local time. (Argentina's time was a further hour behind the local time at Stanley and, as some of the Argentinian forces did not adjust their clocks either, it was probably the first time that two sides met on a battlefield with the clocks set four hours apart.)

There were several incidents during the voyage from Ascension. Nimrod and Hercules aircraft flew from Ascension and dropped urgent stores, mail, the detailed text of important orders and video cassettes. There were gales and more submarine scares; at least one whale had a torpedo aimed at it and suffered the blast of depth charges. Argentinian Boeing 707 military airliners made a series of day and night flights starting on 21 April, attempting to track the approach of the task force. They were successful in plotting the general position of the aircraft-carriers when Sea Harriers had to be sent up to turn them away at ranges varying between 80 and 130 miles. No shots were fired; the Rules of Engagement did not permit the Harriers to attack the unarmed airliners until 25 April after a warning had been sent to the Argentinians. The Boeings then kept their distance. There was one tragedy on the evening of 23 April, at the half-way point to the Falklands, when a Sea King 4 helicopter from *Hermes* was lost. The warships were trying to keep topped up with stores as the task force approached the Falklands. There were plenty of helicopters with the carrier force but only the five commando assault helicopters of 846 Squadron were capable of carrying heavy loads and four of these were working all that day between the warships and the supply ships. The weather and light started to deteriorate and Flight Lieutenant Robert Grundy, an R.A.F. pilot on attachment to the Fleet Air Arm, decided to return to *Hermes* to pick up a second pilot. But the Sea King never reached the aircraft-carrier. The pilot is believed to have become disorientated and the helicopter hit the sea at about 70 knots. The cockpit broke away from the main cabin and Flight Lieutenant Grundy found himself on the surface but there was no sign of his crewman. Grundy could do no more than shout and knock on the side of the main cabin, the tail of which was still floating, but the wreckage quickly sank. *Hermes* realized what had happened when the helicopter failed to arrive, a search was launched and Grundy was rescued. Three ships remained in the area for more than twelve hours but the crewman was never found. Petty Officer Aircrewman Kevin Casey was the first fatal casualty of Operation *Corporate*. (See Appendix 1 for a list of the British dead in the Falklands war.)

CHAPTER FIVE

South Georgia

One of the earliest decisions the War Cabinet had to make after the dispatch of the task force was whether to send a force to recapture South Georgia or to bypass it and wait until the main Falklands operation was completed. The dispersal of one's forces is against normal military doctrine but Admiral Lewin, as Chief of the Defence Staff, wanted an early success 'to convince the politicians that the military could do what they claimed they could do'. He was advised by Admiral Fieldhouse that sufficient forces were available and that an attack on South Georgia would not delay the main Falklands operation. This proposal fitted in well with the War Cabinet's strategy of escalating military action by a series of steps, to convince the Argentinians that Britain was determined to regain all the occupied islands, and this would perhaps induce a voluntary Argentinian withdrawal from the Falklands. The decision had been taken in London by 7 April, only four days after the Argentinians had captured South Georgia! The operation – *Paraquet* – was directly under the control of Admiral Fieldhouse's headquarters at Northwood where a small 'South Georgia cell' was created under Captain Robert Woodard.

The ships allocated were the destroyers *Antrim* and *Plymouth*, the tanker *Tidespring*, which would also act as a troop-ship, and the *Endurance*, which was now coming up from the south after being resupplied by *Fort Austin*. The six helicopters being carried on these ships – two Wessex 5s, one Wessex 3 and three of the tiny Wasp helicopters – were a vital element of the force; they were capable of carrying a miniature landing force of twenty men in one lift. The main landing force would be provided by M Company – the 'Mighty Munch' – of 42 Commando and about 100 special forces men made up of the four troops of D Squadron of the Special Air Service and No. 2 Section of the Special Boat Service. If this was correct, the British force would have a two-to-one superiority in numbers. There was also a party of twelve men from

148 Battery, 29 Commando Regiment, who would act as forward observers and liaison teams for the four 4·5-inch guns – two mounted in *Antrim* and two in *Plymouth* – which could be used to provide fire support for a landing. These units totalled approximately 230 men. These British units were about to take part in one of the most exciting, and lucky, episodes of the war.

The ships from Ascension met *Endurance* on 14 April. *Antrim* manned and cheered ship for *Endurance*, 'a gesture for her having been in the lion's jaws until then'. The 2,800-nautical-mile approach voyage from Ascension was covered in a week. Few men in the force realized that friendly eyes were scouting ahead for them. The submarine *Conqueror* had already been off South Georgia for two days on an anti-shipping patrol but had seen nothing and on 19 April Commander Chris Wreford-Brown brought *Conqueror* in to examine the north coast of the island. The suggestion that *Conqueror* landed an S.B.S. reconnaissance party here is now known to be incorrect. The honour of being first ashore would thus fall to the men of the S.A.S. *Conqueror* would remain near by throughout the South Georgia operation.

The following day, 20 April, the R.A.F. in the form of a Victor K2 tanker aircraft flew down from Ascension on a fourteen-hour reconnaissance flight to make a radar search of the area up to 250 miles out from the South Georgia coast to make sure that no Argentinian ships were in that area. This long flight was made by Squadron Leader John Elliott and his 55 Squadron crew and was the first of three long range reconnaissance flights which would be carried out by the Victors during the South Georgia operation. Each of these sorties was supported by eight other Victor tanker sorties – four each on the outward and return flights – which were early examples of the huge air-to-air refuelling effort which would be deployed during the Falklands war. Squadron Leader Elliott's flight, covering about 6,500 nautical or 7,400 statute miles, was almost certainly a new record for a maritime radar recce flight. No Argentinian ships were detected, only a lot of icebergs. One iceberg, spotted on a later flight, measured thirty-five miles by ten miles and was 500 feet high!

The British landing force arrived off South Georgia on the early morning of 21 April and were ready to start the operation. There had been regular meetings in *Antrim* to discuss the opening moves.

Four diverse characters were involved in the planning. In overall command was Captain Brian Young, a naval officer in the final months of his service. Major Guy Sheridan of the Royal Marines was supposed to be in charge of all the landing forces. But also present were Major Cedric Delves, the S.A.S. commander, who would claim a semi-independent role for his men, and Captain Nick Barker of *Endurance*, with his courteous manner and the appearance of a prosperous businessman but who was the only one with first-hand experience of local conditions. Captain Barker also had the services of Peter Stark, the most senior of the Antarctic Survey men who had remained at their outstations when the Argentinians invaded and who knew the terrain well. *Endurance*'s helicopter had picked Stark up and he was now in *Endurance* but he did not come to *Antrim* for the planning conferences. The command system did not work well and some bad mistakes were made. Few of those present realized how physically and mentally daunting conditions on land were in this part of the world, far more severe than in Norway or Canada, which was where most of the S.A.S. and S.B.S. cold-weather training had been done. There could be rapid changes of the weather; it was, says one officer later, 'a place where you could get all four seasons in a day'.

It was decided that the S.A.S. and the S.B.S. should establish observation posts overlooking the two places where Argentinian forces were likely to be – Grytviken and Leith. It was agreed that the S.A.S. should set up the northern observation posts, around Leith, and that the S.B.S. should take care of Grytviken. Major Delves startled everyone by stating that he wished to land his men on the south-west side of the island and then cross the twenty miles of intervening ground to Leith on foot. Captain Barker was appalled. This was a journey only to be undertaken by the likes of polar explorers; he forecast that the journey would take seven days. He countered with the suggestion that the S.A.S. should land on the north-east coast, a few miles one side or the other of Leith, and that the S.B.S. should land in Hound Bay, six miles from Grytviken. He did not think that the Argentinians would have outlying observation posts and these closer landings would be unobserved. There were lengthy discussions. Delves gave up his plan to trek across the island but would not land as near as Captain Barker suggested. A compromise was reached. *Antrim* and *Tidespring* would land the S.A.S. Mountain Troop – believed to be

sixteen strong, though some reports say thirteen – by helicopter on
the Fortuna Glacier, north of Leith, and *Endurance* and *Plymouth*
would land the S.B.S. men south of Grytviken. Several people
expressed doubts about the wisdom of the S.A.S. landing on a
glacier instead of on the coast; the Antarctic Survey man had
warned that wide crevasses on the glacier would prevent any
movement over it to Leith, but Major Delves persisted and this
part of the plan was allowed to stand.

The operation commenced at 13.00 local time on 21 April when
three Wessex helicopters dropped Captain Gavin Hamilton and
his S.A.S. men on the glacier at the second attempt in poor weather
conditions. This party expected to be overlooking Leith next
morning. At the same time, *Endurance*'s two Wasp helicopters
tried to take the S.B.S. party over a mountain to Cumberland Bay
but were prevented from doing so by a blizzard. The S.B.S. were
later put ashore by Gemini boats further down the coast and would
have to work their way to a point near Grytviken when the weather
allowed.

The S.A.S. men on Fortuna Glacier found that they could not
cross it as fast as they wanted and they were forced to camp on the
glacier overnight. A gale blew up. One tent was blown away. The
gale still blew the next morning and the S.A.S. men had to radio
for help. There followed a rescue operation of the most dangerous
nature. Lieutenant-Commander Ian Stanley, in *Antrim*'s elderly
but radar-equipped Wessex 3, led the two troop-carrying Wessex
5s from *Tidespring* up to the glacier in a blizzard and in conditions
of extreme turbulence because of rapidly changing winds. The two
Wessex 5s held off while Stanley tried to reach the S.A.S. men
1,800 feet up on the glacier. He was not able to do so and, after
three attempts, all the helicopters had to fly back to their parent
ships twenty miles out at sea to refuel. The next attempt succeeded
in finding the S.A.S. men, who were quickly loaded on to the three
helicopters, but the weather clamped down again and, one after
the other, the two Wessex 5s crashed. The third helicopter was
already overloaded and Lieutenant-Commander Stanley had no
option but to fly back to *Antrim* and report what had happened;
not surprisingly, the atmosphere in *Antrim* was 'gloomy'. Back
went Stanley with Major Delves, a doctor and a medical assistant,
blankets and medical supplies. Again the stranded party could not
be reached because of the weather, although radio contact was

made and it was found that no one was injured. After flying back again to *Antrim* to offload the medical men and stores and to refuel, Ian Stanley flew yet again to the glacier. There was a gap in the weather; the S.A.S. men and the crashed helicopter crews were found and all taken safely back to *Antrim* in a very overloaded Wessex.

The greatest of credit is due to Lieutenant-Commander Stanley and his crew – Lieutenant Chris Parry, Sub-Lieutenant Stewart Cooper and Petty Officer Aircrewman David Fitzgerald – whose flying ability had saved twenty lives and averted complete disaster. Stanley was awarded the Distinguished Service Order, a rare distinction for an officer of his rank. The loss of the other two Wessex helicopters was a severe setback to the South Georgia operation. H.M.S. *Brilliant* was hurriedly detached from the main task force and ordered to South Georgia with two Lynx helicopters for the eventual landing. Back in London, Admiral Lewin had been forced to go to Mrs Thatcher and report that the first operation of the task force was going wrong, with two helicopters and possibly an entire troop of S.A.S. men lost on the glacier. It was, he says, the worst moment of the whole war for him.

The Boat Troop of the S.A.S. – under Captain Tim Burls, a former Parachute Regiment officer – were tried next, being put ashore by Geminis on Grass Island, 'right under the Argentinians' noses' the next day. (This was 23 April, the anniversary of the Zeebrugge Raid according to *The Navy Day by Day*, which Commander Angus Sandford in *Antrim* consulted each day and quoted to *Antrim*'s crew when he thought an entry was relevant.) But, yet again, the S.A.S. men encountered severe problems and nearly lost some men. Two Geminis broke down and were swept away in strong winds. The men in one were recovered after a seven-hour search by *Antrim*'s redoubtable Wessex; three men in the second Gemini came ashore further up the coast where they waited until any battle would be over before activating their radio beacons to bring in the rescue team. But the observation of Leith was at last achieved, although the Argentinians also saw the S.A.S. men and would have fired on them if they had possessed a gun with sufficient range. The S.B.S. men further south, however, were not able to achieve their task and were also in trouble, although they never came as near to disaster as the S.A.S. men. The S.B.S. party, landed two days earlier, found they could not move by Gemini across Cumberland Bay; ice coming off the local glacier was puncturing their Geminis. *Endurance*'s helicopter had to pick up the S.B.S. men on the 23rd but they were landed again the next day.

The whole nature of the operation suddenly changed on the afternoon of 23 April when an Argentinian C-130 Hercules appeared after two days of undisturbed presence in the area by the British. The plane approached to within eight miles of the British force, spotting *Plymouth* and *Tidespring* and reporting them as '*Exeter* and tanker'. There were some 'itchy fingers' among the warships' missile crews but the Rules of Engagement then in force did not allow fire to be opened on this unarmed reconnaissance plane. At approximately the same time, information came from London that an Argentinian submarine was approaching the area, a report probably based on signal intelligence. The small British force now considered itself to be in considerable danger, not only from the submarine but from heavier Argentinian warships, if these were at sea, or from long-range bombers dispatched from the mainland. *Plymouth* was ordered to steam out to sea to the east, taking with her *Tidespring* and another tanker, the *Brambleleaf*, which had just arrived and was in the act of refuelling *Tidespring*, an operation that was rudely interrupted by the new orders. *Antrim* followed these ships out to sea; all moved fast, zigzagging, towing decoys and taking cover when possible among icebergs. *Endurance* was left behind; her noisy engines might draw the submarine to the main group. She resumed her old game of hiding among the icebergs off the island and looking after the observation parties which had been left ashore. The submarine *Conqueror* received the report about the Argentinian submarine and came to the southern limit of its patrol area, hoping for action, but was not able to make contact with the enemy.

The British ships remained out at sea for twenty-four hours. Captain Barker of *Endurance* tried to take the initiative while the other warships were away. He believed *Endurance* was in great danger. The Argentinian submarine *Santa Fe* was near by; an Argentinian Boeing 707 flew overhead and spotted *Endurance*, and Antarctic Survey scientists ashore reported that they had seen two Argentinian surface warships. The 'Argentinian ships' report was actually an earlier sighting of two of the British ships but

Barker later found out that Captain Horatio Bicain of the *Santa Fe* had actually had *Endurance* in his sights but had decided not to fire torpedoes; Captain Bicain had been entertained socially on *Endurance* only a few weeks before. Captain Barker consulted S.B.S. officers and an S.A.S. sergeant-major who were on *Endurance*. All were keen to have a go at sailing up to Grytviken and attempting to capture the place with the forty or so S.B.S. and S.A.S. on board who all agreed they had had enough of 'O.P.ing'. Barker could not reach *Antrim* but he could contact Northwood and he remembers with some delight the way in which he asked for the duty staff officer, only to be told that the officer concerned was at lunch. 'Please ask him to ring back as soon as possible' only brought a reply after an hour and a half's delay. Barker's proposal to rush Grytviken was refused.

Brilliant reached the area on the afternoon of the 24th, bringing the welcome reinforcement of two helicopters. The importance of helicopters in the South Georgia operation was a foretaste of what was to come in the Falklands. Captain Young was then ordered by Northwood to bring his three warships back towards South Georgia but to leave *Tidespring* well out to sea. Sunday, 25 April dawned; it would prove to be an exciting day. (Angus Sandford found that it was the anniversary of the first aerial spotting for naval bombardment, off Gallipoli in 1915.) Five helicopters were armed with various weapons and prepared for an attack on the Argentinian submarine.

Lieutenant-Commander Stanley was off first in the only Wessex remaining, hoping to find the submarine quickly with his radar. While the Wessex was up, *Plymouth* heard voice transmissions which suggested that the submarine had landed men at Grytviken (she had). The Wessex soon found the *Santa Fe* on the surface, heading north-west, whether homeward bound or off to hunt the British ships could not be known. The submarine was quite clearly inside a 200 miles exclusion zone around South Georgia, announced earlier to the Argentinians, and within the twenty-five miles danger range to British ships in which the latest British Rules of Engagement permitted instant attack on enemy units. Ian Stanley swept down to less than a hundred feet above the submarine and released two Mark XI depth-charges. These were designed for an attack on a surfaced submarine and were set to explode at a depth of thirty feet. One charge exploded near the

Santa Fe's aft steering planes; the second bounced off the deck and exploded close alongside. The submarine turned back towards Grytviken, steaming normally but leaving a small trail of oil. The other helicopters were called in to attack.

Brilliant's Lynx then appeared and dropped its homing torpedo and the other helicopters attacked with their AS12 rocket missiles. (Lieutenant-Commander J. A. Ellerbeck's crew, from *Endurance*, actually fired the first missile and *Endurance* later signalled Fleet Headquarters, asking if this was the first missile fired by the Royal Navy in time of war; Fleet later confirmed that it was. A reprinted *The Navy Day by Day* may one day contain a suitable entry for this.)

After the rocket hits, Captain Bicain ordered his crew below and, although remaining on the surface, the submarine was steered by periscope back towards the land. Bicain was later court-martialled for being caught on the surface. The submarine's reversal of course was reported to *Antrim* and Captain Young ordered *Plymouth* and *Brilliant* to join him in a rush towards the scene of action. Commander Sandford in *Antrim* remembers this.

We were very anxious, professionally, to ensure that this submarine stayed on the surface; we didn't want him to go to the bottom. War hadn't even been declared but, if we could force him to beach at Grytviken, it would be game, set and match for us. We came out of the mist and saw South Georgia, breathtaking in all its glory, for the first time. It was sheer exhilaration. We had hoisted the battle ensigns and were thundering along at 30 knots. It was a sunny Sunday morning.

The padre came up and asked what time could we have church. He was a lovely man, but he sometimes misread the situation. It was absolutely gripping stuff but very one-sided. No one was shooting at us. The adrenalin was absolutely coursing. Excitement and concentration was absolutely the order of the day; we didn't see the submarine beach, but the helicopter reported it was alongside a jetty near Grytviken, listing badly and with oil coming out. We pretty well realized that we had got him then. We knew they only had four submarines and we had put 25 per cent of those out of action. We really felt quite pleased with ourselves. And there had been no loss of life by anyone at that stage.

There was some rivalry among the various helicopters over which played the greatest part in disabling the *Santa Fe*. The view of the Argentinian captain may help. The success was really the result of shared efforts. The first depth-charges did not seriously damage the *Santa Fe*; her hull was still watertight, although some pipes inside the after end of the submarine were fractured. Captain Bicain did not dive in the interval before the other helicopters arrived in case the Wessex also had a homing torpedo. The rockets fired by *Endurance*'s Wasps did not cause serious damage; they either bounced off the submarine's casing or penetrated the thin skin of the conning tower without exploding, although an unfortunate Argentinian petty officer machine-gunner had his legs taken off by one of the missiles passing through. It was the threat of *Brilliant*'s Lynx, with its homing torpedo, which finally persuaded Bicain not to dive. If he had done so, the 'cavitation' – the extra water noise caused by the rocket holes in the conning tower – would have surely brought a killing hit by a torpedo. It was an exciting action and, because of freak atmospheric conditions, the VHF radio chatter between the helicopter pilots was heard in the ships of the main task force, several hundred miles away.

The submarine crew were seen running towards Grytviken. There was a quick conference in *Antrim* and it was decided that the time for observation-post work and caution was over. Both sides knew the location and approximate strength of the other. With the arrival of the Argentinian reinforcements in the submarine earlier that morning, the Argentinian garrison now numbered well over a hundred but most had only been ashore an hour or so, had not become accustomed to their new surroundings and were probably demoralized by the sudden disabling of the submarine. Major Sheridan proposed that a landing force be prepared from all the troops available in *Antrim* and that these should immediately be put ashore by helicopter to attack Grytviken under cover of fire from the four 4·5-inch guns of *Antrim* and *Plymouth*. To wait for *Tidespring* to come up with the main landing force – the marines of M Company – would entail a delay until the following day with, Sheridan believed, the danger of allowing the Argentinians to reorganize and improve their defences. Seventy-five men could be mustered and formed into three troops. One troop was made up of Royal Marines and a second was mostly S.B.S.; the third was provided by the S.A.S. Mountain Troop. Guy Sheridan would command the force himself; Captain Chris Nunn, the commander of M Company, and Major Delves, the S.A.S. squadron commander, were delighted to act as troop commanders.

There was an infuriating delay of three hours before it could be confirmed that the Argentinian submarine was no longer a threat. H-Hour was set for 14.45. At 14.15, a Wasp landed two Royal Artillery observers. Six helicopters then put seventy-two men ashore in two lifts, all within an hour. *Antrim* and *Plymouth* fired 235 shells, but not at Grytviken; orders had been received from London that the buildings there were not to be damaged if possible. The shelling was really a demonstration to the Argentinians of what might fall upon them if they did not surrender. Major Sheriden established a mortar position and moved his 'infantry' to within a thousand yards of the nearest Argentinians. He then suggested that one of the warships steam into the bay and threaten to fire directly at the Argentinians. *Antrim* did so and called up the garrison on the local radio frequency. A voice was heard replying in English, 'No shoot, no shoot.' There was mention of a man with no legs – the wounded submarine man – which was not understood in *Antrim*. At 17.00, just as it was getting dark, three white flags were hoisted. The Royal Marines and the S.A.S. men had not had to fire a shot. The Argentinians had not fired either. M Company, who had come all the way from England to provide the core of the landing force, was 150 miles out at sea in *Tidespring*. The Argentinian commander at Grytviken, Captain Lagos of the marines, who had only arrived that morning, was later court-martialled 'for contravening Argentina's military code by surrendering without having exhausted his ammunition and without having lost two thirds of his men'.* In his defence Captain Lagos stated that he had been ordered by his superior, Rear-Admiral Busser, not to waste lives in a prolonged defence. The outcome of his court-martial is not known. The *Junta* court-martialled several junior officers to draw attention away from the failure of more senior commanders.

Plymouth and *Endurance* sailed round to Leith the next morning and quietly took the surrender of Captain Alfredo Astiz, who had earlier radioed that he would fight it out with the fourteen marines there. Astiz was an unpleasant character. He was wanted by several countries for the disappearance of foreign nationals while he had been in command of an interrogation centre in Argentina and he tried, unsuccessfully, to lure *Endurance*'s heli-

* *Daily Telegraph* correspondent in Buenos Aires, 28 January 1983.

copter on to a booby-trapped landing pad during the surrender negotiations. In all, 190 Argentinians were taken prisoner in South Georgia, of whom eighteen were members of the submarine crew and thirty-nine were civilian scrap-metal workers. All accounts agree that these last were not fond of soap and water and had a strong odour; fourteen of them later sued the Argentinian armed forces for bringing them into danger and for 'physical and psychological harm'. The only death was that of an Argentinian submarine man, Chief Petty Officer Artuso, who was shot and killed by his Royal Marine guard who thought that Artuso was about to open a valve which would scuttle the *Santa Fe* when the submarine was being moved into Grytviken after the surrender. Most of the Argentinians were transferred to Ascension in *Tidespring* and were flown home on 14 May. Captain Astiz was taken to Britain but his status as a prisoner of war was respected and he followed his men home a month later. Twelve scientists and the two Anglia Television girls, Cindy Buxton and Annie Price, were collected from their outstations and taken to England, together with two live ducks and 'a weird and wonderful collection' of unusual scientific wildlife specimens. A disgusted M Company, which had earlier believed that it might be the only unit to see action if a diplomatic solution was reached after South Georgia was captured, found itself the only unit which did not see action; it stayed at South Georgia as a garrison for the remainder of the war. *Endurance* also remained as guard-ship, while the rest of the ships and the special forces men moved on to take part in the main Falklands campaign.

Britain had retaken South Georgia – 6,580 nautical miles (7,575 statute miles) from Britain – twenty-two days after the Argentinians had captured it. The Argentinians claimed that there was still 'bloody fighting' several hours after the surrender, with Argentinian commandos 'hunting down' the Royal Marines, and they later claimed that a British helicopter had been shot down into the sea and that a British destroyer was ablaze after striking a mine. In London, a relieved Admiral Lewin was able to tell Mrs Thatcher that the debacle of the stranded S.A.S. men on the glacier was now followed by the recapture of South Georgia without any British casualties.

CHAPTER SIX

May the 1st – Black Buck One

Five days after the recapture of South Georgia, this signal arrived at Ascension Island on the evening of 30 April.

From Air Commander

Operation Black Buck

1. Execute op Black Buck 1 AW HQ 18Gp A A A/19F/ K A A 300853Z APR 82.
2. Time on target 010700Z May repeat 010700Z May.
3. Delays in mission launch are acceptable provided that TOT is not later than 010900Z May 82.

This signal set in train a four-day period of air and naval action which would prove to be of the utmost interest and diversity, sometimes violent and often tragic, and yet, historically, it was no more than the opening of the preliminary phase in the campaign to recapture the Falklands.

Black Buck One was the code-name given to the remarkable operation in which a Vulcan bomber flew from Ascension Island to Stanley and back to Ascension, a flight equivalent to an aircraft flying from England to bomb Chicago airport and then flying back to England or, if in an easterly direction, to Western China and back. The origins of Black Buck One were to be found in the R.A.F.'s anxiety to help, in circumstances which had all but excluded that service from the war. The R.A.F.'s long-range bomber force – the V Force – had been reduced almost to a skeleton after the transfer of Britain's nuclear deterrent to the Polaris submarine force, but there were some Vulcans with trained crews still remaining and, if the means could be found to refuel these during flight, they could reach the Falklands or the Argentinian mainland. The R.A.F. suggested that these Vulcans be used to attack the runway at Stanley airfield. Admiral Lewin and Air Chief Marshal Beetham submitted the idea to the War Cabinet. The politicians had to decide whether to escalate the war in this manner and, if the Vulcans were to be used, whether to attack targets in the Falklands or in Argentina. It was quickly decided not to attempt raids against targets on the Argentinian mainland; this would stretch the British justification for military action too far. There was some anxiety over the danger to friendly civilians if the Vulcans bombed Stanley airfield but Air Chief Marshal Beetham was able to reassure the War Cabinet that the people of Stanley, two and a half miles away, would be safe. Permission was given for the Vulcans to be used.

The Vulcan attacks on Stanley's runway and the effect upon Argentinian air operations were later to be the subject of much debate and inter-service comment. It is important to record the exact intention. Stanley's runway, 4,100 feet long, was just on the limits for use by the Argentinian Super Étendard, Mirage and Skyhawk jets but, if the Argentinians did decide to use Stanley as a forward airfield, then the British task-force ships would be faced with the threat of air attack *from any direction* for as long as they remained within Sea Harrier range of the Falklands, particularly if the Argentinians used in-flight refuelling as well. If this happened, the task force would either have to provide an all-round missile and Sea Harrier defence, instead of covering the much narrower cone of threat from aircraft operating only from the mainland, or withdraw 200 miles further east, *which would prevent the Sea Harriers from covering any eventual landing*. But, if the Stanley runway could be hit near its mid point, no high-performance jet could use it until a perfect repair had been made; fast jets can be badly damaged by using an imperfect surface.

If the bombing of the runway was left to the task-force Sea Harriers, these could only bomb with any accuracy from an altitude so low as to be very vulnerable to ground fire. Each Sea Harrier could carry three 1,000-lb bombs; a Vulcan could carry twenty-one bombs and could bomb accurately from a higher and safer altitude which would also give the bombs a much greater terminal velocity and, therefore, penetration power. In short, one relatively safe Vulcan sortie was approximately equal to seven dangerous Sea Harrier sorties and the Vulcan bombs would make bigger craters. Only one Vulcan at a time could be used because of limitations in the number of refuelling aircraft available, but the Vulcan raids could be repeated with some regularity and allow the Sea Harriers to be retained for other work. The old Vulcans, soon destined for the scrap heap, could perform a useful function without robbing the task force of any resources.

The R.A.F.'s remaining Vulcans were all based at Waddington, near Lincoln. Ironically the Argentinians had tried to buy some of the Vulcans being disposed of earlier in the year but the British Government had declined to sell. The Vulcans selected for service in the South Atlantic had first to be converted to receive fuel in flight; this was not difficult because the Vulcan had – thirteen years earlier – operated in this role and it was only a question of refitting fuel probes and lines already available. An inertial navigational set, the *Carousel*, and improved electronic-countermeasures equipment were fitted, the former to help with navigation on long ocean flights and the latter to counter the Argentinian radars believed to be stationed at Stanley. The work was carried out swiftly and efficiently, mostly by R.A.F. ground engineers at Waddington. One Vulcan pilot says, 'If you hand out any bouquets, give them to the engineering guys; they did things in twelve hours that would normally have taken twelve months.' Another officer says that, 'Freed from the dead hand of bureaucracy, modifications were rushed through in one tenth of the time and much cheaper than if normal procedures were used.'

The selection of Vulcan crews for these operations was initially a low-key matter. Five crews from each of the four squadrons at Waddington were ordered to practise formation flying – an essential part of the refuelling preparation – and four of these crews were selected for further training. The feeling was that 'this was just a little bit of show, a little flurry at the end of the V Force's life; we never really thought we would ever bomb seriously'. So low was the initial priority that crews nominated to fly in air displays that summer were removed from consideration and, in one of the four final crews, the regular navigator/plotter was allowed to leave because he had booked a holiday in the United States; this was at the same time that nearly all naval personnel left in Britain after the task force sailed were told that foreign holidays would not be allowed that summer!

The four crews carried out their final training – refuelling flights over the North Sea with Victor tankers and bombing at practice ranges in Scotland and the Isle of Man. The only major problem was caused by faulty seals in the Vulcan's receiving probes, which frequently caused fuel to spray over the aircraft's cockpit. Instructor pilots from the Victor tanking base at Marham were sent to fly as Vulcan co-pilots for the training flights, and it was later decided that these should remain for the operational missions. The regular Vulcan co-pilots were unwilling to be left out and they became reserve pilots and old-fashioned flight-engineers, mainly computing fuel consumptions. Calculations on fuel being used and stocks remaining would dominate every Black Buck flight.

Three of the Vulcan crews flew to Ascension at the end of April; Squadron Leader Alistair Montgomery's crew was the first to leave England.

We came straight off our last training sortie on April 26th, to be met by a service policeman. We flew to Brize Norton by V.I.P. Andover and found a VC10 had been held back for us. We were off to Ascension within four hours of landing from the training sortie.

Before we landed at Ascension my co-pilot and I went to the flight deck to have a good look at the airfield. We naïvely thought that we would be there for one mission. We were looking forward to being the first operational crew but were disappointed; I became the detachment commander and my crew all became the section heads of their own speciality.

Squadron Leader John Reeve and Flight Lieutenant Martin Withers flew two Vulcans out to Ascension on 29 April, being refuelled twice on the way by Victor tankers. On his arrival at Ascension, John Reeve was 'impressed that we were surrounded by professionals. The best of the Air Force was down at Ascension. It was absolutely marvellous to be part of that. I was particularly impressed with the tanker operations people. There was no hanging around and waiting. We got started the next day.'

*

The concentration of interest on the Vulcans in the Black Buck operations is understandable though misleading. These were really combined Vulcan/Victor operations; in fact it can be said that the Black Bucks were really Victor operations on to which a Vulcan bomber was grafted. Both of these fine aircraft were more than twenty years old, being the product of the great leap forward in jet-engine design of the late 1950s. In terms of years of service, it was as though the Battle of Britain in 1940 was fought with Sopwith Camels! The Victors operating from Ascension were all Mark K2s, formerly a bomber but now refitted as a tanker aircraft capable of carrying 123,000 pounds of fuel (nearly fifty-five tons). The Victor crews at Ascension came from 55 and 57 Squadrons and 232 Operational Conversion Unit, all based at Marham, but individual crew identities rapidly disappeared as the detachment immersed itself in a hectic round of operations. Air Marshal Curtiss, under whose command the Victors at Ascension were now operating, says that the ability of the Victors to fly so intensively was the most unexpected bonus in Falklands air operations, especially the way the delicate hydraulics of the *Hudu* fuel-transfer mechanisms stood up to so much use. 'The more you use an aeroplane', he says, 'the fewer problems it usually gives.'

The R.A.F. had never flown anything remotely resembling this complicated operation. The Vulcan bombing plan had been prepared in England but up to a dozen experienced Victor officers took two days to complete the tanking plan at Ascension. Flight Lieutenant Dave Davenall is credited with masterminding the planning process. The main problem was the uncertain fuel consumption of the Vulcan, 'the stranger in our midst'. Squadron Leader John Reeve and his 50 Squadron crew had been told in England that they would be the primary Vulcan crew; Reeve's radar navigator, Flight Lieutenant Mike Cooper, the man who would release the bombs, was regarded as one of the best in the V Force, 'a real scope wizard'. Flight Lieutenant Martin Withers and his 101 Squadron crew would fly the reserve Vulcan. Eleven Victor tankers would take off in two waves. Two Victors would be reserves; the remainder would refuel the Vulcan and each other in diminishing numbers until the Vulcan was finally left with full tanks north of the Falklands. A Nimrod and two further Victors would take off to meet the Vulcan on its return flight. It will be

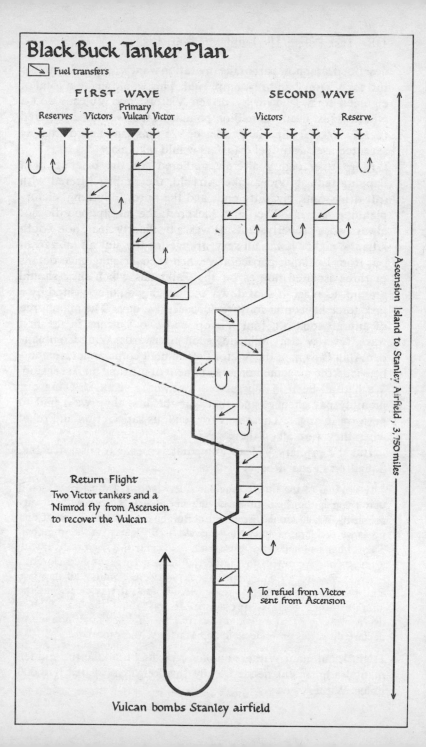

Black Buck Tanker Plan

⬒ Fuel transfers

FIRST WAVE

Reserves   Victors   Primary Victor   Vulcan

SECOND WAVE

Victors   Reserve

Ascension Island to Stanley Airfield, 3,750 miles

Return Flight

Two Victor tankers and a Nimrod fly from Ascension to recover the Vulcan

To refuel from Victor sent from Ascension

Vulcan bombs Stanley airfield

described later how part of the operation went wrong and how two further Victors had to be employed. This would make a total of eighteen sorties flown, by fifteen Victors, two Vulcans and a Nimrod. Just over two million pounds of fuel would be required (nearly 260,000 imperial gallons or 925 tons) and approximately seventeen separate fuel transfers would take place.

More than eighty aircrew gathered for the briefing in a flapping tent at Wideawake Airfield, their tables littered with soft-drink cans, cigarette ends and the notes from long Victor-planning sessions. The lights flickered; the megaphone did not always work properly. Outside was 'a brilliantly clear, hot, South Atlantic night'. It was all very strange for the Vulcan men fresh out from England, particularly when their briefing gave details of three isolated houses on the Falklands which they should attempt to reach if shot down and which would be visited by a task-force helicopter for three consecutive days. The main force of aircraft took off, four Victors and two Vulcans in the first wave, seven Victors following soon afterwards. Wing Commander Alan Bowman, the Victor detachment commander, remembers it as 'the most memorable moment of my time on Ascension. We had all been in the service for twenty years, drawing our monthly pay cheque, and that was the first time we'd had to perform in anger. There was tremendous satisfaction and relief when they were all on their way.'

But the primary Vulcan immediately became unserviceable. Squadron Leader Reeve:

It was one of the little triangular side windows. I must have closed that thing a thousand times during my R.A.F. career without any problems, but as soon as we got airborne the noise went up and up and up as we accelerated away till we could hardly speak on the intercom. The rubber seal had come loose from the frame. We tried to fix it with a polythene bag out of the ration box. Then I opened and closed it several times, to try and get it to seal. We were climbing all the time and, by the time we got to about 16,000 feet, it was clear that the aircraft wasn't going to pressurize. It was about the one fault – decompression – that we couldn't carry on with. I had no option but to declare ourselves unserviceable and Martin Withers took over.

Flight Lieutenant Withers and his crew had to make the sudden mental adjustment needed to fly this long trip of nearly 8,000 miles. Withers says:

The only adrenalin I felt all night was when I found we were taking over. The feeling soon went, we had too much to occupy ourselves with. There were aircraft all over the place. We did one orbit to get into formation and then joined the gaggle going south. I nearly formated on to a particularly bright star; then I formated on to the wrong Victor and had to ask for a red flare from the correct one. He was behind us. It took us thirty minutes to sort everything out but we were moving south the whole time.

It was ironic, after the task-force ships had unloaded all of their overseas exchange officers, that Martin Withers should be an Australian, though with a regular commission in the R.A.F.

One of the Victors also had to turn back; if a third aircraft had been forced to return, the whole operation would have had to be abandoned. But the ten remaining Victors and the Vulcan flew on, the formation being led by the 'tanker lead', always an experienced Victor captain. Squadron Leader M. D. Todd was tanker lead for the first part of the operation, handing over to Squadron Leaders A. M. Tomlin and B. R. Neal for later parts of the flight. The tanking plan unfolded steadily. Some of the refuellings took place in 'CAT' – clear air turbulence. One Victor man says that 'the planners seemed to do a marvellous job of placing a refuelling bracket smack in the middle of areas of CAT; they became affectionately known as the Duty CAT.' But the biggest problem on this flight was that the Vulcan was steadily consuming more fuel than had been estimated. Its best height was 33,000 feet but it had to keep descending to a less economical 27,000 feet to refuel from the heavier Victors.

The most serious incident occurred near the end of the outward flight, when one of the two remaining Victors broke its probe while refuelling in an air of turbulence. The Victor concerned was the one designated to accompany the Vulcan on the final stage of the refuelling operation but it could not take enough fuel from the other Victor. The two aircraft exchanged roles and Flight Lieutenant Bob Tuxford, who should have been returning to Ascension, took back the fuel he had just passed to the damaged aircraft and then flew on with the Vulcan. This setback consumed further fuel. Tuxford gave as much as he could on the final transfer but left himself short of fuel for his own return flight rather than cause the operation to be abandoned. The Vulcan was now carrying an estimated 5,000 pounds short of its planned reserves but Withers decided to press on and accept that shortage; later calculations were to show that he was actually 8,000 pounds short. Withers was later awarded the Distinguished Flying Cross and Tuxford the Air Force Cross.

Three hundred miles out from the Falklands, the Vulcan reached the point known as 'top of descent'. In a long training flight this would be the time when the crew prepared to descend for a landing with the prospect of a bath, a drink and bed. But now it was the moment when Withers and his crew prepared for what might be the most dangerous part of the flight. To stay below the cover of the Argentinian radar, the Vulcan came down to 300 feet, where it was just clipping the top of some sea fog; it was very dark. Then the Vulcan lifted to 1,000 feet at forty miles out and the Falkland hills showed up on radar dead ahead. Flight Lieutenant Gordon Graham, the navigator/plotter, had done a perfect job. (The aircraft did not even have a map of the South Atlantic; the whole operation was plotted on a map of the northern hemisphere turned upside down!) The aircraft climbed hard to 10,000 feet for the bomb run – 350 knots indicated airspeed, actual speed around 400 knots.

The bombing run was almost an anti-climax. Flight Lieutenant Bob Wright, the radar navigator who actually released the bombs, quickly found the echo of Mengeary Point headland, three and a half miles from the runway, which would be used as the 'offset aiming point' for the dropping of the bombs; the runway itself did not show up on radar. The angle of approach was critical. The intention was to get just one bomb somewhere on the runway. The stick of bombs was to be dropped at an angle of thirty degrees across the runway, which was 150 feet wide, the bombs being set to burst at intervals of 100 feet. If any part of the stick straddled the runway, one hit was virtually guaranteed, two were just possible.

The twenty-one bombs were released in just over five seconds while the Vulcan was still two miles out over Port William – the forward speed of the aircraft would carry the bombs to their target. The attack appeared to be carried out perfectly, 'just like a training run'. Flight Lieutenant Withers pulled the bomber round, away from the target. There was no anti-aircraft fire. The air electronics officer, Flight Lieutenant Hugh Prior, had earlier detected the pulses of a Skyguard gun-control radar set and had jammed it.

The bombs were dropped at approximately 3.40 a.m. local time, well inside the two-hour period allowed by the Black Buck signal from Britain. The Vulcan's flight had been plotted by radar in *Coventry*, the task-force ship acting as 'air safety cell', and her packed Operations Room was delighted to hear the code-word 'Superfuze' from the Vulcan; the attack had been carried out successfully. The news was quickly passed on to Fleet Headquarters but someone along the way inadvertently sent the wrong code-word and there was a bad twenty minutes at Northwood until the position was clarified.

The Vulcan crew still had to face the long return flight with low fuel reserves. There were also two Victor tankers still airborne and in trouble, the one which had sacrificed fuel to allow the Vulcan to fly on and another aircraft which had developed a fuel leak. Two Victors which had landed earlier from the first wave were hurriedly prepared and took off to meet the two aircraft in trouble and these were safely recovered. Two further Victors and a Nimrod came out to meet the returning Vulcan, the Nimrod placing itself half-way between the returning Vulcan and the Victor meeting it off Brazil, and guiding the two aircraft to a critical rendezvous. Everyone landed safely at Ascension. The Vulcan had been flying for nearly sixteen hours.

The Vulcan's bombing results were later to be the subject of much unjustified criticism by those who did not understand the true purpose of the mission. The flying of Flight Lieutenant Withers and the bomb aiming of Flight Lieutenant Wright had been near perfect. The bomb release had only been a fraction too late but the first bomb of the stick had struck the southern side of the runway, nearly at its mid point, and produced a large crater. Eighteen of the other bombs had exploded in a long line across the aircraft dispersal and fuel storage area as planned. One Pucará was destroyed and other aircraft were damaged. The remaining two bombs struck rocks and 'deflagrated', the cases breaking open and only part of the explosive igniting in a mild, 'low order' explosion. There were no casualties on the ground. The Argentinians made a rough repair of the Vulcan crater but Royal Engineers who had to make a permanent repair after the war vouched for the severe damage caused, ironically stating that it 'caused us far more trouble than it ever did the Argentinians'. It is debatable whether the Argentinians ever intended to use the runway at Stanley for their high-performance jets. What the Vulcan did was to ensure that the Argentinians never reconsidered this matter, not even in desperation when things were going so badly for them later. The operation was certainly a superb exhibition of airmanship by the Victor and Vulcan crews involved.

A Sea Harrier took a high-level photograph of the airfield and news of the hit on the runway reached Ascension. Among the messages of congratulations was one from the task force, in which Rear-Admiral Woodward said, 'Please can we have a raid like this every night.' Seven days later the Argentinian army authorities in the Falklands published the first issue of a soldiers' newspaper, the *Gaceta Argentina*, in which the editor, described as an army chaplain, promised that 'our first objective will be to tell the truth according to the facts'. The first entry in a 'Summary of Events' was:

1 May 1982. An unidentified enemy aircraft attacks the airfield during the night, dropping two 450-kg bombs.

CHAPTER SEVEN

May the 1st – Fleet Action

Six hours before the Vulcan's bombs were dropped at Stanley airfield, Rear-Admiral Sandy Woodward's battle group entered what until then had been the Maritime Exclusion Zone, the area 200 nautical miles from the centre of the Falklands in which Britain had declared that all Argentinian ships were likely to be attacked after 12 April. It was now 1 May and Woodward's flagship, the *Hermes*, crossed that 200 miles boundary at 01.30 Zulu (actually 10.30 p.m. on 30 April by local Falklands time), his outer ring of Type 42 destroyers being about one hour ahead, *Glasgow* and *Coventry* competing for the honour of being first into the zone. The British ships were approaching from slightly north of due east from Stanley. Twelve British warships entered the exclusion zone. The carriers *Hermes* and *Invincible* each had their immediate Sea Wolf anti-aircraft defence ships, *Broadsword* and *Brilliant*. Not far away were the general-purpose destroyers and frigates, *Plymouth*, *Yarmouth*, *Alacrity* and *Arrow*. Twenty miles out in front, 'up threat', were the Sea Dart anti-aircraft destroyers, *Coventry*, *Glasgow* and *Sheffield*. The supply ships *Resource* and *Olmeda* were also present and *Appleleaf* was not far away. *Glamorgan* had been sent away to the east to transmit radio and radar as a decoy in case an Argentinian reconnaissance aircraft appeared, but she would soon rejoin the others.

Woodward had earlier thought that the entry would be made on 2 May. His ships had been delayed by bad weather and there had been trouble with a shadowing Argentinian trawler, the *Narwhal*, which had eventually been warned off by *Alacrity*. But then the weather improved, better progress was made, and a signal came from Admiral Fieldhouse on the 29th, ordering that the entry into the exclusion zone be made on 1 May. Woodward believed that he was being pushed on by Fieldhouse in case he, Woodward, was 'getting an attack of the wobblers'. Woodward was not 'wobbling' but he did not anticipate an easy victory over the Argentinians. The previous day he had written in his private diary, 'I just hope that the politicians can pull back in time, otherwise a lot of people are going to stay out here.'

The British Government had thrown down a gauntlet five days earlier when a modification to the Maritime Exclusion Zone had been announced. From 11.00 London time on 30 April, this was to be known as the Total Exclusion Zone and aircraft as well as ships within the 200-mile area, of whatever country, were liable to attack if they were carrying reinforcements or supplies to the Argentinian forces in the Falklands. The British plan was to allow approximately two weeks of naval and air action over and around the Falklands in order to draw the Argentinian Air Force and Navy into action and defeat them before the British landing force arrived. The action had already opened with the Vulcan raid; there was to be a further combination of air action, shelling of shore positions and naval manoeuvre which, when seen through Argentinian eyes, might appear to be the prelude to an imminent landing and which, Woodward hoped, would draw the Argentinian aircraft and ships out from the mainland. In addition, men of the Special Air and Special Boat Services would be put ashore each night to find out which areas were clear of Argentinian units and to examine beaches which were being considered for the main landing. The proposed two-week interval would also allow, as with every phase of the war, further time for a diplomatic solution to be reached or for the Argentinians to realize their danger and decide to withdraw from the islands.

The Sea Harriers were to open the action for Woodward's battle group. The aircraft-carriers came in to within 100 miles of Stanley in order for *Hermes* to dispatch twelve Harriers of 800 Squadron – nine to bomb Stanley airfield and three to bomb Goose Green airfield – and *Invincible* to put up six Harriers of 801 Squadron on air patrol. The bombing raids were to take place soon after dawn and had a twofold intention: to start the attrition of the Argentinian aircraft stationed in the Falklands and to provoke the Argentinians into sending out their high-performance jets from the mainland and into combat. The Sea Harrier pilots were told that the airfield defences were strong and many of the Harrier pilots were furious about the Vulcan raid three hours earlier, which was expected to have alerted the Argentinian defences. When four Harriers were detailed to photograph the results of the Vulcan raid, there was so much opposition by the pilots that the photograph had to be taken during a later routine Sea Harrier bombing raid.

The Harrier bombing force took off from *Hermes*, the laden aircraft using their short take-off capability up the carrier's ski ramp. Naval Airman 1st Class Andrew Wroot was a member of the flight-deck party on *Hermes*.

I remember it vividly, that day it all started. This is what we had trained for, practising over and over again, for months and months and months, and this time the bombs were live. We'd come down 8,000 miles and we just knew we were going to wallop the Argies to pieces and this was the day we were going to do it. You ask me whether I was proud to be there – yes, definitely. I think we all felt that – we felt the eye of the world was upon us.

A lot of people came up on deck to try and take photographs, getting in the way. The television crews were there but we didn't mind them; that was a bit of an ego booster.

The first wave of Sea Harriers, led by Lieutenant-Commander Andy Auld, the commander of 800 Squadron, reached the Stanley area in the half light of dawn. The first four aircraft carried out 'toss-bombing' attacks on the Argentinian anti-aircraft positions believed to be situated on Canopus Hill and Mary Hill, slight rises just to the west of the airfield. The Harriers came in fast over the sea, but before reaching the coast, each aircraft lifted its nose at a signal from a computer and released its three 1,000-lb bombs which then curved off towards their targets while the four Harriers pulled safely away. The bombs took about twenty seconds to reach the target area where nine of them burst in the air, triggered by a radar altimeter fuze; two were set to explode on impact and the last bomb was fitted with a one-hour delay fuze. While the gun positions were being thus attacked, five more Harriers came in low and bombed the airfield itself, four aircraft with cluster bombs aimed at parked aircraft and stores and one with parachute-retarded high-explosive bombs aimed at the runway. (The cluster bomb was a 600-lb device containing 147 explosive 'bomblets'; the Argentinians did not like these and complained that they were indiscriminate anti-personnel weapons outlawed by the Geneva Convention. The British replied that the cluster bomb was not in existence when the last protocol of the Geneva Convention was signed in 1977.)

The whole operation at the airfield was over in one and a half minutes. Some damage was caused and fires started in the airport buildings and around positions occupied by the 5th Marine Battalion but there are no details of casualties or other damage, although an Islander aircraft was destroyed and Governor Hunt's private Cessna was damaged some time during this period. The Harriers were too low to cause serious damage to the runway but they did start the process of 'scarifying' the runway with light surface damage which would continue throughout the war. The local Argentinian forces newspaper claimed that two Sea Harriers were shot down and Argentinian television later showed film of the raid and then of the wreckage of a Sea Harrier, but this was an aircraft shot down three days later. The Harriers had been forced to fly through intense light anti-aircraft fire but only the last one was hit, a single cannon shell exploding inside the tail fin of Flight Lieutenant David Morgan's aircraft. Fortunately this was one of the few places in a Harrier where a hit would do no serious damage, only the electrical control wire to the rudder trim gauge being severed, causing the aircraft to vibrate violently at high speed. Morgan throttled back and returned safely to *Hermes*.

Meanwhile, the second force of Sea Harriers was attacking the grass airfield at Goose Green, one of the airfields to which some of the twenty-five Pucará ground-attack aircraft sent to the Falklands had been dispersed. Four Harriers, led by Lieutenant-Commander Frederiksen, caught the Argentinians almost completely by surprise; only a little small-arms fire was seen and no Harriers were hit. Two dropped cluster bombs and the third dropped 1,000 pounders. The Argentinians admitted the destruction of two Pucarás but said nothing about casualties. Eric Goss, the manager at the nearby settlement, has more information.

There had been a gradual build-up of Pucarás during the last few days and these were dispersed around the airfield when the Harriers came. The attack came early in the morning; most of us – ourselves and the Argentinians – were still in bed.

That's what we'd been waiting for. Today it's happened. We were absolutely jubilant. All through April, when people were so despondent, I developed what I call 'Harrier neck'. I had seen a Harrier display in the U.K. and I had lived in a low-flying area in Scotland; I told the people here how the Harriers would come and what they would look like. One of my shepherds who had been outside came and said, 'You were right about the Harriers; they came and went exactly as you described.'

Later, we looked up the airstrip and I think we saw five Pucarás destroyed; the Argies told us three. Seven men were killed, six air force men and one soldier on sentry duty, and they were buried at the far end of the airfield. They buried the soldier apart from the others; even in death their services were separate. An Argentinian officer later told me that another thirty or so men were killed when a bomb hit one of their personnel tents: if that was true, their remains were buried in an unidentified mass grave or dumped at sea.

One of the Argentinian dead was the pilot of a Pucará which was destroyed as it prepared to take off. The Argentinian Press later claimed that the pilot had died heroically while attacking the *Hermes*, which had suffered severe damage. The Goose Green civilians were to suffer much hardship as an indirect result of the Harrier raid, but that will be described later.

The Harriers all returned to the *Hermes* and landed safely less than an hour after taking off. Andrew Wroot was still on duty on the flight deck.

We all stood around, waiting and hoping they would all come back. Then we got the order, 'Stand by to receive two Harriers.' They came over; in the distance we could see a couple more, then others from other directions until the air seemed full of them. We tried to land them on as quickly as possible because we knew they were short of fuel. It was controlled panic. If you were there you would think that everything was going wrong – everyone running about all over the place – but really it was all under control. I think they were all down within about six minutes of the first sighting.

I remember the pilots pulling back canopies, getting out and breathing a sigh of relief; a lot of them had smiles on their faces. Everyone went to have a look at the one with a hole in the tail. Then they all disappeared, as though the curtains had come down, but we had to remain, launching and retrieving the anti-submarine helicopters, one per hour, all round the clock.

The next phase of the day's plan unfolded quickly. Ten minutes after the first bombs fell on Stanley airfield, five warships left the task force and headed towards the land. The ships were in two groups. *Glamorgan*, *Arrow* and *Alacrity* were a gunfire group, on their way to bombard six targets around Stanley. *Brilliant* and *Yarmouth*, together with Sea Kings from *Hermes*, were to steam round to about twenty miles off the coast north of Stanley to carry out an anti-submarine sweep in an area where an Argentinian submarine was suspected.

The anti-submarine group was in action first. As soon as it reached its area of operations, three Argentinian aircraft appeared but they were chased away by two Harriers, an incident which will be described in more detail later. Three Sea Kings of 826 Squadron joined *Brilliant* and *Yarmouth*. These were too large to land on the ship's helicopter platforms but each helicopter winched down a spare crew, refuelled by a pump while hovering, and then commenced a submarine search which lasted the remainder of the day. A contact was soon gained and many depth-charges were dropped by the ships and by the helicopters; a suspected oil slick was seen and there were high hopes that the Argentinian submarine *San Luis* had been sunk. The *San Luis* may have been in this area; she later reported making a torpedo attack on a British ship on this day but there is some doubt about this; she was not resolutely handled during the war and may have been keeping well out of the way of trouble.

Meanwhile, *Glamorgan*, *Alacrity* and *Arrow* had commenced their bombardment. They approached the shore, large battle ensigns flying, and opened fire – an old-fashioned naval sight which few on the three ships ever thought they would see in their service careers. There were six targets. All three ships were to shell the aircraft parking area at the airfield, then switch to individual targets: the road from the airfield to Stanley and suspected radar stations and coastal gun positions north and south of Stanley. There was no evidence that the gun and radar positions were actually there; these were merely the places where such defences might be located and which would be bombarded if a British landing was about to take place. The ships formed a 'gun-line' between 12,000 and 16,000 yards out from the coast and their four 4·5-inch guns – two on *Glamorgan*, one each on the frigates – were soon pumping shells towards the land. *Arrow* claims to have fired the first shot. The ships' radar screens clearly showed a 'conveyor belt' of shells arcing towards the targets. The Argentinian forces newspaper later acknowledged that shelling fell on positions occupied by the 25th Regiment, though without causing serious losses. But the Argentinian troops were to be steadily worn down by naval shelling and a desire by the Argentinian soldiers to stay under cover day and night was born that day and this 'foxhole complex' would intensify as the British shelling and air attack continued.

Two of the helicopters working with the bombardment force had exciting and dangerous brushes with the enemy. *Alacrity*'s Lynx had flown round to Kidney Island, four miles north of Stanley airfield, from where it could correct the fall of shells on the airfield targets. As the helicopter pilot, Lieutenant R. B. Sleeman, climbed away from the shelter of the little island to commence his spotting, he found himself under fire from a launch just the other side of the island. The machine-gunner in the Lynx fired back. Both sides scored hits. The Argentinian gunner was seen to fall; the forces newspaper recorded that he was a Coast Guard man and that he was injured. The Lynx was hit several times, some bullets passing through the cockpit canopy within inches of the pilot's head.

A Wessex gunship helicopter of 845 Squadron, armed with AS12 missiles and with machine-guns, had been attached to the bombardment force in case the shelling forced any ships out of Stanley harbour. It particularly hoped to catch a submarine on the surface, just as *Antrim*'s helicopter had caught the *Santa Fe* off South Georgia a few days earlier. The Wessex was now hovering off Stanley, hoping for a target. Lieutenant David Knight was the pilot; this is the account of Petty Officer Arthur Balls, his missile operator.

We suddenly saw anti-aircraft fire start up around the airfield, very impressive. I saw two Mirages appear, flying south, just inland from Stanley. One of them banked into profile, was hit and crashed into the sea. It was definitely an 'own goal', shot down by their own side.

The firing continued, lots of puffs of brown smoke with bright flashes inside. Then one of the flashes got bigger and I saw it coming towards us, about half a mile from us. I thought it was a shell exploding at first but then I realized it was a missile. Two or three minutes later, I saw another one; this one came high towards us from out of the middle of the anti-aircraft fire. I lost sight of it from the front cockpit but the men in the back reported it coming much closer; they were getting quite excited. So it was time we did something; we turned round and ran away. The missile disappeared into cloud but then appeared again. The men in the back got really excited then; they thought it was going to hit us. It was so near that Sergeant Tommy Sands, our marine machine-gunner, instinctively went as though to put his fingers in his ears, but he'd got a bone dome on and I think his fingers were numb for a while. He told us the story and we had a good laugh about it after-wards.

The missile didn't hit us; we thought it reached the end of its flying time and fell in the sea. They were probably Tigercat missiles. We think the first was visually guided but the flight path of the second one seemed much smoother and we think that one was radar-guided. We were very anxious to get out of it then; the pilot radioed back that he wanted to go back to *Glamorgan* and refuel but the voice came back, cool as a cucumber, 'Sorry, mother is under air attack.'

It should be stated that there were now many things happening at the same time around Stanley. The incident of the Argentinian aircraft shot down by its own side will be described later.

The Argentinian Air Force had come out in force from the mainland, drawn by the three ships sitting off the coast and bombarding the shore. Three Daggers found the three British ships and each aircraft selected a ship for attack. (The Dagger was an Israeli copy of the French-designed Mirage.) The Daggers came round a headland and attacked along the line of ships. *Glamorgan* got off a Sea Cat missile which missed; the other ships could not fire, either because the crossing rate of the approaching aircraft was too fast for their Sea Cat missiles or because other ships were masking their fire. The Argentinian aircraft made one swift attack and then were gone. Two bombs fell either side of *Glamorgan* and two more just astern of *Alacrity*. *Arrow* was hit by cannon fire and Able Seaman Ian Britnell was wounded by a splinter, the first British battle casualty of Operation *Corporate*. It was a very frightening episode and a rude awakening for the British ships to the dangers of operating by day within range of the Argentinian air cover.

The ships stopped firing at the shore targets and Captain Barrow of *Glamorgan* ordered the force away from the coast.

The events around Stanley had drawn Argentinian aircraft into action in considerable numbers and there had been several interesting and significant encounters. The early action was all inconclusive. The Argentinians sent a series of Mirage formations, not to attack the British ships but to fly high over the Falklands and tempt the Sea Harriers into action. The two sides did not meet because the Mirages remained high and the Harriers low, the altitudes at which each aircraft performed best; the Mirages ran short of fuel and disappeared. This was an early action which took place just before the Harriers from *Hermes* took off to bomb Stanley airfield. There had been a major scare in the carrier group while the Harriers were away on the bombing raid. A pair of Harriers from *Invincible*, on routine patrol, were directed towards aircraft which, when spotted, were believed to be a Super Étendard and a Mirage. One of the Harrier pilots suddenly saw a large missile in flight and reported it to the carrier group – about eighty miles away – as an Exocet. This caused a great commotion; most officers had seen film of a trial firing of this French missile in which an old warship had received a massive blow when struck by the missile. Ships turned their sterns to the threat, increased speed and fired off clouds of chaff. But the missile was a Matra missile aimed at the Harriers; it nearly scored a hit, passing only about fifty feet away from the Harrier which had sent the report, leaving a much shaken and embarrassed Fleet Air Arm pilot to return safely to *Invincible* and apologize for the false alarm. This encounter, brief and unsatisfactory as it was, was not only the first air combat of the Falklands war but the first ever for the V/STOL Harrier aircraft.

It was four hours later that the next encounter took place when two Sea Harriers were directed on to three Argentinian planes which had taken off from Stanley airfield to attack *Brilliant* and *Yarmouth* which were carrying out their anti submarine sweep north of Stanley. (This was the incident mentioned briefly earlier.) The Argentinian aircraft were initially logged as Pucarás but they were actually Turbo-Mentor light-attack aircraft of the Argentinian naval air arm. Lieutenant-

Commander Nigel Ward, commander of 801 Squadron, and Lieutenant Mike Watson attacked; there was a brisk burst of cannon fire and some wheeling in and out of cloud until the Argentinian pilots recognized that their slow, bomb-laden, propeller-driven aircraft were no match for the Harriers. They jettisoned their bombs and dashed back to the cover of the Stanley defences. The Harrier pilots had not used their Sidewinder missiles; Lieutenant Watson says, 'It was not considered the done thing at that time to waste an expensive Sidewinder against a cheap, propeller-driven aircraft. It was a pity; I could have become an ace on that first day.' Watson was never again to be presented with such a favourable opportunity. Soon after this, these two Harrier pilots attempted to engage some high-flying Mirages. The Argentinian aircraft came screaming down at a speed which far outmatched that of the Harriers and fired off their missiles. It was the classic single, high-speed pass made by fighters with the advantage of height. The Mirages did not stay to fight; they had expended their missiles and were short of fuel. But the Argentinian missiles were inferior to those used by the Harriers and the Harriers were not hit.

There was again a lull through most of the afternoon but then the Argentinian command on the mainland decided that they had to make a decisive effort to come to the help of the units on the Falklands which were being bombarded. A force of approximately forty aircraft was dispatched – Canberra bombers, Dagger and Skyhawk fighter-bombers and Mirages acting in the fighter role. The first of these aircraft arrived over the Falklands about 4.15 p.m. local time.

The Argentinian aircraft were detected by the forward screen of British ships, and Sea Harriers from both aircraft-carriers were sent into action. Many of the pilots were now flying their third sortie of the day. The first pair of Harriers to make an interception were flown by Flight Lieutenant Paul Barton and Lieutenant Steve Thomas. They were directed by *Glamorgan* on to two Mirages and a missile combat then took place at around 12,000 feet over the northern coast of East Falkland. The result was a clear success for the British pilots. Flight Lieutenant Barton, of *Invincible*'s 801 Squadron, gained the first air success of the war when his Sidewinder struck a Mirage which disintegrated in a mass of flames and fell in two parts. The Argentinian pilot, Lieutenant Carlos Perona, had an amazing escape when he fired his ejection seat and parachuted. He was doubly fortunate to come down in the sea so close to the coast that he was able to walk ashore.

The leader of the Argentinian formation, Captain García Cuerva, was not so lucky. Lieutenant Thomas's Sidewinder chased him into a cloud and the Harrier pilot could only claim a possible. Cuerva evaded the missile for some time but the Sidewinder's proximity fuze eventually exploded the missile sufficiently close to damage the Mirage. Cuerva decided to land at Stanley instead of risking the long flight back to the mainland. He made a slow and careful approach from the west, flying right up Stanley harbour. He was spotted by the Argentinian anti-aircraft gun positions situated on the high ridge behind the town. The aircraft recognition of the gunners was poor and the Argentinians opened fire with gusto and shot the Mirage down and poor Captain Cuerva was killed. There was much dismay among the Argentinians and considerable rejoicing by the Falkland civilians who had seen the incident when the identity of the shot-down plane was discovered. The Argentinian soldiers' newspaper passed the incident off as a collision between the Mirage and a Sea Harrier and stated that the Argentinian pilot survived.

The Mirages engaged by Barton and Thomas may have been trying to clear the way for the main Argentinian effort which came ten minutes later. There were at least two formations of Daggers and one of Canberras. Three of the Daggers found and bombed *Glamorgan*, *Alacrity* and *Arrow* (as described on page 132). A second group of Daggers met the Sea Harrier pair of Flight Lieutenant Tony Penfold and Lieutenant Martin Hale, flying from *Hermes*, the Harriers being at 20,000 feet and the Daggers about 10,000 feet higher. Hale was chased by a missile and only lost it after a steep dive, but then Penfold shot down one of the Daggers with a Sidewinder. Its pilot, Lieutenant José Ardiles was killed; he was a cousin of the Argentinian footballer who played regularly with Tottenham Hotspur.

The slower Canberras, six of them, approached the Falklands at high level and then lost height to get below the radars of the British ships. But they left this manoeuvre too late and were detected. *Invincible*'s operations room plotted the projected course of the Canberras; it led directly to the position of the two British aircraft-carriers. Two further Sea Harriers up on patrol were vectored on to the Canberras and, in a particularly neat operation which reflected the good training of the Harrier pilots, Lieutenant-Commander Mike Broadwater and Lieutenant Al Curtis of 801 Squadron picked up the Canberras on their own radars and intercepted one flight of three aircraft while still 150 miles or so out from the carrier group. Lieutenant Curtis shot down one Canberra, whose two-man crew were seen to eject, and the others turned back. The Harriers could not follow; they were too low on fuel. The two Argentinian crewmen were never found. The other three Canberras were not seen; they did not reach the carrier group. The Canberra shot down was one of a batch sold to the Argentinians in 1970/71 and had previously served with several R.A.F. squadrons.

The Argentinian operation had failed. The Sea Harriers had intercepted three Argentinian formations and shot down three aircraft – a Mirage, a Dagger and a Canberra; a second Mirage had been damaged and then shot down by its own side. Four Argentinian airmen were either dead or would die undiscovered in their dinghies at sea. The Harriers had not received a scratch. The outer ring of the British defences, the Sea Harriers, had succeeded in holding the Argentinians at bay except for the three Daggers which had attacked the bombardment ships. The air defence had been coordinated in *Invincible*'s operations room but most of the early warning and forward control had been carried out by *Coventry*, *Glamorgan* and *Brilliant*. The forward fight controllers on that day were Sub-Lieutenant Andy Moll in *Coventry*, which did much of the distant work, and Lieutenant-Commander Paul Raine in *Glamorgan* and Lieutenant-Commander Lee Hulme in *Brilliant*, the inshore ships which were involved in the air action around Stanley. Commander Chris Craig of *Alacrity* remembers listening to *Brilliant*'s fighter controller: 'He certainly lived up to his ship's name; he was ice cool and clear throughout that first day. I was very impressed. Listening in on the air-defence net, you soon realized which ships' operations rooms were good.'

All of the British units had performed well but the Sea Harrier pilots had emerged with the greatest distinction. Captain David Hart-Dyke, *Coventry*'s captain says:

The Argentinians were clearly testing our resolve and our air defences that day but they came upon some very determined Sea Harriers. I could hear the voices. Our pilots; you could detect the uncertainty and apprehension, going into battle for the first time, it was not surprising. I could also hear some of the Argentinian voices – our Spanish-speaking officers on board were listening. The Argentinians were apprehensive and excitable – even panicky sometimes – and they never gained any confidence in the next few weeks and they were always very apprehensive about our Sea Harriers. But, after that first day, when we came out on top, the voices of our pilots steadied down and their confidence returned. It was then that I thought that we are going to win this war.

The Sea Harriers had done well despite their performance being inferior to that of many of the Argentinian aircraft against which they had fought. The Harrier's 'viffing' manoeuvre, in which the aircraft could suddenly lose speed or vary its direction by swivelling its thrust nozzles, had not been used. The British superiority was achieved through having better missiles – the Sidewinder AIM-9L was superior to the Matra missiles being used by the Argentinians – better 'softwear' systems in their aircraft computers, and much better pilot training. The first two factors quoted above, the poorer quality of the Argentinian armament and equipment, reflects a common failure of many air forces, with a concentration of attention on the aircraft itself to the detriment of its weapons, the delivery of which is the sole purpose of the aircraft's existence. The third factor, the better training of the British forces constantly drilled in the Nato environment, was a feature of the war which would be seen again and again in following weeks.

The burst of air action on that late Saturday afternoon had lasted just thirty-five minutes. It would prove to be the full extent of the air combat phase, in which the fighter aircraft of both sides fought each other, of the whole Falklands war. The Argentinians never came out to fight again, only to attack ships and ground targets when they thought they could avoid the Sea Harriers. Lieutenant-Commander Brian Haigh, of 801 Squadron, says:

We had had lots of training for that day and it had all gone well; for a fighter squadron, that's what it was all about. We felt very elated at the end of it, very pleased with the results. We felt that we had upheld the traditions set for us in the past by the Fleet Air Arm.

Darkness came and the final actions of the day took place. The bombardment ships were ordered in again, to complete the firing programme which had been interrupted by the Argentinian air attack. Captain Barrow, in charge of the group, says:

We went back in again that night, to show that we were not deterred; it was all a question of showing resolve at that stage, right from the start. So, the Three Musketeers were back on the gunline again by 22.40 Zulu – with some trepidation, may I say. I think our names – *Glamorgan*, *Alacrity* and *Arrow* – were the only ones Sandy Woodward knew, we did it so often, but other ships took turns later.

The Argentinians announced that one man was killed and five were injured on Sapper Hill by the gunfire that night but the bodies of five Argentinian soldiers, whose graves were marked '1 May', were later found at Stanley Cemetery. These soldiers were the first fatal casualties suffered in the war by the Argentinian army, exactly one month after the first Argentinians landed at nearby Mullet Creek. The dead men are believed to have belonged to the 25th Infantry Regiment. The bombardment programme was completed in two hours and the three ships sailed back to rejoin the carrier group and to refuel and refill their magazines.

The main group of British warships had been steaming northwest in order to get into position for the next operation, the landing of special forces reconnaissance parties. Details of these operations were kept secret but some deductions can be made. The operations were directed by Colonel Richard Preston of the Royal Marines in *Hermes*, who 'ran' all the special forces in the pre-invasion phase. The helicopter operation was led by Lieutenant Nigel North and it is probable that four Sea Kings of 846 Squadron from *Hermes* were used to land men of G Squadron Special Air Service and from the Special Boat Service. The landings on this night probably took place on the north coast of East Falkland at various places between Stanley and the mouth of Falkland Sound. All went well and the helicopters were back on *Hermes* before midnight local time. The whole task force then turned away to the south-east to resume its position east of the islands, ready for the next day's operations.

It had been an exciting day for the Falkland civilians. The Vulcan and Sea Harrier attacks on Stanley airfield caused the utmost excitement. The civilians had gone to bed the previous night not knowing where the task force was or when any action would commence. The Vulcan raid came without any warning and only those people who happened to be awake actually heard the bombs explode, the salvo of detonations being heard as one long 'WHUMPF'. Most people were awoken by the subsequent shock waves, which were violent enough to twist doors out of alignment in houses at the eastern end of Stanley, more than two miles from the airfield. The faint noise of the Vulcan flying away was then heard, well before the first Argentinian guns opened fire. This was a source of much amusement to the Falklanders, who had been promised at least five minutes' warning by the Argentinians of any air raid. It was assumed that Harriers from the task force had carried out this bombing, until the B.B.C. World Service announced that a Vulcan had flown from Ascension Island. Not many people saw the dawn Sea Harrier raid either; it was over too quickly. The airfield was just too far away from Stanley for the civilians to see much more than a few flashes, to feel the shock waves of the bombs again, after which some smoke appeared. Two ladies, Suzy Packer and Nidge Buckett, found a record of 'Land of Hope and Glory', played it at full blast, called the Argentinian operator on duty at the telephone exchange and placed the phone by the record-player.

The Argentinians were observed to become progressively more nervous as the day wore on. The guns on the high ground behind Stanley opened fire on the slightest pretext and eventually shot down one of their own planes. Soldiers were seen to fire rifles, machine-guns and even revolvers blindly into the air, even when the British ships were shelling targets around Stanley. Myriam Booth, who had manned the British Antarctic Survey office until turned out of it by the Argentinians, remembers her feelings on that day.

Apprehensive and exhilarated – a combination of feelings which I hope never to have to experience again. It was great, but were we going to get out of it alive? Now that the action had started, we thought it would all be over within a few days. I couldn't see how it was going to develop but a lot of people really thought that the end would soon come.

The reaction of the Argentinians at Goose Green, where the Sea Harrier attack on the airfield was particularly effective, was harsh. The entire local population – 114 people at that time – was rounded up and confined in crowded conditions in their community hall, most of them for the next twenty-nine days. The experiences of the Goose Green people during this period will be described in a later chapter.

Some people in outlying settlements, where there were no Argentinians, had listened intently on their radios and heard snatches of conversation from the task force. Richard Stevens at Estancia House heard a ship talking of 'hostiles' and 'caps' and had to wait until after the war to learn about combat air patrols. He heard another voice, probably in *Glamorgan*, saying, 'It's getting hot,' and a later voice saying, 'They are not likely to return; go back in and continue your task.' Immediately after the local doctor's hour, an appeal in both Spanish and English was heard from the task force, asking anyone who was listening to contact the Argentinians. The voice is described as being 'very correct but not colloquially fluent' when speaking Spanish. Two farms in West Falkland were heard acknowledging the messages but it was obvious that the Argentinians at Fox Bay had no intention of passing the appeal on to the headquarters at Stanley.

This situation resulted in an incident of the most bizarre nature. The Spanish-speaking Pitaluga family, at Salvador Settlement on East Falkland, had listened to all the exchanges. Teenage Saul Pitaluga takes up the story.

I suggested to my father that I went by motorbike to Douglas Settlement and try to get through to Stanley by telephone. Father contacted *Hermes* and asked what the message was. They said, 'The Argies should pack it in and surrender to avoid any further bloodshed.' They offered to send in a Sea King at 16.00 to pick up the military governor and the commanders of the three services to discuss terms.

I reached Douglas easily. It was only seventeen miles and good going; the camp was dry that day. I rang from the manager's house to my aunt at Teal Inlet – it was always necessary to make connection by stages – but I only got through as far as Estancia. The Argentinians had cut the line from there to Stanley. I called back on the radio to father and he told *Hermes*. There were more exchanges but the Argies must have been monitoring them all and on my way back home, I was buzzed by a Pucará; they didn't like people travelling between the settlements.

That afternoon, the Pitaluga home suffered a 'mini invasion' by forty Argentinian troops who arrived in two helicopters. Saul and his father Robin were both questioned closely about the morning's events. Their radio was removed and Mr Pitaluga was taken back to Stanley for further questioning at the Police Station. He takes over the story.

I had to write the whole thing out including, verbatim, the message from *Hermes*. There was much verbal abuse and threats, including one gorilla-like army captain, probably from their military police, who never did anything without screaming and shouting. I really did think he was going to shoot me at one stage. He stood me on the sea wall outside the Police Station and put a gun to the back of my head and started clicking the trigger mechanism. Then he pushed me into a trench with two soldiers who were told to shoot me if I tried to escape, but I soon settled down and talked to them in Spanish.

So ended a day of fascinating incident which had started with the Victors and Vulcans of the Black Buck operation taking off from Ascension Island and, through an unbroken chain of events, ended with Robin Pitaluga being threatened with execution on Stanley waterfront. Mr Pitaluga was released the following day but was not allowed to return to his family. He was forced to stay, under supervision, at the Upland Goose Hotel in Stanley. This was not an undue hardship; Mrs King at the Upland Goose was his sister!

CHAPTER EIGHT

Belgrano

The British task force had, temporarily at least, easily defended itself against the best effort which could be mounted by the Argentinian air forces. But what of the considerable Argentinian navy, headed by the most belligerent member of the *Junta*? The British hope was that the Argentinian navy could either be persuaded to stay in port permanently or would come out quickly to be met in battle. It was with this policy of 'contain or fight' in mind that the Swiss Government was asked to convey the following statement to the Argentinian *Junta*.

In announcing the establishment of a Maritime Exclusion Zone around the Falkland Islands, Her Majesty's Government made it clear that this measure was without prejudice to the right of the United Kingdom to take whatever additional measures may be needed in the exercise of its right of self-defence under Article 51 of the United Nations Charter. In this connection, Her Majesty's Government now wishes to make clear that any approach on the part of Argentine warships, including submarines, naval auxiliaries or military aircraft, which could amount to a threat to interfere with the mission of British Forces in the South Atlantic will encounter the appropriate response.

The statement was handed to the Argentinians in Buenos Aires on the afternoon of Friday, 23 April, and the full text released to the world's press the following day. The implications of the statement should have been clear enough. The British were stating that any warship or aircraft was likely to be attacked if it threatened the safety of the British ships approaching the Falklands war zone, *even if the Argentinian ship or plane was outside the 200-mile exclusion zone already announced around the Falklands*.

The world's press did not pay too much attention to this statement, concentrating instead on the recent news from South Georgia, which had just been retaken by the British. This omission was unfortunate and was to lead to much misunderstanding and recrimination around the world over the events of the second day of operations around the Falklands. At the risk of being tiresome, let it be spelt out again what the various statements made by the British to the Argentinians implied. Any Argentinian ship or plane *inside* the 200-mile zone *would* be attacked. Any ship or plane outside the zone *might* be attacked *without further warning* if it was considered to be a threat to the British forces. There is no reason to believe that the Argentinian leaders did not understand exactly what the British statements intended.

Admiral Anaya accepted the challenge. His fleet had returned from the initial operation to the Falklands and South Georgia in early April and exercised for three days in the middle of the month. On 27 April, when the British task force was judged to be approaching the Falklands, the main Argentinian warships sailed again from their base at Puerto Belgrano, making south towards the Falklands. The commander at sea was Rear-Admiral Allara in the aircraft-carrier *Veinticinco de Mayo*. Also dispatched were one cruiser, some seven destroyers, three frigates and three or four tankers. The *Veinticinco de Mayo* – sometimes called *Mayo 25* – was the former British H.M.S. *Venerable*, which served in the Pacific during the closing stages of the Second World War and was then sold to the Netherlands Navy in 1948 to become the *Karel Doorman*. Sold again in 1968, it was now the most important unit in the Argentinian Navy, modernized with an angled flight deck and equipped with eight Skyhawk attack aircraft and some reconnaissance aircraft. The second largest ship, the 13,645-ton *General Belgrano*, was the ex-American *Phoenix*, which was at Pearl Harbor when the Japanese attacked in December 1941 and which saw much subsequent war service. She was modernized and sold to Argentina in 1951 but was now an old ship; she was, however, equipped to carry Exocet missiles, which have a range of more than twenty miles. The most modern of the destroyers were the British-designed Type 42 anti-aircraft ships *Santísimu Trinidad* and *Hércules*, sister ships of *Coventry*, *Glasgow* and *Sheffield*. (One of the Argentinian 42s – the *Hércules* – had been built by Vickers shipyard at Barrow-in-Furness, the other under licence in Argentina. The Argentinians still owed Williams & Glyn's Bank

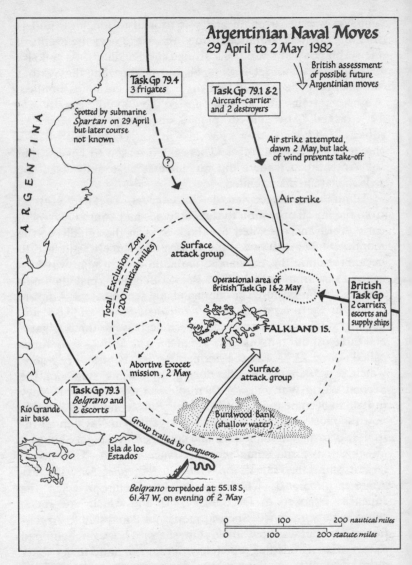

Argentinian Naval Moves
29 April to 2 May 1982

British assessment of possible future Argentinian moves

Task Gp 79.4
3 frigates

Task Gp 79.1 & 2
Aircraft-carrier and 2 destroyers

Spotted by submarine *Spartan* on 29 April but later course not known

Air strike attempted, dawn 2 May, but lack of wind prevents take-off

Air strike

Surface attack group

Operational area of British Task Gp 1 & 2 May

Total Exclusion Zone (200 nautical miles)

British Task Gp
2 carriers, escorts and supply ships

FALKLAND IS.

Abortive Exocet mission, 2 May

Surface attack group

Task Gp 79.3
Belgrano and 2 escorts

Río Grande air base

Burdwood Bank (shallow water)

Group trailed by *Conqueror*

Isla de los Estados

Belgrano torpedoed at 55.18 S, 61.47 W, on evening of 2 May

ARGENTINA

| 0 | 100 | 200 nautical miles |
| 0 | 100 | 200 statute miles |

in London £3·8 million for that deal. If the Argentinians defaulted on the payment, the British Government, under an export guarantee scheme, would have to pay the bank. In the event, the Argentinians quietly paid the sum in full only one month after the war's end.)

The main group of Argentinian ships, centred around the *Veinticinco de Mayo*, was disposed north of the Falklands. The cruiser *Belgrano*, with two destroyer escorts, was sent south. It is important to study the role intended for the *Belgrano*. The Argentinians later claimed that she was never intended to be an attack unit but was screening the area between the Falklands and Tierra del Fuego, ensuring that the Chilean Navy and any British ships or ships sent from New Zealand or Australia did not come round Cape Horn and join Admiral Woodward's task force. She was also, say the Argentinians, acting as forward protection against any British air or naval attacks on the Río Grande air base or the Ushuaia naval base, in Tierra del Fuego. Thus, claimed the Argentinians, the *Belgrano* posed no direct threat to the task force. This all seems reasonable, but other points should be considered. Firstly, the Argentinian *Junta* lied from first to last during the war. Secondly, the British commanders had no way of knowing the Argentinian intention. Thirdly, the Argentinian commanders were free to change their plans and send in the *Belgrano* or her destroyer escorts for an Exocet attack on the task force if the weather clamped down and prevented British air patrols from operating.

The Argentinian ships were all in position by 29 April, two days before the British task force arrived. Admiral Lombardo was told early on 1 May that the British ships were now within Sea Harrier range of the Falklands. The *Veinticinco de Mayo* moved in and, that evening, one of her reconnaissance aircraft picked up the radar echoes of one large and six medium-sized ships; this was the British carrier group, 300 miles to the southeast of the Argentinian ships. The Argentinians closed further, armed their Skyhawks with bombs and prepared for a take-off just before dawn. The British had three submarines in the area, all watching for just such a move as this. *Splendid* and *Spartan* were covering these north-western approaches but neither spotted the Argentinian aircraft-carrier group, although *Splendid* had earlier seen the Type 42s and other ships. But a Sea Harrier, flying a routine night patrol, detected the approaching threat soon after midnight, local time. Flight Lieutenant Ian Mortimer picked up the emissions of the radar sets fitted to the Argentinian Type 42s and then the radar echoes of the Argentinian carrier and her escorts. An intensive series of Harrier patrols was sent out in this direction during the following hours but the Argentinian air strike was unable to go ahead; there was too little wind to enable the loaded Skyhawks to take off with sufficient fuel to reach the British ships, now 180 miles away. Admiral Lombardo turned back to await a better opportunity.

It is ironic that the British had tried to sell Harriers to the Argentinians and had arranged a demonstration when their newly purchased *Veinticinco de Mayo* was sailing through the Channel from Holland in 1969, but the Argentinians had declined. Now the Skyhawks, bought instead of the Harriers, were stranded on the decks of the Argentinian carrier while the Harriers – with their short take-off capability – were able to operate. The Argentinians tried to launch one air attack from the mainland during the day, using two Super Étendards from the Río Grande air base. One was fitted with one of the few airborne Exocet missiles which the Argentinians possessed. But the in-flight refuelling which these heavy aircraft needed to be sure of reaching the British ships failed early in the flight and this operation – the only air operation from the mainland on this day – was aborted.

The *Belgrano*, meanwhile, was steaming to and fro between Tierra del Fuego and the Falklands, keeping outside the 200 miles zone. The British knew all about her; the submarine *Conqueror* had been in visual contact with the cruiser since early the previous afternoon and was reporting its movements both to London and to Woodward's task force. Woodward, on that day (Sunday, 2 May), was unwilling to be driven out of the area by the Argentinian threat; he wanted to keep up the pressure on the Falklands garrison by a further bombardment that night and to land more special forces reconnaissance parties. Fog had now descended on the task force, hampering the Harrier operations, and he did not know the exact position of the Argentinian carrier force. To Woodward and to Fieldhouse at Northwood, the carrier group to the north and the *Belgrano* to the south had all the appearance of a pincer movement against the task force. The rules of engagement then in force did not permit the *Belgrano* to be attacked by submarines; too great a loss of life and thus hostile world opinion might ensue. But, as the daylight hours passed, Woodward was concerned that he was in increasing danger and he asked for the *Conqueror* to be given permission to attack the *Belgrano*. If the cruiser or her destroyers evaded the submarine that evening, came in over the shallow Burdwood Bank, and steamed hard through the night, *Hermes* and *Invincible* might be attacked by Exocets. The members of Mrs Thatcher's War Cabinet were assembling for their daily meeting, due to take place at Chequers after lunch. The request to attack the *Belgrano* arrived just before lunch. There was a quick meeting, standing up in a side room; and a positive decision was taken within twenty minutes. Northwood sent off a general signal to all submarines, authorizing them to attack any Argentinian warships. It should be stated that if *Splendid* or *Spartan* had encountered the *Veinticinco de Mayo* that afternoon, the aircraft-carrier would have been attacked. The Royal Navy would then have been most pleased; this aircraft carrier posed a much greater menace than the *Belgrano* – 'a relatively poor catch'. The fact that further negotiations were taking place between the President of Peru and the American and Argentinian Governments – made much of later by those who condemned the attack on the *Belgrano* – is not relevant. Negotiations of one form or another had been taking place for the past month and no evidence had reached London that the current ones were making any more progress than their predecessors. The Argentinian admirals were certainly not being restrained. One side or the other was likely to deal the other a massive blow at any moment.

The submarine *Conqueror* (Valiant Class, captained by Commander Christopher Wreford-Brown, completed 1971, 4,900 tons, 103-man crew, six torpedo tubes, nuclear-powered) was the oldest of the three British submarines operating off the Falklands at that time. She had been supporting the recent South Georgia operation until 28 April, when she was ordered west. Her actual patrol area was between Isla de los Estados and the Burdwood Bank. The cruiser *Belgrano* had two escorts, the ex-American Sumner-class destroyers *Hipólito Bouchard* and *Piedra Bueno*. Wreford Brown's original orders only allowed him to attack these ships if they attempted to enter the Total Exclusion Zone. He reached his patrol area on 30 April and made long-range sonar contact with a group of ships that evening. There were no regular trade routes in this area, so

Conqueror closed steadily 'down the bearing' of the sonar. The contact remained steady at a position approximately fifty miles off Isla de los Estados.

At dawn the next day, 1 May, *Conqueror* came to periscope depth and found a day of flat calm and perfect visibility, but not yet any sign of the Argentinian ships. She went down again, increased speed, and an hour later came up once more and was rewarded with the sight of not three but four ships. The extra one was a tanker (believed to be the *Puerta de Rosalies*) which was replenishing the *Belgrano*, an event which indicated that the Argentinians were about to set off on a long mission. Commander Wreford-Brown, fourteen years in submarines, was pleased to have made the contact.

I worked my way round on their port quarter. The tanker broke away from the cruiser and set course for the west. The three Argentinian warships formed up into a loose formation, increased speed to about 13 knots and set off to the south-east, I went deep and took up a tracking position astern of them. I had sent the locating report via satellite communications – only a short transmission of a few seconds – and we settled down into an uneventful trail.

Conqueror followed the Argentinian ships all that day and night as they steamed south-eastwards, keeping outside the Total Exclusion Zone but moving steadily behind the potential shelter of the shallow Burdwood Bank. Then, early on 2 May, there was an abrupt change of course as the Argentinians turned west and then appeared to steam an aimless zigzag route, still just outside the Total Exclusion Zone. *Conqueror* was able to keep in touch and sent off further reports to Fleet Headquarters. It was at this time that the British realized that, if the *Belgrano* group split up and steamed fast over or round the Burdwood Bank that night, *Conqueror* would not be able to follow all three and at least two of the three ships could be within Exocet range of the British aircraft-carriers by dawn. Commander Wreford-Brown received a signal permitting him to attack. Wreford-Brown immediately commenced his attack sequence.

I had decided hours before that, if I got the chance, I would attack the cruiser. I had a choice of weapon and I chose the older, closer-range, Mark 8 torpedo because it has a bigger warhead and therefore a better chance of penetrating the warship's armour plating and anti-torpedo bulges – good World War Two stuff; that isn't in the design of ships these days. If I couldn't get within effective weapon range, I could still use my second option of the wire-guided Tigerfish.

I spent more than two hours working my way into an attack position on the port beam of the cruiser. It was still daylight. The visibility was variable; it came down to 2,000 yards at one time. I kept coming up for a look – but when at periscope depth we were losing ground on them – and then going deep and catching up. I did this five or six times. They were not using sonar – just gently zigzagging at about 13 knots. Twice I was in reasonable firing positions but found they had moved off a few degrees.

Listening later to the tape of the attack, it was remarkable how like an ordinary drill it was. It sounded like a good attack in the Attack Teacher at Faslane, everything tidy, no excitement. I'm not an emotional chap and I had been concentrating the whole time on getting into a good position. It was tedious rather than operationally difficult. We eventually got them – ourselves on the cruiser's port beam, with the two destroyers on her starboard bow and beam. I think the escorts were mainly thinking of a threat from the north, while we were to the south. We fired three Mark 8s at 18.57 Zulu, at a range of 1,400 yards. The object was never to hit with all three but to fire a spread to cover any inaccuracies in the fire-control solution. We heard the weapons run and then heard two torpedo hits; we'd got two out of three.

We were still at periscope depth. I think I saw an orange fireball in line with the mainmast – just aft of the centre of the target – and shortly after the second explosion I thought I saw a spout of water, smoke and debris from the back end.

A big cheer went up from the Control Room and only then did I realize how many extra people were crowded into the corners listening. My immediate thoughts then were to evade; in the Attack Teacher at Faslane we would have stopped to have a cup of coffee at that stage.

Six minutes after the torpedo struck, the first 'burst' of sonar was detected as the two Argentinian destroyers started their search and counter attack. Two distant salvoes of depth charges were heard. *Conqueror* dived and evaded but when the depth charges came closer Wreford-Brown decided it was time to leave the scene, 'a decision which, according to the looks on the faces of the men in the control room, met with everyone's approval'. *Conqueror* moved off to the east.

The *Belgrano* had been struck by two torpedoes, the hit towards the stern being the most serious. A large amount of internal damage was caused and, according to Argentinian accounts, most of the Argentinian sailors who lost their lives were killed here. The ship was abandoned, again according to Argentinian accounts, within thirty minutes, and sank fifteen minutes after that. Most of the crew escaped into rubber life-rafts and spent a miserable twenty-four hours in strong winds and high seas before being rescued. One report says that fifteen ships were eventually involved in the search for survivors. The Chileans sent one ship. *Conqueror* returned to the scene two days later, on the 4th, and saw two destroyers, one merchant ship and several helicopters and aircraft on obvious search work. Wreford-Brown made no attempt to attack again.

It is believed that the *Belgrano*'s crew numbered 1,042, of whom 368 were lost. Three sons of admirals were reported dead but most of the men lost were young conscripts. The *Belgrano*'s captain, Captain Héctor Bonzo, was a survivor.

The British cause was to be both strengthened and weakened by the sinking of the *Belgrano*. World support for the British position was definitely shaken by the news that a modern 'nuclear' submarine had sunk an old cruiser outside the Total Exclusion Zone and that one third of the cruiser's crew of young conscript sailors had perished. There was criticism in Britain but most of this was inspired by party politics; Mr Foot, the Labour leader, later confided to the broadcaster Sir Robin Day that Mrs Thatcher 'had no alternative than to order the sinking of the *Belgrano*'. Criticism in South American countries was vociferous; support for Britain in Spain, Italy and France wavered; Ireland turned hostile. Russia said little. Surprisingly, the United States were not concerned too much and most Americans were delighted when their country declared itself for Britain at about that time. Britain's case in the world could have been much helped if the text of the clear warning given to the Argentinians on 23 April, that 'Argentinian warships . . . which could amount to a threat . . . will encounter the appropriate response', had been more fully publicized. (I only realized that this warning had been given when a task force naval captain gave me a copy several months after the war!)

I am trying to be unbiased in this book and to keep my personal feelings to a minimum but I must say that I see criticism of the *Belgrano* sinking order as humbug. The Argentinians had used armed force to invade the Falklands, in clear defiance of a long-standing United Nations declaration that such methods should not be used. The Argentinians had then ignored a United Nations resolution that they should withdraw from the islands. The Argentinians had been clearly warned that their warships were likely to be attacked whether inside or outside the Total Exclusion Zone if those ships threatened the safety of British ships. The Argentinian Navy was not out on a summer cruise; their ships were trying to catch the British aircraft-carriers. The *Belgrano* incident certainly should not be considered in isolation. The cruiser and its escorts were part of a joint Argentinian operation (see p. 144), the intention of which was the destruction of the British aircraft-carriers or at least the forcing away of the British task force from the Falklands. If that operation had been successful, the British bid to liberate the Falklanders would have been ended. If the War Cabinet or Mrs Thatcher or Admiral Fieldhouse had failed to act and if the *Belgrano* and her escorts had come over the Burdwood Bank that night and loosed off a salvo of Exocets and hit *Hermes* or *Invincible* or other ships, the British loss of life would have been enormous, the task force would have been crippled, public opinion would have been baying for resignations and courts martial and the Falklands would still be occupied by the Argentinians.

The Royal Navy had no qualms. Admiral Fieldhouse says:

I have no doubt that it was the best thing we ever did. It cut the heart out of the Argentinian Navy and we only had their Air Force to deal with then. That was a very considerable advantage.

But there was some sympathy. Leading Seaman Geoffrey Lote was in the *Yarmouth*.

There were big cheers but not big cheers – big cheers in myself when I heard that the torpedoes had hit, but then news flashes came through that the ship was heeling over and that many of the Argentinian sailors would be lost. Then no big cheers; they were only boys like our sailors and I remember that night being bitterly cold – calm but very cold with snow outside. They couldn't have lasted long if they'd had to jump over the side. The other chap in the turret with me just shook his head as if to say, 'What a waste!'

The operational benefits to the British were undoubted. The

Argentinian surface warships never again ventured away from the continental shelf along the inland coast where the water was too shallow for the British submarines to operate. The crew of *Conqueror* had not only sunk the second largest ship in the Argentinian Navy, they had neutralized the only aircraft-carrier and many smaller ships. The naval aircraft were transferred from the aircraft-carrier to shore bases and were later involved in hectic action but only at ranges which put them at a disadvantage. Even the small Argentinian submarine force achieved little. Their two small, quiet, German-built type 209s could have been very effective around the Falklands once the British landings took place. They did sail but never pressed home their efforts. The larger submarine, *Santa Luis*, showed the same lack of determination. Admiral Anaya's navy had truly had its heart cut out.

CHAPTER NINE

Sheffield

The 48-hour period following the torpedoing of the *Belgrano* was relatively quiet. The shocked Argentinians made no move to avenge their lost cruiser and indifferent weather kept air activity to a minimum. The British task force was also content to apply only gentle pressure. A coastal bombardment planned for the evening of 2 May was cancelled when news of the attack on the *Belgrano* was received; Sandy Woodward did not wish to reduce the impact of the *Belgrano*'s sinking by losing a ship himself that night. Later in the night, however, a Sea King of 826 Squadron on anti-submarine patrol picked up a small surface-ship contact on its radar off the north coast of East Falkland and was fired upon when it moved in to investigate. Two Lynx helicopters – from *Coventry* and *Glasgow* – were launched, each armed with Sea Skua missiles. *Coventry*'s helicopter made the first attack. Two missiles were launched and, guided by radar, both missiles scored direct hits which resulted in a large explosion. The Lynx's crew – Lieutenant Hubert Ledingham and Lieutenant-Commander Alvin Rich – had carried out the first operational firing of Sea Skua, a weapon which had only been released for use after the task force sailed. The Argentinian ship was an 800-ton naval patrol vessel, the *Comodoro Somellera*, with a crew of approximately fifty men; the number of Argentinian lives lost when she sank is not known. *Glasgow*'s Lynx then came in to search for survivors but was fired upon by a second ship whose presence had not been suspected. This Lynx fired its Sea Skuas and again scored a hit, the bridge and upper works of a second patrol vessel, the *Alférez Sobral*, being smashed. The captain and seven crew men were killed but the ship managed to reach Argentina several days later. The British believed that they had defeated an attempt by blockade runners to bring supplies into the Falklands but the two ships had actually come out from Stanley to continue the search for the two-man crew of the Argentinian Canberra shot down just over twenty-four hours earlier.

The next day, the 3rd, passed almost without incident. An Aeromacchi carrying out a local flight from Stanley crashed in bad weather just east of the airfield and its naval pilot was killed. But, that night, the R.A.F. mounted another Black Buck operation from Ascension. The Vulcan was XM 607, the aircraft flown by Flight Lieutenant Withers and his crew three nights earlier. The primary crew on this occasion was that of Squadron Leader John Reeve of 50 Squadron and his crew, who had been forced to abort on Black Buck One. Reeve's aircraft performed perfectly and the experiences of the first Black Buck operation were put to good use; John Reeve says that the tanker plan 'worked like magic; people were throwing fuel at us from every direction'. The Vulcan reached Stanley without incident and was making what appeared to be a perfect bomb run until the final moments. The Vulcan had an instrument on which two indicators needed to be kept in line on the bomb run, but crews were instructed to ignore any misalignment of these which might occur in the last moments before bomb release, in case the resulting banking of the aircraft threw the bombs too far off the target. Reeve's instruments showed just such a misalignment at the last moment; he resisted the temptation to turn slightly to port as indicated and the bombs were released. They fell just fifty yards west of the runway's end in what would have been a perfect straddle. Reeve, who would never get the chance of another Black Buck flight, would always regret that he did not ignore instructions and give that little touch on his controls which would have rewarded him with a perfect line of bombs across the runway.

The raid had been a good attempt, well backed up by the Victor tankers, but no serious damage was caused. The Vulcan and all the Victors returned safely to Ascension.

The first half of the next morning – of Tuesday, 4 May – passed quietly. Admiral Woodward was planning a day of only minor action. Lynx helicopters flew off at dawn to probe along the coast around Stanley and plot the source of Argentinian radar emissions. The only offensive activity planned for the day was a raid by three Sea Harriers on Goose Green airfield, due to take place in the early afternoon. The morale of the task force was high. Airfields on the Falklands had been bombed four times by Sea Harriers or Vulcans, Argentinian aircraft shot down, shore positions bombarded, a suspected submarine driven off, the *Belgrano* sunk, two patrol boats successfully attacked – and all without a single member of the task force being seriously hurt. All of the task force's weapons systems – except the ships' missile defences – had been successfully tested. But this euphoria was about to suffer a severe blow.

Every man in the task force knew what an Exocet was (the name came from *Exocoetus*, a rare species of flying fish). Some of the British ships were fitted with this French-built anti surface-vessel missile which flew just above the surface of the sea at around the speed of sound. The British ships were not well equipped to counter this low-flying missile; Nato training was mainly directed to defence against high-altitude threats. Exocet could theoretically be knocked down by the Sea Wolf missile or by chance shots from other weapons, but only if the defence reaction was extremely swift. The missile could also be decoyed by clouds of 'chaff', thousands of fibre-glass needles fired into the air by rockets. The best defence was to keep out of range.

The Royal Navy's experience of Exocet was confined to the ship-launched version – the MM-38 – which had a range of twenty-six miles. The Argentinians had recently purchased five of the comparatively new air-launched version – the AM-39 – which, though lighter, had a longer range, partly because of the initial velocity imparted to it by its launcher aircraft – the large, modern Super Étendard, also French-built. The British commanders recognized the air-launched Exocet as the most dangerous weapon in the Argentinian armoury.

Sandy Woodward had to tread the difficult path between keeping his carrier group close enough to the Falklands to conduct operations but far enough away to keep as far out of range as possible of the Super Étendards from the mainland. The possibility that the Super Étendards could be refuelled in the air had been considered but this was not thought to be likely; Woodward hoped that the Vulcan and Sea Harrier attacks could keep the Étendards from using Stanley's runway. The British aircraft-carriers were tied to an arc east of the Falklands by the range of their Sea Harriers and of the helicopters which were putting reconnaissance parties ashore each night. Woodward could only vary his position in that arc by moving to new sectors in it. That morning, the task force was opening the range from the land after the previous night's operations, steaming south-eastwards to take up a location further south than on previous days. The carriers were about 100 nautical miles south-east of Stanley. As always in fine weather, the task-force ships were disposed to give the carriers the best possible protection. There were five layers of defence protecting the carriers – Sea Harriers, the Sea Darts of the Type 42 destroyers, the general-purpose screen, the supply-ship screen and the two Sea Wolf-equipped point-defence ships. Every other ship was a potential decoy which might lure an Exocet away from the aircraft-carriers. It was an operational necessity that *Hermes* and *Invincible* be regarded as indispensable and every other ship as being (in relative terms) expendable.

The Argentinian Exocet unit, the 2nd Naval Attack Squadron, was based at Río Grande, in Argentina's deep south on Tierra del Fuego, 495 miles (430 nautical miles) from the position of the British aircraft-carriers. The Étendards needed good intelligence before they could launch an attack. There was, on this morning, an old piston-engined Neptune aircraft on patrol sufficiently near the British task force to pick up radar transmissions. At approximately 8.15 a.m., local time, the Neptune detected the radars of one of the British Type 42 destroyers. What was believed to be the exact location of the source of the radar emissions was plotted seventy miles due south of Stanley; it could be assumed that the British aircraft-carriers were just to the east of that position. Two Super Étendards took off from Río Grande about half an hour later. Each aircraft had an Exocet under its starboard wing and extra fuel tanks under its fuselage and port wing. This time the refuelling from a Hercules tanker was successful and the two Étendards made an undetected approach. The two aircraft did not communicate with each other but they did receive further reports from the Neptune. When they approached the task force, the two Argentinian pilots – Lieutenant Commander Bedacarratz and Lieutenant Mayora – climbed to 120 feet, located three blips on their radars, one large and two small, set the internal guidance systems of their Exocets

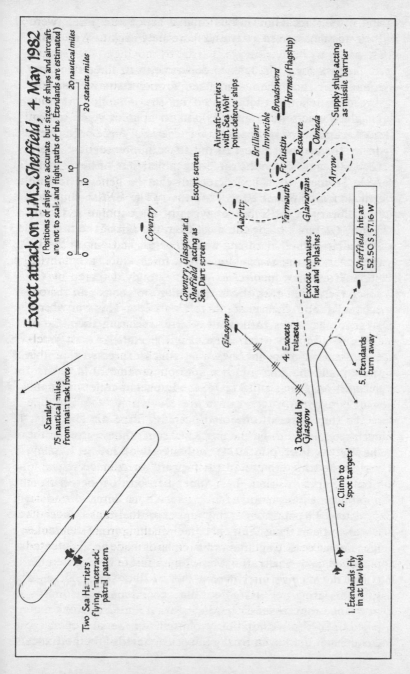

Exocet attack on H.M.S. Sheffield, 4 May 1982
(Positions of ships are accurate but sizes of ships and aircraft are not to scale and flight paths of the Étendards are estimated)

20 nautical miles

20 statute miles

10

10

0

0

Aircraft-carriers with Sea Wolf 'point defence' ships

Supply ships acting as missile barrier

Brilliant
Invincible
Broadsword
Hermes (Flagship)
Ft Austin
Resource
Olmeda
Yarmouth
Glamorgan
Arrow
Escort screen
Alacrity

Coventry

Coventry, Glasgow and Sheffield acting as Sea Dart screen

Sheffield hit at 52.50 S, 57.6 W

Glasgow

Exocet exhausts, fuel and 'splashes'

4: Exocets released

5. Étendards turn away

3 Detected by Glasgow

Stanley 75 nautical miles from main task force

2. Climb to 'spot targets'

1. Étendards fly in at low level

Two Sea Harriers flying 'racetrack' patrol pattern

and, at 11.04 local time (14.04 Zulu or Task Force time), released their missiles, turned away and flew safely back to Argentina.

The task-force air-defence departments had been uneasy for some hours and there had been several tentative reports of Argentinian air activity, based on the detection of real or imagined electronic emissions and on Spanish-voice transmissions. Harriers on routine patrol had several times been directed on to 'threats' which turned out to be non-existent. Suddenly, *Glasgow* came up on the air-defence net. One of her operators, Able Seaman Rose, had detected what he believed to be the emissions of a Super Étendard radar. (The British ships had a tape library of the different types of Argentinian radar emissions.) *Glasgow*, being the most westerly ship of the task force, immediately fired chaff and went to action stations, at the same time alerting the remainder of the force. Soon after this, the Super Étendards themselves were fleetingly detected by *Glasgow*'s radar but, then, both the radar emissions and the radar echoes of the Argentinian aircraft were lost. This was when the Argentinian pilots turned away after releasing their Exocets. *Glasgow*'s reports were received in the Anti-Aircraft Warfare Coordination room in *Invincible*, which expressed doubt over what might be just another spurious contact and asked for verification. It was all too late; the Exocets were on their way – undetected.

The ship nearest to the approaching threat was the Type 42 destroyer *Sheffield*. Displacing 3,660 tons, she was the oldest of the 42s (completed in 1974), with a crew of 286 men. *Sheffield* was at the southern end of the forward air-defence screen. She would normally have had four methods of detecting the approaching danger and of taking some type of avoiding action: detecting the emissions of the Super Étendard's radars (she was closer to them than *Glasgow*); 'seeing' the Super Étendards on her radar; detecting the radar emission of the Exocet once launched; and, finally, the 'Mark 1 eyeballs' of her look-outs. By bad luck, *Sheffield* was denied two of these possibilities. The ship was using her SCOT satellite communications briefly to pass some routine messages during what appeared to be a quiet period and this automatically blotted out the particular radar frequencies used both by the Super Étendards and the Exocets.

There was no avoiding this; messages had to be passed at some time and the period of broken cover rarely lasted more than two minutes. The story published by the *Daily Mirror* in 1986, that *Sheffield* was lost because her captain was 'making a phone call to London when an Exocet attack was expected', was a distressing distortion of the true situation.

The total flight time of the Exocets was no more than two minutes. The missile heading for *Sheffield* could not be picked up by radar because it was too low and it was only in the final seconds of its flight that several people on the upper part of the ship saw the missile approaching. There was no time for any weapons or chaff-rockets to be fired, evasion to be taken or a warning broadcast to the men in the compartments below decks. The Royal Navy had easily neutralized the surface-ship Exocet threat in the shape of the *Belgrano* group but, not for the first time in its history, was caught technically unprepared for a recent development in air power.

The Exocet's internal radar did its work well. The missile smacked hard, at an angle, into *Sheffield*'s starboard side, only slightly forward of the exact midway point of the ship's length; the strike was level with the floor of No. 2 deck, about six feet above the waterline. The exact path taken by the second Exocet is not known but two ships protecting the main group, *Yarmouth* and *Glamorgan*, reported that they had seen or detected a missile. (*Yarmouth* had a 'funny' in its next daily orders: 'The Argentinian missile that passed *Yarmouth* was on a training mission and has returned safely to base.') No other ship was hit and it can be assumed that the second Exocet ran out of fuel and fell harmlessly into the sea, but it had probably approached the western edge of the main group of ships. If this is true, then the Argentinian pilots had not pressed home their attacks hard enough; the three radar blips they had seen were probably those of *Sheffield* (the nearer and larger blip), *Yarmouth* and *Glamorgan* (the nearest ships in the main group). *Yarmouth*, *Glamorgan* and the nearby supply ships – *Olmedu*, *Resource* and *Fort Austin* – had been on the fringes of the danger area but the two aircraft-carriers were well out of range of the missiles. The Argentinian attack was thus only a partial success. Two of their five scarce airborne Exocets had been expended; the *Sheffield* would be lost but would be replaced, and the task force would remain on station to continue operations.

The warhead (165 kilograms or 364 pounds) of the Exocet which hit *Sheffield* did not explode, a fact which saved many lives. But the impact of the missile, whose overall weight was about half a ton and which was moving at nearly 700 miles per hour, was powerful enough to pierce *Sheffield*'s thin side, pass through the starboard passageway and penetrate deep into the ship before it disintegrated and its remaining rocket fuel burst into a huge fire. There were plenty of items inside *Sheffield* to feed this. The fire/explosion took place in or near the galley. Eleven men are believed to have died at once in and around that area, including a young Chinese laundryman who had gone to the galley to brew himself some tea or prepare his own food. There were no survivors from this area but there were some lucky escapes from nearby compartments. Chief Mechanician John Strange, the only man in the machinery space underneath the galley, crawled out, badly burned, the most seriously injured survivor. These are some memories of other survivors. Leading Seaman Alistair Gilchrist was very close to the missile hit.

I was in the main Communications Office and, inside this, was a little office called the Electronic Warfare Office. I was sat monitoring a voice circuit and there was this enormous bang. The piece of equipment in front of me was an old-fashioned one and very sturdy; if it had been a modern one, it would have been much smaller. But it remained intact and protected me. A man the other side of it got badly burnt. Everything flew all around me; a teleprinter came to bits and flew all around me; the door flew off; bookshelves came down.

Able Seaman Mick Morton was in the Operations Room, the next compartment further forward.

I was on a thirty-minute break from the sonar set; I had just sat down. It wasn't a great explosion, just a 'whoosh', and the Ops Room door blew in and hit John Galway, the man standing at a plot near the door, and ended up on the action plot table. Galway was moaning and we found he'd got a big cut in his forehead but it was a bad graze on his buttock that he was moaning about. That was the only casualty in the Ops Room, though there was some shock afterwards.

It went dark instantly except for one small orange light and the room filled with smoke immediately and dimmed that. There were bits of debris and wires hanging, books and charts and dust from all sorts of places in the room all came down. It was just like one of those old comedy films.

Lieutenant Ian Goddard was asleep in his cabin.

I remember not so much a bang as a muffled crump and a shock wave, and my instant thought was that we'd been hit by a 'fish' – a torpedo – because there had been no broadcast warning. I was out of bed, shoes on, life-jacket on, gas mask and immersion suit – which was standard operating procedure; I can't tell how long it took, but it couldn't have been long. I did think to pull on an extra jumper and shoved a torch in my pocket. I was very pleased with these later.

The deck in the passageway outside my cabin was covered with the remains of the deckhead – all the formica panelling, bits of timber and suchlike. As I went aft, the officer in the next cabin appeared and we were going back to the ladders when a steward staggered out from the Wardroom with his head cut open and flopped. We took him out on to the port waist and laid him on the deck and it was then that we heard the 20-millimetre start firing. I ran round the front of the bridge screen and saw a helicopter closing from the starboard beam and our helicopter pilot, who was on the foc's'le, was screaming that it was one of ours – perhaps a couple of expletives in there somewhere. They stopped firing.

I went back to the port waist. Another officer, whose cabin was further aft, asked me if I could let him have a pair of shoes. He had been in his cabin and found a sheet of flame in front of him; he had thought about the shoes for a few seconds then opened the door and ran. The two of us crawled along the floor to my cabin to get this pair of shoes – silly, isn't it? – but it seemed important at the time with the freezing weather and fire all over the place. Shoes were important.

We got him the shoes and crawled back out into the waist. Then I said that we needed bedding for the wounded and set off to start taking blankets off all the beds in the officers' cabins, but the smoke was so thick and so pungent that we had to give up. By now, the smoke was so thick that it was coming out of the door in a great squidge, just like toothpaste out of a tube.

Those of us who were assembling on the foc's'le started laying out the fire hoses, but there was no fire main pressure anywhere in the forward part of the ship. I looked over the side and could see the fire main pumping a stream of water ten feet or so out into the sea through a hole. Obviously, there was a pump running aft but the ruptured main was preventing any water getting forward. I looked over the side and could actually see the Exocet hole and I remember thinking, 'Jesus Christ, I wonder how many guys have snuffed it', because it was a big hole near the Ops Room where a lot of people I knew worked.

The hit had rendered *Sheffield* useless. Damage in the machinery space below the galley put both the main generators out of action and the two back-up generators could not be brought into use; this failure may have been the cause of the ship's main engines losing power. The generator loss, or damage to wiring, caused all power to the ship's weapons to be cut off. There was no main lighting and no communications between departments. The main water supply, which ran right round the ship in the 'Burma Road' passageway, was broken by the missile's initial impact; this starved the main fire-fighting equipment of water. *Sheffield* was in no danger of sinking but she could not steam, could not fight and the crew were having to fight a fierce fire under great disadvantages.

The first attempts to fight the fire were by pathetic bucket-and-rope efforts. One man says, 'We were literally chucking the buckets in the hoggin, passing them along in a chain and throwing them into the fire – where the water just turned into steam.' *Sheffield*'s own Lynx helicopter was able to take off and started lifting casualties from the stern; larger helicopters soon arrived and took the more seriously injured from the forecastle. Portable fire pumps were flown in. *Sheffield*'s Gemini launch took one of these round to the hole made by the Exocet and directed water on to the seat of the fire. The frigate *Arrow* arrived and then *Yarmouth*, although *Yarmouth*'s efforts were diverted by a submarine scare which lasted for some time. *Arrow* was able to pass hoses across but *Yarmouth* could do no more than play water on *Sheffield*'s hull. The crews of the rescue ships were much impressed by the sight of a modern ship, helpless and fighting for its life, but there was some humour. Chief Petty Officer Abi Hirji, on *Sheffield*'s flight deck, remembers, 'Yarmouth's crew was throwing over sweets, chocolates and cans of soft drinks to keep our morale up. There was a lot of shouting, most of it humorous. One of them told us not to make a habit of this; they couldn't spare the nutty again.'

But the battle against the fire was being lost and the fire-fighters were forced steadily back. The Computer Room was a vital position. The six-man watch in it was joined by Lieutenant-Commander John Woodhead and these men worked hard to bring the computers – and hence the ship's weapons – back into action. They remained at this task until their escape routes became blocked by smoke and fire; all were then overcome by smoke and died. Just above the Computer Room was the Operations Room, in which more than twenty men had been waiting for power to be restored; this compartment also filled with smoke but the entire watch on duty was able to escape at the last moment. The ship had now split into three separate parts – the central section, which was full of fire and smoke, and the two ends into which the crew was being driven back. The forward compartment of the ship then started to fill with smoke and the men there were driven out on to the deck. One of the most determined of the men here, Petty Officer David Briggs, made one journey too many in an attempt to retrieve important equipment, was overcome by smoke and died. (Lieutenant-Commander Woodhead and Petty Officer Briggs received post-humous decorations.)

Captain Sam Salt decided that there was nothing further his men could do to save *Sheffield*. Most of his crew were now on the exposed upper deck; it was getting bitterly cold and darkness was approaching; there was a heavy swell. The fire was approaching the Sea Dart magazine and, if this exploded, the rescue ships would also be in danger. To delay longer would risk heavy loss of life. Captain Salt ordered his crew to abandon ship at 17.51 Zulu time; they were all safely away by 18.33, four and a half hours after the Exocet strike. Twenty men were dead and twenty-four injured, four of them seriously; 242 men escaped without injury.

There had been a further, though much smaller, task-force tragedy while the fight to save the *Sheffield* was being lost. Three Sea Harriers had taken off from *Hermes* an hour after *Sheffield* was hit, to carry out a planned bombing raid on Goose Green airfield. Two Harriers were first to make low-level attacks with cluster bombs on any aircraft found at the airfield and the third Harrier was to drop three 1,000-lb bombs to crater the grass airstrip. But the Argentinian anti-aircraft fire hit one of the first two aircraft while it was approaching over the sea and the Harrier crashed, carving a long furrow up the beach and in the grass just short of the airfield. The pilot was killed, probably by the burst of fire which hit his aircraft. He was Lieutenant Nick Taylor, a former R.A.F. pilot who had transferred to the Fleet Air Arm. This was the task force's first Harrier loss. The

Argentinians gave the dead pilot a military burial at the far end of the airfield, not far from the graves of the seven Argentinians killed in the first Harrier attack three days previously. The burial was televised throughout the world.

The events of 4 May forced some important changes to the task force's operations and to the attitude of the men in the British ships. Sandy Woodward – a former captain of *Sheffield* – confesses himself to have been 'very frustrated and down; I talked to the Commander-in-Chief and that bucked me up a bit'. Woodward's first action was to withdraw his ships further east to keep away from the Super Étendards; he knew the Argentinians had three airborne Exocets left. The new position for his carriers was about 160 nautical miles east of Stanley, further out in clear weather, sometimes closer in if the weather was bad. The theoretical line beyond which the carriers were considered to be safe from the Étendards became known in the task force as 'the Yellow Brick Road'. There was some sneering comment over this decision but Woodward knew that he could lose the war at a stroke if he lost one of his carriers, and the move was a prudent one. The changes affected the Sea Harrier pilots most. They were thus forced to make longer flights over the sea every time they attacked a target on the islands and a much heavier programme of routine patrol flights was instituted in an effort to minimize the lack of airborne early warning which had been highlighted by the Exocet attack. The loss of Lieutenant Taylor's Harrier over Goose Green led to direct attacks on defended land targets being temporarily abandoned; only the relatively inaccurate, three-miles-out 'toss bombing' was continued for the time being.

Much thought was given to countering further Exocet attacks and a nervous condition known as 'Exocetitis' would linger among many ranks in the task-force ships until the end of hostilities. The greatest strain was felt in *Sheffield*'s two remaining sister ships, *Coventry* and *Glasgow*, which now had to maintain the forward defence screen alone until other Type 42s arrived. Captain David Hart-Dyke commanded *Coventry*.

There is no need to tell you of the sobering and demoralizing effect that swept through the ship. People didn't talk very much for about twenty-four hours. But I must say that, after that twenty-four hours, there was no question; people wanted to go to war with even greater determination to get back at the Argies – it actually stiffened our resolve and morale came back up to a very high quality.

The following story is told about a Sea Dart operator on one of the 42s:

On the way down, this chap had been saying that he would never be able to press the button when the time came; he could never kill someone like that. Later, after his first Sea Dart firing, someone asked him how it felt. 'Bloody marvellous.' That was the post-*Sheffield* feeling.

Another ship to be much affected was *Hermes*, to which the *Sheffield* casualties were brought. Andrew Wroot was a member of the *Hermes* deck party.

The first one came in by helicopter, arms covered in plastic bags, face all black. He seemed confused. I had never seen anything like it before – horrific. There were many more after that and a lot of hard work to be done. We all thought *Sheffield* would be saved; we never thought a single missile could knock out one of our most modern ships.

Later, when they piped that *Sheffield* had been abandoned and should be considered lost, all of us in the junior ratings dining hall became completely hushed, perhaps the odd whisper and a murmur, but mainly general disbelief. That was a really black day for us.

The news caused a great wave of sadness to sweep through Britain, whose population had to relearn the blow-and-counter-blow aspects of warfare not experienced for thirty years. The next meeting of the War Cabinet was a sombre affair. Only Mr Pym and Mr Whitelaw were old enough to have served as junior officers in the Second World War and most of the other members present were badly shaken, both by the casualties and by the fact that *Sheffield* had been lost to one missile launched by an unseen aircraft. It fell to Admiral Lewin to steady the meeting. He stated that ships were inevitably going to be lost and that smaller warships always had to be regarded as expendable in times of war. He quoted the example of the *Pedestal* convoy to Malta in the Second World War, which lost much of its escort and two thirds of its merchant ships, but history showed that the remaining ships delivered enough stores to prevent Malta from falling. The War Cabinet accepted Lewin's reassurances and the political resolve remained steady.

*

Sheffield remained afloat for a further six days. Helicopters put parties aboard to inspect the ship's condition and retrieve some equipment. For four days the hulk was left drifting as a lure for Argentinian submarines, anti-submarine Sea King helicopters hovering near by. But no submarine appeared and *Yarmouth* was sent to tow *Sheffield* out of the area to meet a tug which would take the destroyer back to Ascension. *Sheffield*'s hull was still in sound condition and *Yarmouth* towed the empty ship all through 9 May and for most of the next night, making about 6 knots. But *Sheffield*'s movement through the water caused her to take in increasing quantities of it and she started to list heavily to starboard. Just after dawn the following day, 10 May, the look-out at the stern of *Yarmouth* realized that *Sheffield* had sunk. No one saw her go, because of fog and because *Yarmouth*'s radar had been switched off for concealment purposes. The tow-line was immediately released. Captain Salt, who had been in *Yarmouth* throughout the tow, followed his old ship down by sonar until he was satisfied that she was finally at rest on the bottom. *Sheffield* sank about 135 nautical miles east of the Falklands, of whose islands she had never been within sight. The ship became a war grave for the bodies of the nineteen men still left aboard; one of the dead had earlier been buried at sea. *Sheffield* was the first ship of the Royal Navy to be sunk by enemy action since the Second World War. A missile costing just over £100,000 had sunk a destroyer which would cost about £120 million to replace.

The *Sheffield* survivors were put aboard the *British Esk*, a tanker returning to Ascension, and then flown back to England on 27 May, in the first of many emotional homecomings for men of the task force.

CHAPTER TEN

Life in the Falklands

The loss of the *Sheffield* marked the end of that first, violent series of actions which followed the arrival of the task force off the Falklands. The next phase would be a quieter one, with the Argentinian Navy confined to its bases, with the Argentinian Air Force refusing serious action, with the British Sea Harriers being conserved, with the British carrying out reconnaissance and minor naval action in preparation for the arrival of the landing force. It is an opportune moment to look at what was happening on the Falklands themselves.

Some of the outlying settlements on the island never saw an Argentinian soldier throughout the war and life there continued almost as normal, though with avid listening to the B.B.C. World Service and to the scraps of news available from other settlements on the restricted inter-island radio-telephone service. But the Argentinians did come to some of the medium-sized settlements. Alan Miller, the manager at Port San Carlos, received an early visit.

They came by helicopter on that first Sunday, about 9.30 a.m. As it was coming into land, I started to walk out to meet it but soon stopped and put my hands in the air when a door slid open and a dozen fully armed little green men jumped out and spread out among the houses. An officer, who it later transpired was the infamous Major Patricio Dowling, the Argentine Gestapo chief, ran down to where I was standing, shoved his gun barrel into my stomach, pushing me back against the wall of the building behind me. He then turned me round and banged my head against the wall with his gun butt – charming fellow! He then ran off towards my house, leaving me under the guard of a terrified young soldier, who kept looking petrified around himself, in between peering over his rifle sights straight at me, with his finger on the trigger. I can assure you that I kept very still. Eventually, some of them came back, bringing the menfolk from some of the houses. It appears they were looking for Bill Luxton – another settlement manager who was very outspoken and was on Dowling's wanted list; they picked him up later in the day somewhere else.

Eventually I told my guard that I was the *administrador* of the *estancia* and he took me to what appeared to be a very senior air force officer who had come from the helicopter. He asked me what I thought about 'this mess' and I gave it to him straight. I told him they'd made an even bigger mess of it all by coming in like this. He just smiled and made some non-committal remark.

Then my eighteen-year-old son Philip came out. He had been in bed because it was Sunday and, as soon as he heard the helicopter, he'd thrown his clothes on and come out to see it. Of all the things to wear, he had chosen his combat jacket and the whole lot made a bee-line for him and there was about ten minutes of rapid-fire questions.

We carried on with our normal work after they had gone away; we knew Maggie was coming.

Major Dowling was the most feared and hated of the Argentinian invaders, well known in Stanley as the head of military intelligence. It is believed that his part-Irish blood may have been responsible for his unusually hostile attitude to the 'British' Falklanders.

The civilians at Port Howard – a settlement on West Falkland where there was a permanent Argentinian garrison – held a meeting at which they discussed their attitude to the war. They decided that they would rather leave the Falklands and settle in another part of the world than have the lives of British servicemen lost getting rid of the Argentinians. This view was personal to the Port Howard people and it probably did not become known in Britain until after the war. No other community expressed this feeling in such definite terms but the general attitude of the Falklanders was of the dual horror of the Argentinian occupation and of young British soldiers, sailors and airmen losing their lives to get rid of the Argentinians.

The people of Goose Green had a grim time after the Sea Harriers carried out their first attack on Goose Green airfield on 1 May. These notes were made during an interview with Eric Goss the farm manager.

The Argentinians came running down from the airfield area to the settlement within fifteen minutes of the Harrier attack. They moved two Chinooks and some smaller helicopters right into the settlement among the houses. They put all the local people into the Recreation Hall – 114 people, most of them for twenty-nine days – and moved their own men into the civilians' houses so they wouldn't be bombed again. Sanitation in the hall was grim. We ran out of water on the third day; the two toilets were blocked and there was some dysentery. We persuaded the Argentinians to bring sea water in barrels for the toilets; an old chap, Mike Robson, did sterling work keeping them going.

Two young men, Bob McLeod and Ray Robson, both radio hams, found an old broken radio, part of the club equipment, in a junk cupboard. They made this work and we listened each evening to the B.B.C. World Service; the others made a noise at the windows to cover the crackling of the broadcast and we were never discovered.

Our dogs had no food and were not cleaned out for nine days the first time, then another week, then every two days. Human rights are fairly low on their list of priorities; animals don't reckon at all.

The people of Stanley had the Argentinians around them from first to last. Initially, the occupiers had moved in an administration whose purpose was to oversee a 'transitional phase', pending full incorporation into Argentina. Brigadier-General Menéndez tried hard to make friends. He was handicapped by his inability to speak English and left most civil affairs in the hands of Air Commodore Carlos Bloomer-Reeve and Captain Barry Melbourne Hussey – air and naval officers but acting mainly as civil administrators. There was a public meeting at which various plans for the future were proposed; but the task force was on its way and the local people felt confident enough to reject every Argentinian suggestion. The Argentinians started a dual Spanish and English language television service; an Argentinian dealer provided television sets on hire purchase and several Falkland families became the owners of such sets on which only two payments were made before the Argentinians left.

Most of the civilian services in Stanley had been quickly restored after the initial occupation but others never were. The pilots of the local air service all refused to fly under Argentinian control. The teachers were pressed to reopen school classes. They agreed, provided no armed troops were present and there was no Argentinian interference of any sort; the Argentinians would not agree and the school remained closed. Harold Rowlands, Financial Secretary to the original Falkland Government, was one of the most senior officials left after the Argentinians had deported Governor Hunt and his staff; he describes the introduction of Argentinian administration in his department.

We all had to decide whether to carry on or not. Our Governor had said to us just before the invasion that the civil servants were expected to 'crack down' under the new masters and carry on for the benefit of the population. I felt it was my duty to do this. The ones that I dealt with were gentlemen. I told them I wouldn't work with them but I would work for the people of the islands. They accepted this for what we both referred to as the interim period, after which there was no way I would live under their rule and, if they were staying, I would leave and take up residence wherever someone would allow me to live, preferably Britain. I think that was the general reaction of the local administration.

I was ordered to continue normal operations, like raising taxes, but I decided not to. The Treasury Account at the Government Savings Bank, which paid the local public salaries and all pensions, had a good credit balance, so there was no problem. There was a panic in the first few days; the depositors at the Savings Bank started a run on the bank but I advised them that everything was all right – we had big reserves in the U.K. – and most of the money started coming back – also the pesos which the Argentinians were spending. They ordered that the Falkland Islanders must accept pesos from Argentinians; they fixed a rate of 20,000 pesos to the pound and we soon had millions of them circulating. I was ordered to change the Government Account over to pesos but I told them I was too old to learn and, anyway, our columns were too narrow for all the noughts; I refused and they didn't force it.

(The Stanley Savings Bank held 1,600 million pesos at the end of the war, which the British Government Claims Officer eventually took away in exchange for a cheque for £80,000; the rate of exchange against the peso had quadrupled in the interval, so the British Government lost £60,000 on the deal – about a fifth of the cost of one Sea Dart missile.)

Bill Etheridge was the Postmaster.

One of their colonels invited me formally to stay on as the senior postal officer in the Falklands but employed by the Argentinian postal service. I told them I would only keep the postal system intact and running until I left. But, when I heard that the task force was being formed, I knew the Argies would eventually be thrown out and, providing I was alive at the end of the battle, I decided not to go when some of the others went. I wanted to stay and see the thing out.

We reopened for business but they stopped us selling Falkland stamps.

There was never any Malvinas stamps; they regarded the Malvinas as an integral part of Argentina. Those stamps which were overprinted, 'The Malvinas are Argentinian', were issued in Buenos Aires for propaganda purposes and were not, despite what many collectors in the world believed, on sale in the Falklands.

They moved their own postal people in – three of them, under a Señor Caballero, who was a very senior, very efficient man, a man of very high quality; in other circumstances, we would have found it very easy to work with each other. The Argentinians ran the Argentinian mail and my staff continued to run the civil mail. Most mail from Britain stopped and most Falklanders wrote fewer letters, assuming that the Argentinians were stopping the mail but, as far as I know, there was no systematic censorship or curtailment of the mail. I sent trial letters to a friend in the U.K. and it was acknowledged by a code message in the B.B.C. World Service.

Anton Livermore was one of the local policemen.

Their Major Dowling interviewed me; he told me I'd got to collaborate. I said, 'No, cooperate perhaps.' I was willing to do that so that the local people didn't get into difficulties with the Argentinians. I said that I was willing to carry on working for the local people and he agreed with that. Our police chief appointed me sergeant at once because he believed, rightly, that I would need more authority. He went back to the U.K. and the woman police constable went out to the camp, leaving me the only one. But I never received the sergeant's pay and I lost my stripes straightaway on liberation. I'm still a P.C.

The Argentinian military police moved into the station and I got on fairly well with them, professionally all the time and personally when their officers were not present. They were very good really and kept strict discipline among their own army people but Major Dowling was a problem and eventually there came an incident which led to my finishing working in uniform. They sent me, under armed threat, to arrest a civilian and I refused to do it again, only to help out of uniform with that type of problem. They didn't like that answer and threatened me, but Monsignor Spraggon sorted that out; I have a lot to thank Monsignor for.

Monsignor Daniel Spraggon was the Tyneside-born missionary priest ministering to the Catholics in the Falklands. With the common faith he shared with the Argentinians, his title of Prefect Apostolic of the Falkland Islands and his ten years of service in Stanley, he became one of the principal go-betweens during the occupation and obtained many concessions for the civilians. A large number of Argentinian army padres appeared in Stanley; the senior one, Padre Mafezzini, was much admired by Monsignor Spraggon for his fervour and for his care for young Argentinian soldiers, but the younger Argentinian priests were less popular. They flooded Stanley with Spanish-language religious-political pamphlets, attempting to justify the occupation. Monsignor Spraggon was very annoyed at this and considered complaining to the Vatican, but he realized the priests were probably acting on military orders and let the matter rest. The army padres were among several Argentinians who stayed at local hotels. There were seven priests at the Upland Goose, where Barbara and Alison King thought they were

. . . a poor lot, mostly too old. We often argued with them and had nicknames for them: 'Little Italy' – a small Italian, 'Old Sanctimonious' – always blessing his food, 'The Gaucho' – he wore a poncho, 'The Priestling' – a young one, 'The Walking Wounded' – he had a bandage over one eye. They used to take the soldiers upstairs to give them showers. We switched the hot water off; that was our passive resistance – we said kerosene was short.

The population of Stanley declined in numbers as the war progressed. Many people moved out to the settlements as soon as the Argentinians arrived, the husband often sending wife and children out to stay with friends. The Argentinians deported several of the senior officials and their families with Governor Hunt on the day of the invasion and others were sent out later. The Argentinians publicly announced that everyone was free to leave the islands and the procedure for doing so was quite simple. People went to the LADE airline office, bought a ticket and were flown to the mainland and then allowed to fly on to any part of the world they wished. Unfortunately, the ordinary Falklanders, holders only of British Colonial passports, were forced to apply for an Argentinian passport as they were regarded as residents of the 'Malvinas', an integral part of Argentina. Only a few people with pressing personal reasons took this step, giving the Argentinians only a small propaganda success. An offer of full Argentinian citizenship was only taken up by one man, a champion sheep-shearer. This man went to live in Argentina but for some reason fell foul of the authorities later in the year and finished up in prison.

But the 'expatriates', the holders of full British passports, were free to go without any formality. This freedom was the cause of immense personal anguish and sometimes resulted in bad blood between those who left and those who stayed and saw the war out in Stanley. There were heated discussions between husbands, who wanted to stay, and young wives who were not prepared to risk their children in the middle of a dispute which, they felt, had nothing to do with them. Most of the people involved were the families of British officials serving in Stanley under government contract. At least one wife threatened divorce if her husband did not leave with the entire family for Britain. The exact number of people who left in this way is not known; a figure of between sixty and one hundred has been mentioned. Some had been doing key jobs in Stanley and their skills would be missed. A move was made to resolve the problem by sending a letter to Britain, signed by many of the Stanley officials, asking that a protecting power appointed under the terms of the Geneva Convention make arrangements to evacuate temporarily the entire civilian population. The British Government was horrified; to evacuate all the civilians would play right into the hands of the Argentinians and remove part of the justification for the dispatch of the task force. Strenuous efforts were made to show that the signatories of the letter did not contain any elected representatives of the local people and the matter lapsed.

On 27 April, two days after the Argentinians lost South Georgia, a big round-up took place of people whom the Argentinians considered to be potential resistance leaders. Velma Malcolm was the outspoken and formidable part-owner of the Rose Hotel, a well-known opponent of any transfer of sovereignty to the Argentinians.

Three men – I think it was a corporal in charge – banged on the side of the house and said that I was going to the camp and they were going to search my house. I told them to go ahead and help themselves; I had nothing to hide. They told me to pack and I started to do so but I found him following me round with a pistol. I told him he didn't need that, but he said, 'We know you don't like Argentinians.' I told him that was why, because we didn't live with guns.

Gerald Cheek had been a member of the local defence force.

I was at my parents' when my wife phoned and said there was a soldier at the front gate but he could not speak English. I went home quickly and found that a vehicle had arrived and there were several men, revolvers in hands, I believe under an N.C.O. They burst into the house and insisted that my wife and two daughters and I should all pack. I insisted that, if I was going, I wanted my parents to come, but they said no. They told me I could leave my wife and children behind. 'You will only be gone two or three days,' they said.

It was a very difficult decision to take. They came up to the bedroom, threatening my wife and daughters, hurrying us to get packed. My daughters were crying and my wife was on the verge. We talked very quickly and decided on the spur of the moment that it would be better to leave them. It was the most terrifying moment of my life. I didn't expect to see my family again and they didn't expect to see me again. I didn't know where we were going and asked if I needed my passport, but they said it would be 'internal'.

Eventually, fourteen people found themselves at the airport, not far from a Hercules transport aircraft with its engines running. They were convinced they were off to Argentina. Velma Malcolm could understand Spanish.

I heard the corporal in charge of the guard say to the others, 'I'd like to kill them all, the fat ones, the thin ones, the ugly ones, the lot.' It all coincided with the time that the Argentinians started getting jittery and also when the purely military people started to dominate the officers sent in to run the government.

The fourteen deportees did not go to the mainland but were taken by helicopter to Fox Bay on West Falkland, where their arrival was unexpected by the local Argentinian commander. The party stayed in the Fox Bay area for the remainder of the war, mostly confined to the house and garden of Richard Cockwell, the manager at Fox Bay West. The day after their arrival, Griselda Cockwell courageously went on the doctor's radio programme and asked Stanley if there was any mail for . . . and then read out the fourteen names so that relatives in Stanley would know the whereabouts of the deportees. The Argentinians were furious and she was never allowed to use her radio again.

The deportation party included Doctor Daniel Haines and his wife, also a doctor. Haines had been an outspoken opponent of the occupation and had tried to send military information – he was a former R.A.M.C. officer – to the task force by carefully phrased answers to medical questions on his daily radio schedule.

The departure of the two Haines doctors left medical affairs in Stanley in the hands of Doctor Alison Bleaney, who had retired to look after her young family, and newly qualified Doctor Mary Elphinstone, who was visiting the Falklands when the war started. One of Alison Bleaney's friends, Veronica Fowler, says:

There was a big gap when the Haines were taken away. The Argentinians wanted the whole hospital. Alison went to see Commodore Bloomer-Reeve and demanded vigorously, while breast-feeding her five-month-old daughter Emma, that some civilians should stay in the hospital. She was not embarrassed; Alison would breast-feed anywhere. She found it to be a good tactic; the Argentinians were very fond of babies. They agreed and Alison became the acting Senior Medical Officer, something she had never envisaged. Mary Elphinstone, a slip of a girl, became her assistant; it was a great strain on such young people. These two carried the civilian population through the war medically; it was a big boost to us in Stanley and an invaluable lifeline to the camp through their daily doctor's link.

An official Argentinian census showed that the population of Stanley declined from an original strength of about 1,500 people to 450 at the end of the war. Alison Bleaney was only one of several natural leaders who emerged when so many specialists left or were forced to leave. Many old people remained, a common reason being 'if we are going to die, at least we will die in our own beds', but it was the younger married people who often emerged to provide leadership and example to the dwindling population of Stanley. It would be wrong to mention names, because of the difficulty of compiling a full list, but I came away from my visit to Stanley after the war with the greatest admiration for those people who stayed on and led their community through the war. The people of Stanley will know their names.

Conditions became more tense after the British retook South Georgia and a curfew was imposed but the arrival of the task force in the area brought a huge boost to morale. The Harrier and Vulcan raids did not worry the Stanley civilians but the naval gunfire which took place most nights was a great trial, the navy's shelling was usually accurate but many of their targets were only just outside Stanley. Radio became more than ever a lifeline. When news was received that South Georgia had been retaken, announcer David Emesley made his first record on the local radio 'Georgia on My Mind', but the Argentinian on duty quickly ordered it taken off the turntable. Several presenters were able to play records with double meanings but Steve Whitley, the local vet, is the best remembered for his resourcefulness. He calmly sent greetings to 'Liz' on the Queen's birthday and played 'Happy Birthday to You'. The local radio was taken off the air on 29 April. The B.B.C. World Service was then listened to even more avidly, particularly an evening programme 'Calling the Falklands'. Barbara King at the Upland Goose was one of many regular listeners.

Before the war, it had only been on Sundays but now it came on every night. It gave us all the news and told us what support there was for us in the U.K. There were messages from families that we knew. It really made you feel good, because you knew someone cared. There was no other way; there was no mail and we knew the British task force was here but we couldn't hear them or know exactly where they were. The priests staying at the hotel were telling us the Virgin Mary and God were on their side and that didn't leave much left for our side, so that programme was a real lifeline.

We used to sit huddled round the radio, everyone 'shooshing', and we knew people all over the Falklands were listening. To start with, the B.B.C. used two frequencies but then the Argies started jamming so they used two more as well. Sometimes they managed to jam them all. There was nothing more frustrating then; we used to tape it and try to decipher it afterwards. There was one message from a child in England to her pen-friend here, telling her that her daddy was on the way with the task force to save her and her family. That made us all cry.

When he signed off each night, he always said, 'Heads down, hearts high. Nighters.' He wanted to end on a cheery note and the least we could do was cheer up a bit.

CHAPTER ELEVEN

Reinforcements

As the days and weeks passed without a voluntary Argentinian withdrawal or a negotiated settlement, it was judged that 3 Commando Brigade and its supporting ships and aircraft could not alone complete the successful reoccupation of the Falklands and preparations proceeded for the dispatch of further British forces. Everyone was conscious that the original task force had no reserves within 8,000 miles of the Falklands unless more help were sent.

The most urgent need was believed to be for more aircraft; the task force was hoping to draw the Argentinian air force into action and more Sea Harriers were expected to be lost. Both the Fleet Air Arm and the R.A.F. hurriedly prepared to send help. A new naval squadron – 809 Squadron – was formed at Yeovilton; reserve aircraft were brought from store at St Athan and from other places; former Sea Harrier pilots were brought back from all over the world. Eight further Sea Harriers were made ready by the end of April and they flew out to Ascension Island early in May, being refuelled up to seven times *en route*. The R.A.F. added a contribution from No. 1 Squadron at Wittering, which was equipped with the GR3 version of the Harrier. This type of Harrier normally operated as close army-support, ground-attack aircraft but it was now designated as a replacement fighter aircraft for the Sea Harriers which were expected to be lost in combat off the Falklands. The squadron hurriedly converted its aircraft to carry Sidewinder missiles and to operate from aircraft-carriers, practised ski ramp take-offs at Yeovilton and intensive air combat. Nine of the R.A.F. Harriers then flew down to Ascension.

The other major aircraft requirement was for load-carrying helicopters, an item specifically requested by Brigadier Thompson when he met Admiral Fieldhouse at the Ascension conference on 17 April. The Fleet Air Arm threw every available helicopter into the campaign and three new operational squadrons were now formed. (They were not really new squadrons but training squadrons which renumbered, because squadrons in the 800 series automatically had the 01 priority in the computerized issue of spares, while squadrons in the 700s only had an 02 priority.) 847 and 848 Squadrons, equipped with Wessex helicopters, were formed at Yeovilton, which was a hive of activity throughout this period. Culdrose provided the third new load-carrying squadron when ten Sea King anti-submarine training helicopters of 706 Squadron were stripped of their submarine-detection equipment and became 825 Squadron, crewed by instructors and pupil crews who had never flown commando-type operations, 'flung green into the field' as one officer says. The R.A.F. contributed five of their huge Chinooks from 18 Squadron at Odiham. It was later calculated that nearly 200 helicopters operated with the task force, all but the American-made Chinooks being manufactured in England by Westlands.

The transportation of the Harrier and helicopter reinforcements presented a problem. The navy had no aircraft-carriers left and, although the Harriers might just be able to reach the task force from Ascension, their delivery by air in this way would be a risky project. Again the STUFT operation saved the day when the Cunard container-carrying ships *Atlantic Conveyor* and *Atlantic Causeway* were chartered. These large ships, each nearly as big as H.M.S. *Invincible*, had proved uneconomical in their planned North Atlantic run. The *Conveyor* was lying idly in the River Fal and was immediately available in perfect condition. She moved to Devonport, was loaded with a mass of army stores below decks, and her open upper deck became an aircraft park, lined on either side by a weather shield of piled freight containers. Six Wessexes of 848 Squadron and the five R.A.F. Chinooks were loaded and a Sea Harrier arrived to provide a very public display of 'flying on'; the Ministry of Defence did nothing to correct press reports that twenty Harriers were also loaded but the ship sailed on 25 April without the Harriers, which caught up when *Atlantic Conveyor* reached Ascension. One Chinook was off-loaded for work at Ascension and eight Sea Harriers and six R.A.F. Harriers were flown aboard. There was no room for the remaining three R.A.F.

Harriers and these were left at Ascension temporarily. The *Atlantic Causeway* sailed later with more stores and helicopters.

Hard on the heels of the additional aircraft followed a large Army reinforcement. A request by Admiral Fieldhouse for more troops led to a War Cabinet decision, taken on 25 April, to provide a further full brigade of infantry with its supporting arms. But there were problems over the selection of the units required. It was decided not to rob Britain's front-line defences in the British Army of the Rhine. The 1st Infantry Brigade, based at Salisbury Plain, was the Army's 'Ace Mobile Force' and most of it was available but it was a Nato-designated unit – although 3 Commando Brigade had also been so designated – and one of its three battalions was from the Territorial Army. The 5th Infantry Brigade at Aldershot was the 'out-of-Nato-area' unit earmarked for just such a task as this, but two of its three battalions – 2 and 3 Para – had already been attached to the original task force. The Army had already decided that, if more troops were needed, 5 Brigade was to go; United Kingdom Land Forces Headquarters had already replaced the two missing battalions and had sent the brigade to carry out intensive training in Wales.

The only original battalion in 5 Brigade was the 1st/7th Gurkha Rifles, based at Church Crookham. This battalion contained four full rifle companies, one more than the average British battalion, and, with 720 front-line infantrymen, would be the largest unit to serve in the Falklands. To replace the two battalions already withdrawn from 5 Brigade, the Army decided to allocate two Guards battalions from London District. The choice fell upon the 1st Welsh Guards at Pirbright, whose current tasks were mainly to provide ceremonial guards at Windsor Castle and being on stand-by for 'terrorism and hijack duty' at Heathrow Airport, and the 2nd Scots Guards at Chelsea Barracks who were providing the Buckingham Palace guard. The Scots Guards were preferred to the 2nd Grenadier Guards, also at Chelsea, because it was felt that the Scots Guards 'were ready for a break' after a longer period of duty at Chelsea than the Grenadiers.

It is difficult to imagine a greater contrast than mounting guard at one of the royal palaces and operations in the Falklands and doubts were expressed over the selection of the two Guards battalions. Some thought that at least one of the battalions in 1 Brigade should have been provided; the 1st Queen's Own Highlanders in particular were specially equipped for operations in a cold climate and were at a peak of training and fitness. The two Guards battalions realized that there were doubts about their suitability but reckoned that their sound basic infantry training and their regimental tradition and discipline would make up for any recent lack of field training. Both battalions benefited from two weeks of hard work in the Sennybridge training area in Wales, on terrain similar to that of the Falklands. 5 Brigade would encounter many problems not of their own making in the Falklands and would find some parts of their campaign to be an unhappy experience. There would be ill-informed and unjustified criticism. It should be borne in mind that this was a new formation with an inexperienced head-quarters, with some deficiencies of equipment, that two of its original battalions had already been lost to 3 Commando Brigade, that the replacement battalions and some of the minor units were only gathered together three weeks before the brigade embarked and that some of those in authority believed that the war would be over before 5 Brigade reached the Falklands.

The strength of this second wave of troops numbered approximately 3,200 men, making the potential landing force 10,500 troops strong, stretched out between the S.A.S. and S.B.S. men now hiding in their observation posts on the Falklands to the rearmost units of 5 Brigade still training in Wales. Of all these units, approximately thirteen and a half companies of infantry and some of the special forces units – perhaps 1,300 men in all – would come face to face with Argentinian forces; many of the remainder would be bombed and shelled, and all would get wet and cold and suffer other privations. The eight infantry battalions constituted the largest British force sent into conventional military action since the Second World War, just exceeding the force maintained in Korea during the war there.

The provision of a reinforcement brigade made the appointment of a new land-forces commander essential. The eight battalions of the enlarged force was almost equivalent in strength to a division. The Royal Marines had no such thing as a divisional organization and, with five of those eight battalions coming from the Army, it might have been thought that the Army would provide an experienced divisional commander and staff. But this did not happen; the Task Force was still under the control of the Royal Navy. Major-General Jeremy Moore, Commander Commando Forces on the outbreak of the war and then military adviser to Admiral Fieldhouse, was now designated Commander Land Forces Falkland Islands. Lieutenant-General Richard Trant, General Officer Commanding South-East District, became the military adviser to Northwood. The appointment of Jeremy Moore was a reasonable move. The Army did not have a spare divisional headquarters and Moore's old staff at Commando Forces Headquarters had become almost redundant after it had prepared 3 Commando Brigade for sailing. Moore was given an army deputy, Brigadier John Waters, but most of his staff officers came from his old headquarters at Plymouth.

Fifty-three years old Jeremy Moore had served thirty-five years in the Royal Marines, winning a Military Cross as a troop commander in Malaya and a bar as a company commander in Sarawak; he had commanded 42 Commando in Northern Ireland. Ironically, it was the actions of the I.R.A. which brought Moore to this new appointment; he should have retired before the Argentinians invaded the Falklands but a serious I.R.A. bomb injury to Moore's superior, Lieutenant-General Pringle, extended Moore's service by the few months necessary to give him this Falklands command, undoubtedly the most interesting for any Royal Marine officer since the Second World War.

There was some frustration in the Army over Moore's appointment; one senior officer who had to send units to the Falklands says:

The Army took a greater and greater role but the normal chain of command was cut and we had no direct line to our units in the field. Everything went through Northwood; it was a little clumsy sometimes and there was the feeling that Northwood played it too close to the chest. The Navy was determined to hang on to it.

It worked, but only because a lot of 'entrepreneurs' were fixing things; often people who should have been informed and involved were not done so. The war was certainly not fought, as far as we were concerned, as it had always been planned.

Fair comment maybe, but Major-General Moore would carry out his part in the campaign very effectively.

To transport 5 Brigade and the other units of this wave, the government requisitioned the *Queen Elizabeth 2*, another example of the almost breathtaking scale of commitment to the campaign. The cost of taking over this great liner and its crew was approximately £1 million per week (*Canberra* was costing about £700,000 per week). The guns and heavy equipment were loaded into two Stena roll-on roll-off ships, the *Baltic Ferry* and the *Nordic Ferry*, which sailed on 9 May, and the *QE2* followed from Southampton three days later with much public display. It would only take 5 Brigade three weeks to reach the fighting, compared to nearly six weeks for 3 Commando Brigade. Major-General Moore and his staff flew down to Ascension and were helicoptered out to *QE2* as it sailed past the island.

The Royal Navy had also been preparing reinforcements. A third aircraft-carrier, the new *Illustrious*, was the most important ship to be designated for the South Atlantic, but she would not reach the Falklands until after the war. The loss of *Sheffield* on 4 May hastened the sailing of what may loosely be called the 'Bristol group' of ships which sailed between 10 and 12 May. The large destroyer *Bristol* was the only ship of the now cancelled Type 82 Class and was fitted with an advanced form of American-designed communications system which would make her an alternative flagship if anything happened to *Hermes*. The *Bristol* group made a fast voyage south and arrived off the Falklands within two weeks. *Avenger* claimed the fastest U.K.–Falklands sailing of the whole war – fifteen days with an average speed of 29½ knots, beating her sister ship *Active* by one hour. The *Bristol* group was joined by more ships as it sailed south, *Exeter* giving up her task as guard ship off Belize and *Cardiff* being released from the Arabian Gulf patrol. The provision of *Cardiff* was made possible by an offer from the New Zealand Prime Minister, Mr Muldoon, to provide a ship for duty somewhere out of the war zone to relieve the Royal Navy. This offer was made at a 10 Downing Street dinner for Mr Muldoon during a conference of Commonwealth ministers. Admiral Lewin, also present, promptly accepted and this led to the eventual relief of *Cardiff* in the Gulf by the New Zealand ship *Canterbury*. *Cardiff* joined the *Bristol* group via Gibraltar, three of *Cardiff*'s crew being married in the local Register Office to fiancées flown out from England. The public announcement of the valuable New Zealand help embarrassed the Australian Government, whose only action so far had been to insist that Australian exchange officers be taken off task-force ships. Mr Fraser, the Australian Prime Minister, then offered not to enforce the agreed sale of *Invincible* after the war; the British promptly accepted this, to the dismay of the Australian Navy, who thus saw their naval air arm without an aircraft-carrier for the foreseeable future.

The naval reinforcements totalled nine surface warships: the *Bristol*, the Sea Dart destroyers *Cardiff* and *Exeter*, the Leander Class frigates *Andromeda*, *Minerva* and *Penelope* and the Type 21 frigates *Active*, *Ambuscade* and *Avenger*. Three more submarines also sailed south: the *Courageous* and the *Valiant*, and the smaller *Onyx* which would be used, among other duties, for putting S.B.S. parties ashore on the Falklands without having to surface. The number of warships now committed to the South Atlantic was three aircraft-carriers, twenty-four destroyers and frigates, and six submarines. The total number of ships of all kinds was more than one hundred and the number of sailors, soldiers and airmen in the Task Force under Admiral Fieldhouse's command was more than 25,000. There would be no significant further reinforcement before the end of the war.

CHAPTER TWELVE

Interlude

It has taken four chapters of this book to describe the first four days of task-force operations; the following two weeks will require only one short chapter. Argentinian surface warships remained completely absent; the Argentinian air force came out only once or twice; the British landing force was not yet ready. Admiral Woodward's ships remained off the Falklands during this period, constantly on guard against attack and continuing with the tasks of reconnaissance and of isolating and wearing down the Argentinian troops in the islands. There were occasional bursts of minor action. The crews of the British ships felt very isolated so far from any friendly force or base; Sandy Woodward says 'there was a certain overlying menace to the period'.

The most constant activity was that of providing anti-submarine protection, a task which went on day and night. The main burden fell on 820 and 826 Squadrons in *Invincible* and *Hermes* respectively. 826 Squadron lost two Sea Kings during this period. The first suffered a mechanical failure, 'pancaked' on to the sea and sank during the afternoon of 12 May. The second was lost during the night of 17 May; its radio altimeter failed and it crashed into the sea. The crews of both helicopters were rescued. There was certainly no let-up for the Sea Harrier pilots during this seemingly quiet period; after *Sheffield* was hit the number of standing air patrols was increased in an attempt to make up for the lack of airborne early warning. The average pilot in 801 Squadron, which bore the brunt of the air-defence duty, would fly fifty-seven sorties before the war ended, an average of more than one each day. To this must be added time spent on cockpit-standby and other states of readiness. There were no tours of duty and no one knew how long the war would last. No encounters took place with Argentinian aircraft in the period between the first hectic day of operations on 1 May and the arrival of the landing force, but two Sea Harriers of 801 Squadron and their pilots were lost. Soon after dawn on 6 May, Lieutenant-Commander John Eyton-Jones and Lieutenant Al Curtis were on routine patrol in misty conditions, with visibility down to half a mile or so. A possible radar contact to the south of the task force was reported by one of the ships and the two Harriers, flying independently of each other because of the bad visibility, both made towards it. Their own radar blips – still well apart – then disappeared as they lost height and the Harriers were never seen again. The Argentinians made no claims and it can only be assumed that the two aircraft collided in the murk. Eyton-Jones was an 'old hand' at Harrier work, having been recalled from a training squadron for the Falklands operation; Curtis, who had earlier flown both with the New Zealand Air Force and the R.A.F., was on his first Sea Harrier tour and was one of the successful pilots from the first day of action only five days earlier, when he had shot down an Argentinian Canberra.

The Sea Harriers flew few bombing raids in this period, in order to conserve this small force of aircraft which had now suffered three losses. Another Vulcan raid from Ascension, Black Buck Three, was prepared for the night of 16/17 May, with Flight Lieutenant Withers and his crew designated for the flight and Squadron Leader Reeve's crew to fly the reserve Vulcan. But the operation did not take place. It was later stated that a forecast of strong head winds caused the cancellation, but this was not the full story. It is said that the Vulcans' parent formation in Britain, 1 Group, had intervened with a signal urging that the Vulcans should not fly unless a further 1,000 pounds of fuel could be passed at certain refuelling transfers. This was a reasonable safety measure but the tanker planners at Ascension could not provide the extra fuel with the Victors available and so Black Buck Three was cancelled. The Vulcans had to be temporarily withdrawn from Ascension now, to provide parking space for other aircraft required for more urgent operations.

The task force was particularly passive in the immediate aftermath of the *Sheffield* being hit but a more active policy was adopted on 8 May; Sandy Woodward wrote in his diary, 'Finally decided to get on with my war – air activity is not getting us anywhere . . . *Sheffield* set me back and it took me this long to shake the last effects off.' Woodward decided to commit four of his ships to offensive action that night. *Brilliant* was sent 'to terrorize the north end of Falkland Sound'. The frigate steamed round to this exposed position, deliberately showing lights, but the Argentinians were not tempted and she returned to the main group the next day. *Alacrity* was sent alone to the gun-line off Stanley and fired ninety shells at suspected Argentinian positions near the racecourse. She too returned without drawing any response. A more ambitious move was to send a '42–22 combination missile trap' to appear at dawn off Stanley as a deliberate bait to Argentinian aircraft. The Type 42 destroyer's Sea Dart missiles would attack any raider approaching at medium range; the Type 22's Sea Wolf would engage any aircraft which penetrated to close quarters. *Coventry* and *Broadsword* were the two ships dispatched on this hazardous duty. Admiral Woodward's diary reads, 'If this combination does not work, we are out of business . . . I'm worried for them, putting two surface ships off an enemy airfield in clear daylight, but it is essential.'

The 9th of May proved to be an interesting day. Two Sea Harriers were launched to bomb Stanley airfield as a further prod to the Argentinians to send up their own aircraft, but the Stanley area was covered with low cloud. The bombing raid was abandoned and the two Harriers carried out a patrol during which the Argentinian trawler *Narwhal* was picked up on radar. This was the second encounter with this ship; *Alacrity* had shooed her away on the evening of the task force's entry into the Falklands operational area ten days earlier. The Harriers fired cannon ahead of the ship, but this was ignored so they dropped the bombs they were carrying. One bomb struck the trawler but it did not explode because it was fuzed to be dropped from 5,000 feet. Admiral Woodward then ordered that *Narwhal*, which was unarmed, was to be captured and Sea Kings from *Invincible* embarked a boarding party of S.B.S. men. Before these men could be put aboard, however, two further Harriers arrived on the scene and requested permission to attack. *Coventry*, the Harrier control ship in the area, was unaware of the boarding plan and gave permission. The Harriers attacked with cannon fire and caused severe damage and some casualties. The S.B.S. boarding party then arrived and the trawler's crew submitted without a fight. The trawler was not considered salvable and was left to sink. One wounded Argentinian prisoner died. The Argentinians protested at these attacks on a 'defenceless ship', but one of the captives turned out to be a naval officer, and equipment and documents found on board confirmed that the *Narwhal* was being used as a shadower.

Coventry was soon involved in further action. She and *Broadsword* had taken station only ten miles off Stanley and they picked up radar echoes of five aircraft approaching the Falklands from the west or north-west. These were a Hercules transport and four escorts. The Argentinians must have considered themselves safe, still being forty miles or so from Stanley. *Coventry* fired three Sea Darts at extreme range – the first Sea Dart missiles fired during the Falklands war. *Coventry*'s computer assessed that the second Sea Dart had exploded and was a likely hit. *Broadsword*'s doppler radar, which could track aircraft among land returns, indicated that the aircraft emitting the two smaller echoes could have been destroyed and this was confirmed by the radio chatter of the pilots of the other Argentinian planes. The two aircraft destroyed were Skyhawks; both pilots were killed. *Coventry* picked up another target three hours later, much closer this time, and another Sea Dart was fired. A Puma helicopter, on its way to help the *Narwhal*, was hit and destroyed; its three crew men died. These successes by *Coventry* brought an end to the 9 May actions.

The next day, 10 May, Admiral Woodward ordered another of his ships to go in to provoke the Argentinians and to carry out some important reconnaissance. The Type 21 frigate *Alacrity* was chosen for an operation which demonstrated the Royal Navy's traditional skill in small-ship work. No British ship had yet ventured into any part of the area in which *Alacrity* was being sent. Her captain was Commander Chris Craig.

My orders were to probe the southern garrisons along both East and West Falkland and transit Falkland Sound from south to north, to carry out interdiction against Argentinian resupply. We were also to find out if the Sound was mined; this was not mentioned but when I said, 'I suppose you would like me to go up one channel and back down the other if there was time and see that they were clear' the task group commander gave a hollow laugh. There was no mention of mines but we both knew that's what he would have liked. We had no mine-detection gear.

After reconnoitring the rocky natural harbours on the south coast of both East and West Falkland that afternoon with her Lynx helicopter, *Alacrity* entered the narrow southern straits of Falkland Sound just before midnight, covered by low cloud, mist and rain. The Lynx was again launched to reconnoitre Fox Bay and to act as a diversion, although flying conditions were appalling. *Alacrity* then detected a moving radar contact six miles ahead in the channel and increased speed to close the gap and identify the contact. A star shell was fired in the hope of achieving the identification but this was not possible and fire was opened, the first dozen rounds being fuzed for air-bursts in the hope of stopping the ship without causing it serious damage. The firing was stopped after two minutes to assess the effect but the target was seen, on radar, to be fleeing for the shelter of the land. High-explosive shells were then fired and at least three hits were observed through the heavy rain, followed by a large orange explosion; the radar contact faded from *Alacrity*'s screen two minutes later. The ship was later identified as the Argentinian naval transport *Isla de los Estados*, which was carrying 325,000 litres of aviation fuel and military vehicles. The number of casualties on the Argentinian ship is not known. This was the only action between surface ships during the entire war. *Alacrity* was met by her sister ship, *Arrow*, at the northern end of Falkland Sound and the two frigates returned safely to the carrier group.

The next thirty-six hours were quiet, but then another burst of activity erupted on the afternoon of 12 May. *Glasgow* and *Brilliant* were now the 42–22 combination off Stanley. *Glasgow* shelled targets on shore during the morning and eventually the Argentinians on the mainland decided to send out aircraft; at least eight Skyhawks were committed to action. The first four arrived as the Sea Harrier air patrol covering the two ships was being changed. The four Argentinian planes came in low across the Falklands but were detected by *Brilliant*'s radar. The targets were initially handed over to *Glasgow*'s Sea Dart system to engage, but the Skyhawks were too low and too fast and they were not 'acquired' until the range was too short for Sea Dart. Captain John Coward of *Brilliant* describes how his ship took over. It was the first operational firing for her Sea Wolf close-defence missiles.

After their Sea Dart failed to fire, *Glasgow* held steady while we fought the battle. We hoped to be able to fire Sea Wolf fully automatic. We took our hands off and let it fire itself to see how it behaved, which it did perfectly. Two missiles were fired and they took out the first two aircraft. They both disintegrated in balls of flame about a mile away. A third aircraft ran straight into the disintegrating parts of these two and it crashed into the sea as well. Its engine catapulted right over our flight deck. The fourth aircraft passed overhead and her bombs hit the water and skipped over the stern of the ship. We asked the flight deck crew if they had seen the markings on the aircraft; they said they had even seen the markings on the bombs!
We were very happy with the Sea Wolf.

The Argentinians had indeed lost three out of the four planes and all three pilots were killed.

The second group of four Skyhawks came in twenty minutes later, straight out of the sun. This time, *Glasgow*'s Sea Dart system acquired the targets at twelve miles range but a frustrating malfunction in the missile system occurred and again it would not fire and again *Brilliant* had to take over. Captain Coward resumes:

It was then that I realized how lucky we had been with Sea Wolf the first time. The system is designed to identify fast, incoming missiles, selecting the one which is definitely going to hit the parent ship and rejecting the others. On this occasion, however, our Sea Wolf rejected the aircraft as 'non-missile' targets and would not fire. All the planes dropped bombs. After this, we realized that Sea Wolf had a problem when faced with a group of aircraft instead of missiles and a rapid exchange of messages with Marconi took place. It was settled by an adjustment to the computer. We did fire Sea Wolf again, later, but we never had the same clear opportunity as on that day.

Glasgow had been hit by one bomb but this passed clean through the ship's main engine room without striking any substantial obstacle and went out the other side without exploding. The entry and exit holes were almost exactly level. A third Argentinian raid appeared to be approaching but two Sea Harriers were now arriving and no further attack took place.

It had been a success for the 42–22 missile trap, with three Skyhawks destroyed, much learnt about the Sea Dart and Sea Wolf systems, one ship slightly damaged and no British seamen hurt. There was further grief for the Argentinians when the Skyhawk whose bomb had hit *Glasgow* flew too near an area around Goose Green airfield which was prohibited to Argentinian aircraft. Argentinian anti-aircraft gunners shot down the Skyhawk and the pilot was killed. *Glasgow*'s damage was not serious. The holes were soon sealed, but damage to her electrical system later caused problems and she left the task force before the end of the war; she would be the first warship to return to Britain from the Falklands. The task force now had only one Sea Dart ship fully operational, *Coventry*, so the missile trap and gunfire groups were temporarily withdrawn from the Falklands coast in daylight – a limited success for the Argentinians.

The intended landings of the marines and paras was now less than ten days away and the main interest switches to the more active preparations being made for the landings and to a campaign of deception being carried out. Parties of Special Air and Special Boat Service men had been in and out of the Falklands by helicopter for the last two weeks; now, a major raid was planned. The target was the grass airstrip on Pebble Island, off the north coast of West Falkland, where a considerable part of the Argentinian Pucará force was located. The Pucarás would be a major threat to British ground forces when the landings took place. The plan called for *Hermes*, *Broadsword* (as air-defence ship for *Hermes*) and *Glamorgan* (to provide bombardment of the shore) to steam round the north of the Falklands to land an S.A.S. raiding party. The operation was planned for the night of 13/14 May but there was a postponement of twenty-four hours owing to a delay in the sending back of intelligence by the S.A.S. patrol which had been earlier landed by canoe.

The three warships approached the land after dark on the 14th, being much buffeted by heavy seas and a gale. *Broadsword*'s Sea Wolf became defective and she could not keep up. *Glamorgan* went in to seven miles off the coast to give covering fire and to be available if helicopters had to ditch. The more valuable and more vulnerable *Hermes* closed to within forty miles of the land, much closer than planned, to give the Sea Kings of 846

Squadron a shorter flight in the strong winds. Thereafter, everything went according to plan. Forty-five (or forty-eight, accounts differ) men of D Squadron, S.A.S., who had earlier encountered so much trouble on the glacier in South Georgia, were landed together with an artillery observation team. Every plane at the airstrip was destroyed, either by S.A.S. demolition charges or by *Glamorgan*'s gunfire; a ton of ammunition was blown up; a radar station may have been hit by *Glamorgan*'s shelling; a half-hearted Argentinian counter-attack was repulsed; only two S.A.S. men were slightly injured. The aircraft destroyed were six Pucarás, four Turbo-Mentors (also a ground-attack aircraft) and a Skyvan transport. The grass airstrip was also badly cratered.

Helicopters returned for the raiding party five hours later, and *Hermes* and *Glamorgan* turned east and steamed as fast as the state of the sea would allow. They were in considerable danger if caught by Argentinian aircraft or submarines, but they returned safely to the main group. The enraged Argentinians at Pebble Island confined settlement manager Griff Evans and twenty or so civilians to the manager's house for the remainder of the war.

The general reconnaissance task of the S.A.S. and S.B.S. continued throughout this period. Little information has been released about the work of these secretive units. The Argentinians probably knew they were operating; the helicopters of 846 Squadron came in most nights and must sometimes have been detected, but none of the reconnaissance parties was discovered by the Argentinians before the main landing. The frigate *Alacrity* went back into Falkland Sound on the night of 16/17 May and landed two teams of S.B.S. men and an artillery observation officer who went up on to Sussex Mountain which overlooked part of the selected landing area. The Geminis which had landed these men returned to *Alacrity*, bringing with them a piece of 'Falkland rock' for Commander Craig's birthday – a totally smooth piece of stone weighing eighty pounds and claimed to be the 'first piece of the Falklands recovered by British forces'.

Glamorgan was given an interesting task during the period; her captain, Mike Barrow, the most senior captain in the task force, was an old term-mate of Admiral Woodward at Dartmouth. He describes his task.

All of our orders were simple and clear. We were *'on our own, to produce such deception activities as we could with our resources, to make the enemy believe that the amphibious landing would take place in the south of East Falkland'*. We were given a particular area – from the north side of Choiseul Sound entrance to Fitzroy.

We received a daily sitrep of shore conditions and positions, which I presume came from islanders via the S.B.S. or S.A.S. We went in each night and shelled Stanley and then moved along to the deception area. We put ourselves in the position of the amphibious planners and tried to choose the locations of good landing positions. We used a lot of high power, a lot of fuel, a lot of ammunition and lost a lot of sleep. I met Major Keeble, 2 Para's second-in-command, after the war, on holiday in Denmark. He told me he was surprised to find some of the Argentinian defences facing the wrong way, so we must have done some good. We also did some communications spoofing, talking to non-existent landing craft.

This story of deception now leads on to one of the war's best-kept secrets – the story behind the British Sea King helicopter which went to Chile. The helicopter crash-landed near the Chilean port of Punta Arenas, 400 nautical miles west of the Falklands and well over 500 miles away from the task force's normal helicopter launching area. The three crewmen gave themselves up to the Chileans and were soon returned to Britain. The Ministry of Defence in London announced that the helicopter had become lost in bad weather while on a reconnaissance flight – blatant cover story.

In the original version of this book, I floated the suggestion that the helicopter flight was a hoax operation, that no S.A.S. men were landed in Argentina, but that it was hoped that the possibility of S.A.S. having been landed would cause the Argentinians to fear another Pebble Island-type raid and withdraw their strike aircraft from the forward airfields and out of range of the Falklands – all this *on the eve of the British landings on the Falklands*. I based that supposition on my belief that the British War Cabinet would never escalate the war to the mainland.

I now think that the hoax scenario was wrong. There has still not, at the time of writing this edition, been any authoritative statement, but several fragments of information have come my way. They all lead to the view that the Sea King did put down S.A.S. men who were to operate near those forward mainland

1. Air activity at Ascension Island – a VC-10 troop transport, a Hercules freighter, a Task Force Sea King and assorted stores. Also present would have been many Victor tankers, Nimrods and, sometimes, Vulcans

2. Victor tanker crews at Ascension wait for a briefing.

left 3. The runway at Stanley after the first two Vulcan raids; the craters made by bombs falling in soft ground have caved in and appear much smaller than those in the harder ground near the runway.

right 4. The runway at the end of the war. Some of the Vulcan bomb craters are still unfilled (in the foreground) and the temporary repair to the crater beyond the mid-point is also visible. Nine captured Aeromacchis are on the airfield.

5. A Sea Harrier of 800 Squadron lands on *Hermes* after a combat air patrol, its two Sidewinder missiles unfired. Messages chalked on the 1,000-lb bombs: 'Biff' and '*Accipe hoc!*'

6. An R.A.F. Harrier GR3 makes a dusk landing on *Hermes*.

7 and 8. Exocet. *Sheffield* and *Atlantic Conveyor* after being struck by the missiles. Both ships were destroyed by fire.

9. *Hermes* in the operational zone, on her deck a Lynx, a Sea King, a mixture of Sea Harriers and GR 3s and many stores stacked forward.

10. Bravo November, the one Chinook to survive the sinking of the *Atlantic Conveyor*, delivers bombs to *Hermes* from a supply ship. There was an air alarm soon after this picture was taken and the deck crews all ran to the rear of the flight deck to be away from the bomb stores.

13. Men of C Coy, 40 Cdo, dig in at San Carlos.

top left 11. D-Day. Heavily laden men of 2 Para in the Continental Lounge of *Norland* just before transferring to landing craft. These are mostly men of D and Support Coys. The man facing the camera on the left, Cpl Paul Sullivan, was killed at Goose Green.

bottom left 12. Part of 45 Cdo, feeling exposed by their late landing in daylight, moving to Ajax Bay. *Fort Austin* is in the background.

14, 15 and 16. Bombing. *Ardent* (top), *Antelope* (bottom) and *Sir Galahad* (right), all hit with heavy loss of life.

17 and 18. The Falklands terrain. Marines on the move near Mount Kent and guns of 7 Battery, 29 Cdo Regt, in action in front of the hills around Stanley. The top photograph shows the type of rocky outcrops in which the Argentinian defences were positioned to cover the open ground around.

19, 20 and 21. The dead. British dead are buried in body bags at Ajax Bay after the Battle of Goose Green and bombing attacks on Ajax and San Carlos. The body of Lt-Col. Jones is at the far end. The newly arrived Maj.-Gen. Moore stands at attention on the left; the ITN camera team film on the right. Shown below are dead Argentinians on Mt Harriet.

22 and 23. Lt-Col. Vaux of 42 Cdo gives orders to his officers before the brilliant attack on Mt Harriet and talks with some of his Marines after the battle.

24 and 25. Prisoners of war. Cpl Geoff Saxton, crewman of a 3 Bde Air Sqn Gazelle, arranges the loading of prisoners near Mt Harriet, and (below) Argentinians captured by 2 Para are escorted away from Wireless Ridge.

26 and 27. 'It's all over!' 7 Platoon, G Coy, Scots Guards on Tumbledown and men of 3 Para on the outskirts of Stanley.

28 and 29. Tired Marines of 4 Troop, Y Coy, 45 Cdo, led by L/Cpl Wiggington, enter Stanley after 'yomping' right across East Falkland, while the Argentinians march out of Stanley and are disarmed by J Coy, 42 Cdo, on the airport road.

30 and 31. *Canberra* brings 3 Commando Brigade home, the greeting at Southampton. Below, no caption needed.

air bases, probably in a covert manner, by installing devices which could detect the take-off of Argentinian aircraft, possibly even using miniature television cameras. The main targets were the all-important Super Étendards capable of carrying Exocets at Río Grande airfield which was directly on the way to Punta Arenas. The story goes on that some of the S.A.S. men were detected and engaged by the Argentinians, that one S.A.S. man was killed, at least one badly wounded and some taken prisoner. This may not have happened until the main task was completed; the S.A.S. men are said to have done well and there is talk of an 'unpublished D.S.O.'.

An obvious question should be asked: Why did the Argentinians not make capital by displaying their prisoners? The answer given me was that the prisoners were secretly exchanged for Lieutenant-Commander Astiz, the naval officer captured at South Georgia and taken to England and who was wanted by several countries over the disappearance of their people in the 'dirty war'. There is no confirmation, as I write this, on any part of this account, but I do now believe that there was a link between the 'Chile helicopter' and S.A.S. men operating on the Argentinian mainland.

There were no further major incidents before the landings, but some minor action. A pair of Sea Harriers caught two Argentinian supply ships in Falkland Sound on 16 May. One ship, at Fox Bay jetty, was attacked with cannon fire. This ship was the well-known naval auxiliary *Bahía Buen Suceso*, which had been needling the British since the 1950s, when she fired on a British Antarctic Survey ship; more recently, she had taken Señor Davidoff and his scrap-men to South Georgia and set off a train of events which led directly to this war. She was beached soon after the Harrier attack, the Argentinians saying she had drifted ashore, the Fox Bay civilians that the crew deliberately beached her so that she would not have to go to sea again. The second ship was far enough away from the settlement to be bombed, and it was sunk. The following day, two Super Étendards attempted another Exocet attack on a British ship reported by the Argentinians off Stanley – the *Glamorgan* – but failed to find their target and took their scarce missiles back home.

The first reinforcements arrived for the task force. *Uganda* reported her presence as a hospital ship on 13 May but remained separate from the warship group. The submarine *Valiant* arrived three days later, making a total of four submarines now in the area of operations. The large container ship *Atlantic Conveyor* then arrived and transferred its fourteen Harriers to the two aircraft-carriers. The eight Sea Harriers of 809 Squadron were split evenly between *Hermes* and *Invincible*, where they were easily absorbed into the existing squadrons; the six GR3s all went to *Hermes*. The R.A.F. pilots found that they were not now required for air-defence work, because so few Sea Harriers had been lost, but were to revert to their original role of ground attack. There was some dismay over this after the effort put into their air-combat training, effort which could have been used in sharpening up the ground-attack techniques which would now be required. The R.A.F. aircraft were soon in action. The day after his arrival in *Hermes*, Wing Commander Peter Squire took off with one of his pilots on a familiarization flight and found himself directed on to a suspected Boeing 707 shadower 200 miles away, but no contact was made. The next day, Squire and his two flight commanders, Squadron Leaders Jerry Pook and Bob Iveson, attacked some Argentinian fuel dumps near Fox Bay.

On 19 May, the first contact was made with the force of amphibious troop-ships coming down from Ascension.

CHAPTER THIRTEEN

Plans and Preparations

After as much training as possible for the military units and a vast transfer and restowage of stores, Commodore Mike Clapp's amphibious ships had sailed from Ascension in various groups between 30 April and 8 May. There were thirteen troop-carrying or supply ships and three escorts. The voyage south was without major incident, although Russian reconnaissance aircraft operating from Luanda found and tracked the force. The British ships were within range of Argentinian Canberras and possibly other aircraft operating from the mainland but, as has been said, the Russians were not passing information and the Argentinians were unaware of the British approach. The ships of the landing force finally made contact with Woodward's battle group about 200 miles north-east of the Falklands on 18 May. Marine Steve Oyitch of 45 Commando, in *Intrepid*, did not see the warships until the next day.

My mate and I thought we would come up and get a bit of fresh air before breakfast. We went out on to the big flight deck aft. It was dismal weather, drizzle and a big swell, overcast, a fairly strong wind, but everywhere we looked – right to the horizon – there were ships. You didn't think Britain was capable of putting a fleet together and there it was, spread out before your eyes. I don't know exactly how many. I have never seen anything like it and I don't think I ever will again. We couldn't say anything, we were just awestruck. *Hermes* was right next to us – she was covered in rust and the paint looked old, she had been at sea for so long. But it was beautiful to see her, just beautiful. She was just turning into the wind, causing an arc in the water, and we watched two Harriers take off on a routine combat air patrol.

To plan a landing on an enemy coastline is one of the most interesting but most difficult of military exercises. Most of the detailed work was done at Ascension by a team of officers from 3 Commando Brigade, with Commodore Mike Clapp and his staff also being involved. It was quickly decided – both in London and at Ascension – that every effort should be made to avoid an opposed landing and that the whole of the reinforced 3 Commando Brigade should be put ashore in an undefended area with sufficient support and stores to fight, initially, a defensive battle. Once that was achieved, the next phase of operations could develop. The eventual target was always acknowledged to be Stanley; whoever controlled Stanley controlled the Falklands. The landing would be known as Operation *Sutton*.

The planners had to face the three usual questions: Where? When? How? The 'When?' was the easiest of these to answer. The only advantage in delay would have been if the Argentinian air forces were being whittled down by Woodward's battle group, but this was clearly not happening; Woodward had warned 3 Commando Brigade, as early as 6 May, that he could not guarantee air superiority for the landings. These had to take place as soon as possible; any delay would risk a land campaign being prolonged into the winter and would stretch the ability of the supporting warships to remain at sea. That was why there was no attempt to wait until 5 Brigade in QE2 arrived. An early proposal to land on 16 May had to be abandoned and a new one substituted – the early hours of Thursday 20 May.

The 'Where?' presented the most difficulty. Everyone knew that the Argentinians had concentrated their forces around Stanley and everyone agreed that an opposed assault on that area should be avoided. But this led to a conflict between the requirements of the land forces, who needed a short approach march to Stanley and short lines of communications, and the naval requirement, which wanted an anchorage as secure as possible against bad weather and Argentinian air attack. An early plan to land on West Falkland was quickly ruled out; its only merit was that it would establish a symbolic reoccupation which might lead to a settlement, but it would place the British forces nearer to the Argentinian mainland air bases and further from the ultimate objective of Stanley and would require a second amphibious landing across Falkland Sound to East Falkland. There followed much poring over the charts and studying of reports. Major Ewan Southby-Tailyour, a former commander of the Royal Marine garrison at Stanley, produced the notes he had made during his yachting tours of the Falklands coast; various Falkland civilians who happened to be in Britain

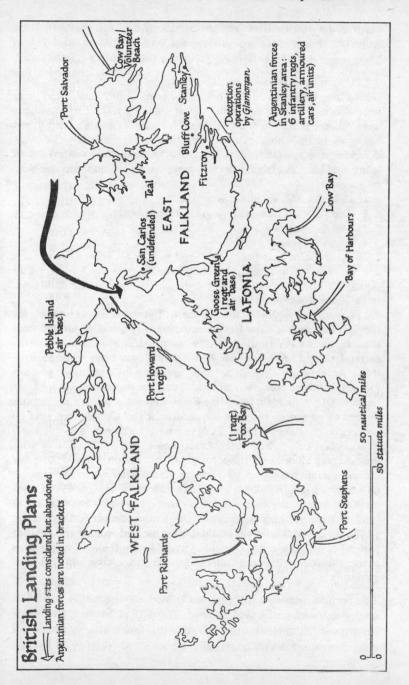

British Landing Plans

⇐ Landing sites considered but abandoned
Argentinian forces are noted in brackets

Cow Bay/Volunteer Beach

Port Salvador

Stanley

Bluff Cove

Teal

Fitzroy

Deception operations by Glamorgan

(Argentinian forces in Stanley area: 6 infantry regts, artillery, armoured cars, air units)

EAST FALKLAND

San Carlos (undefended)

Goose Green (1 regt and air base)

LAFONIA

Low Bay

Bay of Harbours

Pebble Island (air base)

Port Howard (1 regt)

(1 regt) Fox Bay

WEST FALKLAND

Port Stephens

Port Richards

50 nautical miles

50 statute miles

contributed information and the S.A.S. and S.B.S. parties ashore were obtaining more up-to-date information. Brigadier Julian Thompson describes the long process of selecting a suitable site.

Commodore Clapp and myself had frequent consultations with charts spread over the cabin floor of one or other of us in *Fearless* and there were also regular meetings with the full planning team. First questions were, 'Where is the enemy, and where is he not?', and we tried to put ourselves in the shoes of our Argentinian counterparts if we were defending the Falklands from invasion. We then asked ourselves how many suitable, undefended landing places were left and came up with four answers:

1. Cow Bay and Volunteer Beach.
2. The Port Salvador area – a great inlet of water.
3. The San Carlos area.
4. Various bays in Lafonia.

This appreciation had been completed before we reached Ascension and was handed to Jeremy Moore at the *Hermes* conference on 17 April. We said we could not refine the choice any further until more intelligence was given.

Later, when Georgia fell to us so easily, there was talk of us going for the Salvador and/or Cow Bay areas; there was a feeling around that, though it was risky landing so close to Stanley, a quick Argentinian surrender could be engineered. But it would have been a bad solution if the Argentinians decided to stand and fight, as they did, and I am pleased that those areas were not imposed upon us. Jeremy Moore came down to talk to us again and Mike Clapp and I said that, of the options we had listed earlier, the Cow Bay, Salvador and Carlos areas were all acceptable on different balances of closeness to Stanley and suitability of anchorages, though privately Clapp and I both preferred the Carlos area. Jeremy Moore took this information to London, together with our outline operational orders for all three areas. After we left Ascension, he told us on secure voice telephone that we were to go for San Carlos and the formal order from Fleet was dropped to us by Hercules on 12 May. Actually we were all ready for San Carlos, having had the feeling all the time that this would be the best area, and we had our detailed plan all ready. I gave my formal O Group to the C.O.s of major units in the wardroom of *Fearless* on 13 May – it was a tense and expectant gathering.

The conflict between military and naval requirements had been settled in favour of naval factors. The Royal Marine Brigadier Thompson, as is clear from his account, quite accepted this; an army brigadier might not have done so so readily. The San

Carlos landing provided a number of suitable beaches, two fine natural anchorages, rising ground on all sides which could be used as the initial infantry perimeter and as the location of the Rapier air-defence missile posts and, most important, a virtually Exocet-proof area because the radar of these missiles could not operate near land. The drawbacks were a known Argentinian defensive position on Fanning Head which dominated the entrance to the anchorage, the presence of a large Argentinian garrison and some Pucará ground-attack aircraft at Goose Green, sixteen miles away, and, above all, the fifty-mile distance from Stanley across whose almost trackless expanse the British troops would eventually have to move. The long and arduous march of many of 3 Commando Brigade's units and the later coastal leap-frogging of the 5 Brigade units – which led to the *Sir Galahad* tragedy – were direct results of the choice of the San Carlos landing area. But, barring last-minute Argentinian moves, the landing would be unopposed. It would still be fascinating, however, to speculate on how a landing in the Port Salvador area would have progressed.

The exact 'How?' of the landings posed some interesting tactical problems. British forces had not carried out an amphibious landing at brigade strength for nearly twenty years, yet now it was required that five battalions and their supporting units should be put ashore, admittedly in an area which was not heavily defended, but certainly without local air superiority. The constraining factor would be the numbers of landing craft and helicopters available. The 3 Commando Brigade planners were quickly told that neither *Hermes* nor *Invincible* could come inshore to be used as a helicopter platform but must remain out at sea to provide the Harrier cover. This led to the decision to make the initial landing in the dark, using landing craft only, so that four of the infantry units and all the available tanks could be established ashore before dawn; the artillery, the air-defence units and stores were to follow as quickly as possible by helicopters and by whatever landing craft were available. (The map on page 209 shows the exact disposition of units.) The initial plan was for Blue and Green Beaches (Port San Carlos and San Carlos) to be taken simultaneously by 45 Commando and 40 Commando respectively but, when it was heard that Argentinian troops with helicopters might be located *north* of Darwin, all landing craft were allocated to the putting ashore of 40 Commando and 2 Para at San Carlos, with 2 Para moving off as quickly as possible to secure the Sussex Mountain line and block any Argentinian move from Darwin.

The plan to put nearly 5,000 men ashore on the Falklands was finalized, sent to London and approved without dissent by the full Cabinet in London (the only time the full Cabinet was involved). Rear-Admiral Woodward would cease to have full control of operations around the Falklands as soon as the landings commenced; his role for the remainder of the war would be to support the amphibious shipping and the forces ashore. Commodore Clapp would control all shipping in 'the Amphibious Area'; and Brigadier Thompson would become 'Commander Land Forces' as soon as the first of his units was ashore. His latest orders from Major-General Moore, who would become land commander when he arrived later, ran (according to Jeremy Moore's memory) thus:

> To establish a beach-head and secure it so that the next brigade could land into it for the repossession of the islands; to seek information to assist in the planning of that and, where the opportunity offered, to close with the enemy and to seek a battle which we were going to win in order to establish our moral superiority over the enemy.

The main significance of these orders is that a full break-out from the beach-head was not envisaged until 5 Brigade arrived.

It would be the first amphibious landing by British forces since the operation to repossess the Suez Canal in 1956. There were many thoughts back to that unhappy campaign, when Britain had been forced to retire by world opinion. Others, with a more historical bent, harked back to the Normandy D-Day landings in 1944 or even as far back as Gallipoli in 1915 when a British landing force on beaches far from home failed to break out and became penned into a horribly costly beach-head until forced to evacuate. Could the Argentinians yet secure the help of world opinion as had the Egyptians in 1956, or would they fight as well as the Germans in Normandy or the Turks at Gallipoli?

The final move into the Falklands and the landing at San Carlos had been planned for the night of 19/20 May, but a delay of twenty-four hours had to be imposed on this plan. The main reason for this was an order from London, probably originating with Mrs Thatcher, that *Canberra* should not approach the coast while still carrying three major army units and that two of these units should be transferred to other ships so that there should not be a huge casualty list if *Canberra* was hit. The transfer took place during the daylight hours of Wednesday, 19 May. It was a day blessed by remarkably calm weather and most of the programme was completed safely. 3 Para was moved by landing craft from *Canberra* to *Intrepid* and 40 Commando to *Fearless*. These transfers would leave the five infantry units each with their own ship; 42 Commando remained in *Canberra*, 45 Commando was in *Stromness* and 2 Para in *Norland*. Marine Michael Spence of 40 Commando describes his move from *Canberra*.

We transferred through the big loading doors in the side of *Canberra*; we had to jump down into a landing craft which was moving up and down four or five feet. We were fully loaded, with much extra ammunition, and one man in C Company missed and fell in the sea between the landing craft and *Canberra*. He was in serious danger of sinking or being crushed. All the men in the landing craft held the two ships apart with rifles and the man was able to keep afloat and swam around until he was pulled out. It brought home to us that this was how it was going to be; we were on active service and there would be none of the safety facilities normally present.

We were very cramped in *Fearless*; the whole of my company was in the Wardroom. We spent much of the next day and a half asleep, but we did watch *Grease* on video. That was good; it kept our full attention, kept our minds off what was coming.

Captain David Pentreath, of *Plymouth*, had the opportunity to see the whole gathering of ships present that day.

It was perfect weather. The ships were widely dispersed. I flew in my helicopter to *Fearless* for the final briefing and it was very impressive. It wasn't the biggest force I'd seen but, considering that it was the other side of the world and that they had been assembled in such a hurry, it was very impressive and a truly memorable achievement. I was thrilled to see such a gathering and it was obviously a very positive sign of our intent. It was one of my abiding impressions of the war that, despite all the political and diplomatic manoeuvrings going on, there was never any question of the ultimate aim not being achieved. There was no question of being deflected from that aim.

There were also some minor moves taking place by helicopter. One of these was the transfer of the special forces organization which had been housed in *Hermes* while the carrier group had been operating off the Falklands. Four Sea Kings of 846 Squadron and all of the S.B.S. and S.A.S. men present were being transferred to ships of the landing force from which they would operate for the remainder of the campaign. The very last transfer of the day – from *Hermes* to *Intrepid* – took place as darkness was falling. It met with a tragic accident. Sea King ZA 294, with a crew of three, was loaded with twenty-seven other men and much equipment. As the helicopter approached *Intrepid* from directly astern, the pilot was concentrating hard on his controls and the second pilot was temporarily working the navigation computer; no one was keeping a visual watch. The helicopter crashed into the sea. Rescue work commenced immediately. The two pilots of the crashed helicopter and seven of the passengers were rescued, but twenty-one men were lost. It is said that one survivor, trapped by a piece of harness around his ankle, felt a hand from below cut through the restraint, allowing him to escape, but his saviour did not. A dead albatross found on the surface was also collected as possible evidence of the cause of the accident. It was widely believed that this large bird had struck and shattered the anti-icing shield over the engine intakes, causing pieces to be sucked into the engine, but later examination of the albatross showed it to have been long dead, its flesh mostly nibbled away by fish. The cause of the tragedy was the combination of an overloaded helicopter being flown by a tired crew.

Eighteen of the dead were S.A.S. men – two warrant officers, fifteen N.C.O.s and one private soldier. They were all members of D Squadron, which had already operated on the South Georgia glacier and carried out the raid on the Pebble Island airfield. It was later said that most of these were not 'badged' S.A.S. men but attached signallers. The full list of the previous regiments of the S.A.S. casualties does not support this; there were four men from the Guards, two from the Parachute Regiment, two from the Royal Green Jackets, five from other units and only four from the Royal Signals. Two of the other dead were a forward air control team working with the S.A.S. – Flight Lieutenant Garth Hawkins of the R.A.F. and Corporal

Douglas McCormack of the Royal Signals. The final casualty was Corporal Aircrewman Michael Love of the Royal Marines, who had been the cabin crewman in the helicopter; his body was recovered and later buried at sea. Love had already flown on seven insertion sorties with the S.A.S. and S.B.S. and was awarded a posthumous Distinguished Service Medal for this work. These casualties brought the number of British dead by enemy action or accident in the Falklands war to forty-five (compared to an Argentinian figure at that stage of approximately 470). Among the dead in that helicopter were the first fatal casualties of the war for the Army and the Royal Marines and the R.A.F.'s only death of the war.

The accidental loss of so many men caused despondency throughout the task force. It also affected future operations because these men are believed to have been earmarked for landing by helicopter at three places on the south coast of East Falkland as part of the deception landing plan which *Glamorgan* had been carrying out in this area on recent nights. The three S.A.S. parties would have broadcast signals to simulate the landing of a major force. The lost S.A.S. men were not replaced until three weeks later when new men parachuted into the sea off the Falklands and were picked up, coincidentally, by *Glamorgan*.

When the transference of units was finished, the ships intended for the actual landing formed up into convoy. There were twelve amphibious landing ships: the command ship *Fearless*, with Brigadier Thompson and Commodore Clapp on board, with its sister ship *Intrepid*; the smaller landing ships *Galahad, Geraint, Percivale, Tristram* and *Lancelot*; the R.F.A. ships *Stromness*, acting as a troop-ship, and *Fort Austin*, acting as a helicopter carrier; and the civilian ships *Canberra, Norland* and *Europic Ferry*, which would sail right into a potentially enemy-held bay, a venture never envisaged when these ships sailed with such fanfares from England. Admiral Woodward detached seven warships as escorts; the control of these would pass to Commodore Clapp for as long as they remained in the Amphibious Area – or afloat. The choices of the destroyers *Antrim* and the frigates *Ardent, Argonaut, Plymouth* and *Yarmouth* were natural ones; these general-purpose ships were ideally suited. But Woodward was concerned with anti-aircraft defence, especially during the daylight hours of 20 May when the convoy must steam along the north coast of East Falkland to reach its jumping-off area for the landing that night. He decided to risk adding *Brilliant* and *Broadsword*, his two new Type 22 Sea Wolf destroyers which normally provided close protection for the aircraft-carriers. Unknown to the men in the convoy, the hospital ship *Uganda* was sailing independently into the nearby 'Red Cross Box' negotiated for her and, later in the day, a Nimrod flown by the 206 Squadron crew of Flight Lieutenant David Ford would fly from Ascension and carry out a long-range reconnaissance flight between the convoy and the Argentinian mainland naval bases.

The convoy set off during the night of 19 May. The daylight hours of the next day were a most anxious time but, again, the most favourable possible conditions for the British cause appeared. A Force 6 gale may have caused many men to be seasick but the mist and low cloud which accompanied the weather front successfully concealed the convoy all day. Naval officers were delighted at the safe passage so far. Captain Bill Canning of *Broadsword*:

> We had that lovely, foul weather all day. I thought the submarine threat was the greatest, with the spectre of two torpedoes going into *Canberra* – that would have knocked the whole operation on the head – and the nearer we got to the islands, the more this spectre haunted me, bearing in mind that those little German submarines the Argentinians had were in their element in inshore waters. Nightfall came with considerable relief.

Captain Kit Layman of *Argonaut:*

> We had expected the Argentinians to concentrate a great effort at that stage. We took them by surprise and, if I was writing a history book, I would judge that to be a major success, that we had been coming out from England for six weeks with the world's attention on us and we still surprised them. There were no scares on the way in.

Captain Ian Gardiner of 45 Commando had thoughts more for the morrow.

> There was no more to be said. All plans had been made, all briefings finalized, all kit checked. All that remained was to issue ammunition.

So the last fifteen hours of peace for me passed very calmly. I slept a bit and ate a lot, not being entirely sure how reliable food resupply would be ashore. I wandered round the ship speaking to my men. Contentment was what I identified. We had set them up in every way to go and do a job – and now they had been given the job. The process was complete.

Down in the mess deck that evening, the atmosphere was not unlike a rugger changing room before the match, each man going about his business preparing himself, dressing, packing, loaning magazines. Quiet determination was evident everywhere – and humour. I saw one man examining his face closely in the mirror. 'Keith!' 'Yes, Mate?' 'I can't possibly go ashore.' A concerned 'Why not?' out of the red gloom. 'I've got spots.'

The weather changed again, for the third time providing ideal conditions – a calmer sea, clear visibility; there was no moon. *Antrim* and *Ardent* went ahead to carry out special tasks. The landing force formed itself into single file and, led by *Plymouth*, slipped through the narrow entrance to Falkland Sound.

CHAPTER FOURTEEN

D-Day and Bomb Alley

The first necessity in the main landing area was to remove, by one means or another, the threat of the Argentinian position on Fanning Head, a towering headland nearly 800 feet high which overlooked the narrow entrance into San Carlos Water. A small party of Argentinian troops from 25 Regiment had taken up position there about eight days earlier but had suffered such privations that Alan Miller, manager of Port San Carlos Settlement six miles away, had felt obliged 'for humanity's sake' to rescue them by Land-Rover two days later and revive them with hot soup. A larger Argentinian party from 12 Regiment had then arrived and the combined group of sixty-two men then based itself in Mr Miller's community hall and school, with approximately one third of its strength providing the garrison on Fanning Head by turns. Argentinian helicopters lifted two 106-mm anti-tank guns and two 81-mm mortars on to the headland. This threat had to be removed.

Brigadier Thompson's staff devised a plan which, it was hoped, would induce an Argentinian surrender with a minimum of bloodshed. The destroyer *Antrim*, two Wessex helicopters, about thirty-five Special Boat Service men, one Royal Artillery naval-gunfire officer, one Royal Marine officer-interpreter, and some old- and new-fashioned equipment were allocated to the task. One of the helicopters was *Antrim*'s own Wessex 3, equipped with radar and flown by Lieutenant-Commander Ian Stanley and his crew, who had done so well in the recapture of South Georgia; the second was a Wessex 5 flown by Lieutenant Mike Crabtree of 845 Squadron. The helicopters made their first flights when *Antrim* was still forty miles out to sea. Stanley's helicopter landed a small S.B.S. party to guide the main group in later. Crabtree's helicopter was carrying a 'thermal imager'; this piece of modern kit mapped the whole headland in a series of sweeps and identified the exact location of the Argentinians from their body heat.

About three hours later, when the landing force was nearly up to Fanning Head, the two helicopters ferried the main S.B.S. party to the landing zone on the landward side of the headland, out of sight and sound of the Argentinians who were on the seaward side. *Antrim*'s two 4·5-inch guns gave covering fire. The interpreter was Captain Rod Bell, who had lived for many years in South and Central American countries because of his father's work as a United Nations official. Bell's equipment consisted of a loudspeaker borrowed from *Fearless* and a heavy, 24-volt dry-cell battery. The S.B.S. 'killer group' established itself on high ground overlooking the Argentinian position and Bell placed his loudspeaker about a hundred yards to the right. It was still dark. The S.B.S. had a hand-held thermal imager and found that the Argentinians had moved about 200 yards from their original position because of the recent shell fire. Rod Bell broke the silence and announced himself as a representative of the British forces, whose commander wanted to avoid bloodshed, pointed out that British troops surrounded the position, and called upon the Argentinians to surrender. There was no response but the thermal imager showed that the Argentinians were moving towards the British, whether to attack or seek better shelter or in confusion was not known. The S.B.S. opened fire, the targets being identified by the thermal imager. After some time, Rod Bell asked the S.B.S. officer to cease fire.

We were massacring them and I wanted to talk again. Our ships were moving in now just under Fanning Head and our object was already achieved; the Argentinians were not in a position to engage. The S.B.S. officer said he wasn't prepared to risk his men but I could go down to talk to them on my own if I wished. Two S.B.S. men offered to come with me; I was very grateful to them. There had been no Argentinian return fire yet but they did so now for the first time. The S.B.S. men and I took cover and our main position opened fire again. I said to myself, 'That's all I'm going to do; I've given them every chance.' It was getting daylight now and the Argentinians were running up to the highest ground. It was a duck shoot; you could see them falling. I called out again and they waved three white flags and six men came in; there were three others injured. But we couldn't see any bodies; I was amazed and lost all faith in our weapons for a few weeks. One Argentinian N.C.O. even said that he had gone back to sleep during the fighting. But I heard that 40 Commando searched the area later and they found ten or eleven bodies in the rough ground.

We all sat around then and admired the landing scene below.

Two other diversions were being carried out as the landings commenced. The main purpose of all these operations was to simulate a Pebble Island-type raid and mask the main landing for as long as possible. *Glamorgan* was in Berkeley Sound, shelling positions north of Stanley, and the S.A.S. were carrying out a major raid around Darwin/Goose Green assisted by gunfire from *Ardent* with the aim of containing the considerable Argentinian garrison there. The raiding party was provided by approximately forty men of D Squadron, S.A.S., under Major Cedric Delves, who were landed north of Darwin by helicopter. An initial bombardment by *Ardent* was not carried out because the artillery observation officer ashore was lacking a certain piece of equipment and could not correct *Ardent*'s fire by night, but the S.A.S. men provided a fine demonstration of fire-power with anti-tank launchers, mortars and machine-guns. *Ardent* came into action at dawn, shelling the airfield and claiming hits on three Pucarás; she later spotted what was believed to be a column of vehicles moving along a track from Darwin towards the landing area. Fire was opened on this target, but then hurriedly stopped when the vehicles turned out to be a line of grazing cows which 'leapt about in all directions'.

The Argentinians at Goose Green were heard reporting to Stanley that they were being attacked by 'a battalion of British troops' and were successfully prevented from moving towards the main landing area fifteen miles away – a major contribution by the S.A.S. to the success of the day's operations. Eric Goss, manager at Goose Green, 'marvelled' at the accuracy of *Ardent*'s shelling. When an Argentinian officer warned him of the danger of walking from his home to the community hall where most of the civilians were still imprisoned, Goss replied, 'I'm all right; you're the one in danger. They're not shelling me. Admiral Woodward must have got up in a bad mood this morning.'

These actions enabled the major British units to land safely. The landings were delayed by seventy minutes because of the difficulty encountered by heavily burdened men in moving through the narrow passageways and exits in ships, particularly by a 2 Para man who fell and injured himself in *Norland*. Zero Hour for the first touchdown should have been 06.40 Zulu (3.40

a.m. local) but was set back one hour; this still left three and a half hours of darkness remaining for the infantry units to establish themselves ashore. The troop-carrying ships anchored in the dark outside the entrance to the anchorage. 2 Para, from *Norland*, and 40 Commando, from *Fearless*, moved first. Their

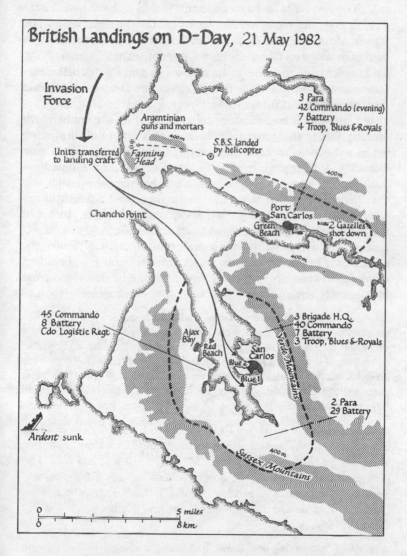

British Landings on D-Day, 21 May 1982

Invasion Force

Argentinian guns and mortars

Units transferred to landing craft

Fanning Head

400 m

S.B.S. landed by helicopter

3 Para
42 Commando (evening)
7 Battery
4 Troop, Blues & Royals

400 m

Chancho Point

Port San Carlos

Green Beach

2 Gazelles shot down

400 m

45 Commando
8 Battery
Cdo Logistic Regt

Ajax Bay

Red Beach

Blue 2

San Carlos

Blue 1

3 Brigade H.Q.
40 Commando
7 Battery
3 Troop, Blues & Royals

Verde Mountains

2 Para
29 Battery

Ardent sunk

400 m

Sussex Mountains

0   5 miles
0   8 km

sixteen landing craft moved quietly under Fanning Head, where the S.B.S. were in action, and into the bay. The frigate *Plymouth* was the only large ship to accompany the landing craft, ready to give direct gunfire support if needed – but she would not be required. The landing craft ran in over a smooth sea; it was very cold. The two units had agreed that it would be better for 2 Para to land first; the paras needed to set off fast for their objective, Sussex Mountain. Major John Crosland's B Company were the first men ashore. There were no Argentinians about and the paras quickly came ashore, turned south and started the four-mile climb to the 700–900-ft high ridge-line facing the possible Argentinian threat from Goose Green.

40 Commando landed seven minutes later on the small beach directly facing the settlement houses, this amphibious landing being a historic moment for any marine unit. Major Shane Cusack's A Company landed first and established a defensive position half-way up the smooth green between the beach and the manager's house (exactly where the San Carlos Military Cemetery is now located). Two Scorpions and two Scimitars of the Blues and Royals were also landed. Captain Andy Pillar's C Company then came through and moved tactically into the settlement. Pillar decided to knock up the manager. This first meeting between liberators and liberated can be described from both sides. Captain Roger Williams of 40 Commando:

> Our chaps surrounded the manager's house and Andy Pillar knocked on the door to ask if there were any Argies about. There was some delay and then Mr Short appeared and said something like, 'Oh, you've come then' – a common reaction throughout the settlement. There were no garlands of flowers or kisses on both cheeks. If we had been grass–skirted Chinamen with daggers in our teeth, the reception would have been the same.

All encounters with the Falklanders, from this one at isolated San Carlos all the way through to the final liberation of Stanley, would be of this low-key nature. Many British soldiers would be disappointed at this but the quiet Falklanders are not emotional people. Their inner gratitude to the British servicemen who came to liberate the islands was deep and genuine. This is Pat Short's memory of the event.

> About 1.30 a.m. my son Derek had heard shelling – we later found

out it was on Fanning Head – and had woken us up. Then we heard engine noises out in the bay. I went outside but couldn't see anything. There was a pup in the porch and it woke up, wagged its tail and hit a tin. It made me jump and I went back in again; I thought it might have been a shot somewhere. I went back to bed, waiting and wondering what was happening. Then, at 4.00 or 4.30 a.m. there was a hefty knocking. I tried to unlock the door but dropped the key and kept them waiting while I scrambled about on the floor to find it. When I opened it there were all these blokes with paint on their faces and camouflage. I asked them if they were British and one of them said, 'Hello, Pat, do you remember me?' I couldn't pick anyone out that I could recognize and he told me it was John Thurman. I had met him when he'd been out here a year or so ago and had visited us at the settlement. I shook his hands, a few times I think, and they told us to go back to bed and forget all about it until the morning.

Most of 40 Commando fanned out and climbed up to establish their defensive perimeter positions on Verde Mountains, not quite as high as 2 Para's Sussex Mountain. Both units dug in along the reverse slopes and established observation posts on the crests. The peat and stone sangars and trenches they constructed that morning would be their homes for the next week.

The landing craft returned to the troop-ships and picked up the next wave of troops, 45 Commando from *Stromness* and 3 Para from *Intrepid*, these ships still being outside the entrance to the anchorage. 45 Commando was to land at Ajax Bay and 3 Para a mile west of Port San Carlos Settlement, thus broadening the beach-head on either flank. But difficulties were accumulating. Captain Ian Gardiner was waiting on *Stromness*.

Eventually the time came early in the morning for us to muster at our assault stations. I think this period was one of the worst of the whole campaign. Hitherto, it had been academic. We had been conveyed in considerable comfort and good cheer from Ascension and the most we had done was talk about it. Now we were about to make the transition between comfort and discomfort, safety and danger, certainty to complete uncertainty. But the moment for embarking passed and no boats had appeared. It was two hours late that we climbed over the guardrail and down the scrambling net. Further delays waiting for the fourth landing craft. No sign. It had broken down somewhere. Back alongside to cram the remaining people into the three landing craft! Further delay.

The sky was getting grey as we sailed. Had there been any opposition at all, we would have taken a hammering. The greater part of 45 Commando was crammed into three small boats, sailing up a sea loch surrounded by unsecured land. It needed one machine-gun or one wideawake pilot and things would not have gone well for us. The landing went as planned. We secured our objectives unopposed and quickly deployed to the hill. For the first, and not the last time, fortune had smiled on us. It was a beautiful morning. We tramped off feeling a great deal better. We were away from our vulnerable moments and were now in our own element.

3 Para was delayed even further, firstly by the difficulty of loading from *Intrepid*, then when one of the larger landing craft ran aground and its company had to be transferred to smaller craft. The old rivalry between the paras and the marines erupted and 'there was much comment, not always polite', directed at the landing-craft crews. 3 Para's account says: 'It seemed a long journey to Green One Beach as we anxiously scanned the clear skies of a beautiful dawn, but at this stage only the gulls, the ducks and the Upland Geese were noisily airborne. It was good to be on land, and safe – though we hardly felt we had made history.'

There were no civilians at Ajax Bay, where 45 Commando landed, only a deserted meat-packing plant, but Port San Carlos again provided a meeting between the troops and civilians. Alan Miller was the manager.

We had heard the gunfire on Fanning Head and hoped that it was the prelude to a landing. I went up on the hill behind the house at first light and saw the *Canberra* three or four miles away and some landing craft. I took one look and turned round to go back and inform the others. I had a system that, if anyone wanted to call all the seven houses in the settlement at once, they gave one long cranked handle ring. They all came up and I told them, 'The Fleet's in.' Myself, my son Philip and another lad, Ron Dickson, ran down to meet the first landing craft in a little bay about a mile to the west. I wanted to tell them that there were Argies about but what we didn't know was that they already knew this; the S.A.S. or S.B.S. had been there watching us for three weeks.

About three quarters of the way down to the beach – I was running, waving a white handkerchief for safety – a major in the Parachute Regiment stood up from behind a gorse bush and said, with a big grin on his face, 'Ah. Good morning, Mr Miller.' That absolutely floored me, that he knew my name. The detailed planning and knowledge of the Task Force was incredible. He introduced himself but I never caught his name; I would have liked to have known who he was.

The major was David Collett of 3 Para's A Company; much of his information came from Alan Miller's wife, who was in England at the time of the Argentinian invasion and who had provided much help to the Ministry of Defence 'desk' which had been set up in London to gather local information. Alan Miller warned Collett that the Argentinian rear party from the Fanning Head garrison was at the eastern end of the settlement. The paras speak very well of the civilians of this small community. Some men recovered hidden rifles and were prepared to march with the paras, each lady produced between fifteen and twenty gallons of tea that day; small boys helped to dig trenches and, in the words of Alan Miller, 'all our vehicles – which had been broken down for the Argentinians – became repairable and worked non-stop for the next several days and nights hauling equipment and ammunition from the landing craft to wherever it was needed.'

The next phase of the landings had started as soon as dawn broke. Helicopters were lifting the 105-mm guns of 29 Commando Regiment and the Rapier air-defence missile launchers of T Battery to their allocated positions ashore. It was an appropriate date for 29 Regiment, the twentieth anniversary of its becoming a 'Commando' unit and earning the green beret of the marines. The lift of the guns and missiles was carried out by eight Sea King 4s of Lieutenant-Commander Simon Thornewill's 846 Squadron, which was operating from *Fort Austin* and various other ships. This was the start of a flat-out helicopter operation which, because of the lack of roads in the Falklands, would continue unabated until the end of the war. The Sea Kings would lift 407 tons of guns and stores and 520 troops before the end of the day. The smaller Gazelles and Scouts of 3 Brigade Air Squadron were also being used, but mainly as escorts for the large helicopters and for reconnaissance work. The guns were quickly brought ashore but the lift of the Rapier launchers did not go so well. These had been stowed below decks in the ships, to protect their delicate equipment, and there were delays in getting them ashore. Later in the day, the beach-head would be well protected from a ground threat which did not appear but ill prepared for air attack which did.

The first British setback of the day occurred soon after the helicopters started to bring their loads ashore. 3 Para's delay in reaching Port San Carlos Settlement had not, apparently, been reported to the helicopters. The paras were still clearing the settlement when a Sea King and a Gazelle suddenly swooped overhead, flying at low level in the direction of the area just east of the settlement into which the Argentinian party from Port San Carlos had moved. The Sea King was carrying an underslung load of mortar ammunition and the Gazelle, from the Brigade Air Squadron's C Flight, was reconnoitring the area to establish if the Argentinians were still present. The Argentinians opened fire with rifles and machine-guns. The Sea King unhooked and dumped its load and swept out over the bay at high speed. The Gazelle also banked, but in a wider turn. It was hit in the tail rotor and engine, and crashed into the water fifty yards off the end of the settlement jetty. It sank at once. The crew, Sergeants Andy Evans and Eddie Candlish, escaped but an Argentinian machine-gunner opened up on them in the water. Local men put out in a small motor boat, picked up the survivors and brought them ashore. Sergeant Evans was found to have a bullet wound, believed to have been received while in the water. He was taken to the settlement bunkhouse where Thora Alazia, the cook for the unmarried workers, and Suzanne McCormick, the settlement schoolteacher, tried to stem the flow of blood with towels, but Andy Evans died. Sergeant Candlish was unhurt.

The loss of this helicopter was compounded when another Gazelle arrived on the scene. This one was piloted by Lieutenant Ken Francis, one of the youngest but keenest pilots in C Flight. His radio was not functioning properly but, while flying in his allotted sector to the south of the landing area, he is assumed to have heard the original call for a reconnaissance of the Port San Carlos area and he came across to help. His Gazelle arrived about fifteen minutes after the first shooting down, was likewise fired upon and hit, and crashed into a hillside less than a mile east of the settlement. Lieutenant Francis and his crewman, Lance-Corporal Brett Giffin were both struck by bullets and killed. News of these events had still not reached the command ship when yet another Gazelle flew into the same area. Captain

Robin Makeig-Jones and Corporal Roy Fleming were greeted by more shots and their helicopter was hit by about ten bullets but Makeig-Jones was able to turn away and escape.

The loss of two Gazelles and the damaging of a third, in the first use by British forces of light support helicopters in serious action, was to bring a hurried change of policy. The tasks of observing for artillery shoots and of armed offensive action on the battlefield were largely abandoned for the remainder of the war and the Gazelles and Scouts were used primarily for communications, for casualty evacuation and for the moving of light loads over already secure territory.

3 Para now made a determined effort to clear the Argentinians from this area. Mortars were used and 79 Battery across at San Carlos fired the first artillery rounds of the war but the Argentinians slipped away to the north. The troop of Blues and Royals tanks attached to 3 Para were disappointed not to have been called upon to pursue the Argentinians but official opinion was that their tanks, light as they were, could not operate on the peaty open ground away from tracks.

The Argentinian side of this story is to be found in a report submitted later by the officer in charge of the Argentinian party to his superiors; the report was later captured by British forces. The officer was Lieutenant Carlos Daniel Esteban of J Company, 25 Regiment. Esteban had forty-one men under his command but only one section were front-line soldiers; the remainder were cooks, signallers and administrative men who had provided the base in Port San Carlos Settlement for the Fanning Head post. He had seen 3 Para (mistakenly reported by radio to Stanley as 'marines'), coming in by landing craft and had taken up position on the high ground east of the settlement 'to prevent the enemy encircling the settlement'. Esteban's men shot down the two helicopters (identified as 'Sea Lynxes') as described, Lieutenant Francis's Gazelle crashing a few yards from the Argentinians. Lieutenant Esteban moved his location after each incident, to avoid retaliation, and he reported that the British small-arms, mortar and shell fire was never nearer than 500 metres away although one of the helicopters shot down, almost certainly Lieutenant Francis's, had attacked and knocked out the Argentinian party's *Instalaza* rocket launcher.

Lieutenant Esteban and his men reached Douglas Settlement

after a four-day march and were evacuated from there to Stanley by helicopter. We shall meet him again.

With the coming of daylight and the news that the main beachhead area was secure, twelve troop and supply ships filed through the one-and-a-half-mile wide entrance to the anchorage and distributed themselves among the individual moorings allotted to them by Commodore Clapp's staff. The larger ships had to keep to the deeper part of the anchorage but the smaller '*Sirs*' and the *Europic Ferry* were able to tuck themselves into the southern end of San Carlos Water. The individual moorings had been carefully selected so that Argentinian aircraft approaching from over the hills would experience difficulty in lining themselves up on any of the ships. *Norland* claims to have been the first of the ships to anchor. It was a picturesque scene, the bright colours of the chartered civilian ships contrasting with the naval grey; the huge white bulk of *Canberra* drew every eye. There was no room in the anchorage for the naval escorts and these had to remain outside, one of them always being posted in the entrance – the 'cork in the bottle' – and the remainder disposed in Falkland Sound where there would be at least some space for radars and weapons to engage Argentinian aircraft attacks.

The unloading programme proceeded at full pace, any craft capable of running up to the beach now, like the helicopters, being committed to an all-out effort. The troops of the landing units had taken two days' rations ashore with them, so the major unloading effort for the remainder of D-Day was for ammunition, some being delivered to unit sites but a large central reserve being built up by the Commando Logistic Regiment at Ajax Bay which would be the main supply base for the rest of the campaign. Also at Ajax Bay would be the joint marine and para medical dressing station.*

Across the bay, at San Carlos Settlement, the command vehicles of brigade headquarters were landed and parked among the settlement houses. A Union Jack was run up on Mr Short's flagpole with just a handful of civilians and marines present – 'no speeches, no cheers, very quiet really, but a good feeling – lots of smiles'. When Brigadier Thompson came ashore in the late

---

* Well described by Surgeon-Commander Rick Jolly in his book, *Red and Green Life Machine* (Century, 1983).

afternoon, he ordered the brigade vehicles to be moved out of the settlement, into a series of deep bunkers gouged out of the gorse by bulldozer.

The remainder of the D-Day land activities can be disposed of quickly. The ridge lines overlooking the beach-head were all secured without opposition. Much digging-in was done. Patrols were sent out. It was a completely successful operation, owing much to the careful reconnaissance work and the diversions carried out by the S.A.S. and the S.B.S. Most of the landing area – San Carlos Settlement, Sussex Mountain, the Verde Mountains, Ajax Bay and the extensive peninsula between these places and the sea – was well within the single San Carlos property, which extended more than ten miles away to the south-east, to include Mount Usborne; this all belonged to the firm of David Smith & Co. Ltd, wool merchants of Bradford. The Port San Carlos area, in which 3 Para was established and where the two Gazelles were shot down, belonged to an Irish family, the Camerons of Beaufort, County Kerry.

Admiral Woodward had brought his depleted battle group in much closer to land than in recent days; *Hermes* and *Invincible* were now less than a hundred nautical miles north-east of Stanley and about 130 miles from the landing area. The first Sea Harriers were sent out on air patrol at dawn; the R.A.F. GR3 Harriers were sent up to provide close ground support soon afterwards. The GR3s were soon in action. They owed their first success to a four-man team from G Squadron, S.A.S., commanded by Captain Aldwin Wight, who had joined the S.A.S. from the Welsh Guards and who was now in a hidden position near Stanley, watching and reporting on Argentinian movements. It was a duty that Wight and his team would carry out for an unbroken period of twenty-six days and for which he would receive a Military Cross. The S.A.S. party had established that the Argentinians had a night dispersal area for helicopters ten miles west of Stanley, in a saddle between Mount Kent and Mount Estancia. The helicopters were taken out here on most nights to avoid being shelled by British warships or subjected to a Pebble Island-type raid. The helicopters were returned to the cover of the Stanley defences by day but the S.A.S. advised that they could be attacked if caught at first light. The GR3 pair of

Squadron Leader Jerry Pook and Flight Lieutenant Mark Hare was directed on to this choice target and attacked from the north with their 30-mm cannons. They found the helicopters well spread out and had to make four or five passes. A Huey helicopter was already on the move when the Harriers arrived and it escaped, but a Chinook and two Pumas were hit and set on fire. Hare's Harrier was hit by three bullets but he returned safely to *Hermes*. Later in the day, an Argentinian officer came to Mr Richard Stevens at Estancia House and demanded a Land-Rover, saying that there were many casualties at the helicopter site.

The next GR3 action was not so successful. Two more aircraft took off to provide support in the landing area an hour later, but the leader's aircraft could not retract its undercarriage and had to return. Flight Lieutenant Jeff Glover carried on; it was his first operational flight over the Falklands. He contacted his forward air controller on arrival at San Carlos but there were no targets for him in the immediate landing area. He was sent, instead, to look for targets of opportunity around Port Howard, twenty miles away on the other side of Falkland Sound. Glover made one run over the area but could see no targets. The controller then asked him to photograph the area. Glover allowed a quarter of an hour for the Argentinians to settle down and then flew over the sea past the settlement at low level so that his camera could take side shots of the area. The Harrier was hit by three shots, part of the wing came off and the aircraft rolled over. Glover waited for the roll to complete itself, then pulled the ejection handle; the canopy automatically shattered and the seat rockets fired him clear of the aircraft. Glover's shoulder was badly injured in the process. He was rescued from the sea by a civilian and some soldiers in a boat, well treated, taken to Argentina and returned home after the war. The GR3 pilots quickly learned that one pass over a defended area was enough and, also, that single aircraft should not proceed on a mission if one of a pair had to return with mechanical trouble. The Argentinian forces newspaper in the Falklands made the most of the incident; they claimed the Harrier's destruction three times over.

The Argentinians' own air reaction was slow. Only the units based in the Falklands made any attempt to intervene during the first hours of the landings. Pucarás, probably from Goose

Green, were first on the scene. The first of these appeared at San Carlos about two hours after dawn and made half-hearted cannon attacks on various ships but was deterred by the mass of fire put up by the British weapons. Another Pucará had earlier flown unwittingly over the S.A.S. men tramping back from the Goose Green diversion and was promptly shot down by a shoulder-held Stinger missile. The pilot ejected and walked back into Goose Green. The next attack was made by a lone Aeromacchi flying from Stanley on what may have been a routine dawn patrol around East Falkland island. Lieutenant Crippa came in from the north and suddenly found himself among the British warships outside the anchorage. Captain Kit Layman of *Argonaut* describes the resulting attack on his ship.

He came in over the hill, flying very well, nap of the earth stuff, and strafed us with cannon fire and rockets. We had some minor damage and one hole blown in our 965 radar aerial but this didn't affect its work. He was too close for us to fire our Sea Cat but the Bofors and small arms on the upper deck opened fire. We had three men hurt, including one who lost an eye, and Master-at-Arms Francis who had a piece of shrapnel one inch above his heart.

Lieutenant Crippa fired off all his ammunition and returned to his base at Stanley; he was awarded a decoration for this attack and survived the war.

The locally based Argentinian aircraft never were a serious problem to the British ships but, at around 10.30 a.m. local time, there began a series of fierce attacks by Argentinian aircraft flying from the mainland that would last for nearly five hours. The first of these attacks was made by Mirages but the main effort was by A-4 Skyhawk and Dagger fighter-bombers; the older Canberras and the modern Super Étendards were not used. The Argentinians units involved were Grupos 4 and 5 (Skyhawks), Grupo 6 (Daggers), Grupo 8 (Mirages) and the 3rd Naval Squadron (Skyhawks); these aircraft were flying from the Río Grande and Río Gallegos airfields.

The British defences would be severely tested by these attacks, hampered as always by the lack of airborne early-warning aircraft and by the fact that *Hermes* and *Invincible* had to stay back because of the Exocet danger. The limited number of Sea Harriers available and the distance from the carriers meant that complete patrol cover could not be provided, and there were problems over the forward control of the Harriers. The ship of the amphibious group designated as air-defence controller, *Antrim*, was inside Falkland Sound and her primary duty was to coordinate the local defences in the landing area. But Antrim's radar view of incoming raids was hindered by the land surrounding her. She could do little to help the Harriers intercept incoming raids but would be able to give the Harrier pilots details of raids taking place and of Argentinian aircraft leaving the area.

The direct defence of the landing area was entrusted to the ships at San Carlos and the shore defences; that defence would rely mainly on the warships which guarded the anchorage. Much praise was later heaped upon what became known as 'the May the Twenty-First Gunline'. These ships were *Antrim*, *Ardent*, *Argonaut*, *Brilliant*, *Broadsword*, *Plymouth* and *Yarmouth*. They were equipped with a variety of weapon systems ranging from modern Sea Wolf missiles to old Bofors guns and to machine-guns lashed to upper deck railings. The units ashore had their Rapiers, Blowpipes and small arms although, as has been described, the Rapiers were not yet fully positioned. The entire local defence was hampered by two factors. Firstly, the area of Falkland Sound and the San Carlos anchorage was too extensive for complete, overlapping protection. Secondly, the Argentinian approach would always be so low and so often masked by land that ships' radars rarely acquired targets quickly enough for the missile systems to be used properly. But the Argentinians had their difficulties as well. Faced with missiles and gunfire for the first time, the Argentinian pilots invariably used the low, fast approach to attack. This frequently prevented their bombs from exploding. The standard Mark 17 1,000-lb bomb (British-made) which the Argentinians were using required more time in the air for the spinner, which was activated on release from the aircraft, to arm the bomb. All the theorists agree that the Argentinians committed a major error on 21 May and in ensuing days by not being prepared to fly a little higher, accepting heavier losses, but sinking more ships. It did not appear so simple to the Argentinian pilots facing the British fire in and around San Carlos.

*

The Argentinian air attacks can be split into two phases, the first of which lasted from 10.30 a.m. to about 1.0 p.m. (local times). There was no coordination; the Argentinians approached in flights of four or six aircraft, possibly flying twenty or more sorties in this period. Most came in over West Falkland and burst upon the warships in Falkland Sound. It is possible that the Argentinian mainland intelligence had not yet realized that there was a major force of landing ships inside the anchorage. The British warships initially were not deployed in a purely air-defence role and were well scattered. Some were standing by to give gunfire support to the army units ashore; others were on anti-submarine watch.

Antrim, *Ardent*, *Argonaut* and *Broadsword* were all attacked in a series of scattered actions during this phase. *Broadsword* was only hit by cannon fire and had four men injured but *Antrim* was seriously damaged. Cannon fire scored many hits on her upperworks; one man was blinded and another seven were injured. One man describes the upper deck as 'looking like a battlefield'. The faithful old Wessex helicopter 'Humphrey', which had performed so well in the recapture of South Georgia and had helped with the Fanning Head operation that morning, was riddled with splinters. (This twenty-two-year-old helicopter never flew again and is now on display at the Fleet Air Arm Museum at Yeovilton. Chief Petty Officer Terry Bullingham, the man blinded in the attack on *Antrim*, also worked at the museum after the war.) *Antrim*'s most serious damage, however, was from a bomb which penetrated the stern, passed through a pyrotechnic locker and the Sea Slug magazine – fortunately being deflected by a steel pipe from two live missiles each containing two and a half tons of explosive – and then 'just flopped down' without exploding in the heads (toilets), causing 'some concern' to one man who was inside but no casualties. A second bomb bounced off the forecastle and exploded when it hit the water. But *Antrim* was able to remain on duty, helping with the control of the air defence and firing a further 350 4·5-inch shells in later air attacks.

The Argentinians lost four aircraft during this series of raids. *Broadsword* shot down a Mirage just south of the entrance into San Carlos with a Sea Wolf missile in the very first air attack. There were later to be many claims and counter-claims for aircraft shot down but *Broadsword* captured the whole engagement and the destruction of the enemy aircraft on cine film on this occasion. Three Sea Harriers shot down a Pucará with cannon fire during this period but then met with no further success until right at the end of this phase when they were directed on to four Skyhawks which had just attacked *Ardent* and shot two of them down with Sidewinders after a long chase across West Falkland.

The next wave of attacks started in the early afternoon and lasted for little more than an hour. The Argentinians were now improving their tactics; it is possible that their pilots returning from the morning attack had been debriefed. There was more concentration of effort; at least one attack came in at a more suitable height to allow bombs to explode; at least one attack penetrated the anchorage to attack the richer targets of the troops and the supply ships. The British defence also improved. Argentinian submarines were judged not to be a serious threat and there were no calls from the land for gunfire support, so the warships in Falkland Sound were brought together in a tighter group. The control of the Sea Harriers improved. *Brilliant*, with its doppler radar which could pick out targets from among land returns and which had a very experienced 'old-fashioned fighter-control officer', Lieutenant-Commander Lee Hulme, was brought into use more. *Invincible*, whose operations room had overall control of the Sea Harrier department, was asked to provide more sorties and to place the patrol areas further forward. Lieutenant-Commander Duncan Ford, who was on duty for most of the day in *Antrim* as the official 'Amphibious Group Anti-Air Warfare Co-ordinator' says that '*Invincible* was always very responsive to suggestions'. *Hermes* and *Invincible* were able to provide as many as ten Sea Harriers on patrol at one time at the peak of the action.

The Sea Harriers responded by shooting down a further six Argentinian aircraft out of possibly twenty or twenty-five arriving in the operational area. Five of the successes were achieved with Sidewinder missiles, the sixth by cannon fire. There was much luck as to which Harrier pilots found themselves in favourable positions at the right time. Some never saw an Argentinian aircraft all day; indeed, some were never to do so all through the war while more fortunate colleagues were able to gain several successes in favourable circumstances.

But again the 'May the Twenty-First Gunline' ships had to face the brunt of the attacks and their strength was steadily reduced. *Argonaut* was hit first. This is Captain Layman's account.

We were the cork in the bottle off the mouth of the anchorage but, at the time, we were trying to get rid of three earlier strafing casualties by helicopter to *Canberra* and were steaming north, to give the Wessex pilot a headwind and so that he could see us out of his left-hand seat.

We saw six Skyhawks coming over West Falkland. We had received warning from our own radar that a new raid was coming in. We knew they were about but not exactly where, because we were losing them over the land. But we were all ready to go when they suddenly appeared over the hills on the other side of the sound. They were spread out in pairs, in an extended line abreast. I have a clear recollection of counting the six aircraft and you could tell the exact moment when they were committing themselves for a bombing run on H.M.S. *Argonaut*. And in they came. We fired at them, of course, but it's jolly difficult to take on six. We certainly got one. I don't know what we hit him with but he piled into Fanning Head with a very satisfactory thump. We may have discouraged one or two others who saw all this stuff going up; we were deliberately using a lot of tracer. They all had two bombs and I think they all released – there were certainly a lot of splashes, bangs and thumps and two bombs hit us.

The door of the forward magazine hoist blew off and smoke started coming out of it but it was a long time before I realized the two bombs hadn't gone off. The bomb forward had caused an explosion in the Sea Cat magazine – we found out later that two missiles had exploded and probably some Bofors ammunition. There was such a scene of devastation down there. I knew something else had gone wrong because we were losing all power. I found out later that we had an unexploded bomb in the boiler room.

Two seamen were killed in *Argonaut*'s Sea Cat magazine. The ship could not steam and was later towed into San Carlos Water by three of *Fearless*'s landing craft. Her crew, together with some visiting bomb-disposal experts, would spend an exhausting week defuzing and removing the bomb in the boiler room and then cutting away large chunks of the decks above the second bomb and eventually dumping it at sea. The damage was temporarily repaired but *Argonaut* was destined for an early return to England.

Brilliant was also attacked at about this time, being raked by a burst of cannon fire. Several cannon shells pierced the thin side-skin of this modern warship and exploded inside the ship. A mass of wiring associated with the Sea Wolf system and other computers was cut; the sonar was put out of action and three men were injured in the Operations Room. But most of *Brilliant*'s damage would be repaired within twenty-four hours.

The ship to be most badly hit was the Type 21 frigate *Ardent* (3,250 tons, crew of approximately 200), a ship which had done more work that day than most, having been off Goose Green giving gunfire support to the S.A.S. diversion there all morning. She had been twice attacked by Argentinian aircraft during this period but not hit. The task at Goose Green completed, *Ardent* was ordered to join the main group of warships just outside the landing area. The crew had a meal and a little rest; morale was very high after the satisfactory action at Goose Green. The frigate was given a one-mile 'beat' with *Yarmouth* at the southern end of the line of warships. Initially all was quiet, but then the storm broke. Most of the Argentinian attacks were now coming up Falkland Sound from the south-west and *Ardent* was one of the first ships they encountered. Skyhawks and Daggers carried out three further separate attacks on this ship. Three Daggers caught the frigate with an inexplicable fault in its Sea Cat missile system and with its 4·5-inch gun unable to bear. Only 20-mm Oerlikon cannons and the recently 'lashed-up' machine-guns could engage the aircraft. They were not enough. The Argentinian pilots made their attacks sufficiently carefully for the bombs they dropped to explode properly. Commander Alan West provides this account.

The first aircraft dropped two bombs, both of which hit the ship. One burst in the hangar with a big explosion and there was a secondary explosion from one of the torpedoes in our tubes. The second bomb went into the after auxiliary machinery room, but it did not explode. The rest of the three aircraft dropped six bombs but they all missed; one hit the water and bounced between the masts.

Up forward it felt as though a giant was holding the ship by the stern and whacking it down on the sea. There was a great deal of smoke going up about a hundred feet into the air. I saw our Sea Cat launcher, on top of the hangar, go straight up into the air and fall back on to the top of the flight deck and, sadly, on to the top of my supply officer, Richard Banfield. Also killed were John Sephton and Brian Murphy, the Lynx helicopter pilot and observer who were both seen firing at the attacking aircraft – John with a Sterling sub-machine-gun, real cowboy stuff, and Brian had a Bren gun; he was last seen firing straight up into the air at the plane whose bomb killed him. John Sephton had sent the two young men of his flight-deck crew to take cover and they reported this later.

Besides the three officers mentioned in Commander West's account, at least one more man in the flight-deck crew was killed in this attack. Lieutenant-Commander John Sephton was awarded a posthumous Distinguished Service Cross for his attempts to defend his ship.

Ardent was now ordered to steam north, to take cover among the other warships, many of whom were also in a damaged condition by now. But the frigate never reached that protection. She was attacked twice more and hit on the stern again in each attack. It is believed that a total of seven bombs hit *Ardent* and exploded and two further bombs remained inside the ship unexploded. The total number of men killed rose to twenty-two, most of them being the after control party who were killed by bomb explosions while they were fighting the fires in the stern of the ship. Alan West, who was on the bridge, sums up these further attacks.

The noise of each explosion was incredible, the whip effect enormous; the deckplates just jumped up into the air and people who had dived on to the deck were lifted three foot up into the air. It could only have been two minutes between the first and last of those seven bombs but it seemed an eternity.

Commander West received reports which showed that the fires in the stern were now out of control and that the after engine-room bulkhead was in danger of collapsing under the pressure of the water behind it; if this happened, the ship would sink at once. He ordered *Ardent* to be abandoned. It was a timely decision. *Yarmouth* placed her stern against *Ardent*'s bow and most of the survivors transferred in this way. At least three men had been blown into the water by the bomb explosions. Lieutenant Stephen Rideout, the ship's doctor, was picked up by another ship; the other two were picked up by Lieutenant Mike Crab-tree's medical evacuation helicopter, Surgeon-Commander Rick Jolly twice going down on the winch-line without having being trained in this work and without the correct clothing. *Ardent* burnt herself out and sank the next day. Five members of her crew were awarded decorations in the post-war honours list, more than in any other warship of the Task Force. Among those decorated was the ship's Naafi manager, John Leake, whose machine-gun fire damaged one Skyhawk so severely that it later crashed while attempting an emergency landing at Stanley airfield.

Only one of the air attacks had made a serious attempt to enter the anchorage to bomb the troop and supply ships, but none of these ships had been hit and no casualties had been suffered by any of the landing ships or the units ashore. There had been much wild firing by the landing ships and the soldiers ashore had been more in danger from stray British rounds falling among them than from the enemy.

So ended a momentous day. The all-important landings had proceeded according to plan. Three thousand troops and a thousand tons of supplies were safely ashore. The carrier group out at sea had seen no action but the seven destroyers and frigates in Falkland Sound had sacrificed themselves to protect the landing ships and forces ashore. *Ardent* was sinking; *Antrim* and *Argonaut* were out of action with unexploded bombs lodged inside them; *Brilliant* and *Broadsword* had been damaged. Only *Plymouth* and *Yarmouth* were unscathed. From a strictly military point of view, these losses were not serious; all of these ships could be replaced. In fact three fresh ships – *Exeter*, *Ambuscade* and *Antelope* – would join Woodward's battle group the follow-ing day. Captain Kit Layman of the *Argonaut* gives the profes-sional view of the Argentinian air attacks. 'It looked impressive – aviators love it – but they were too low. If they had been higher, they could have picked out *Canberra* – the "great white whale" – and our landing ships. When I heard the reports that *Ardent*, *Brilliant*, *Antrim* and later ourselves had been hit, but no reports of hits on the amphibians, I was absolutely elated.' But three British naval officers and twenty-one ratings were dead and a further twenty-five injured, some seriously; one man would never see again. The total British dead for D-Day numbered twenty-seven.

The Argentinians had launched approximately fifty to sixty sorties from their mainland air bases and a small effort from the Falkland-based units. Post-war research shows that thirteen Argentinian aircraft were probably shot down – one Mirage, five Skyhawks, five Daggers and two Pucarás. The Sea Harriers had probably gained nine of the successes (five by 800 Squadron pilots and four by 801 Squadron pilots); the ships shot down three aircraft and the S.A.S. one. It is believed that five Argentinian pilots died. Three Argentinian helicopters had been destroyed on the ground by Harrier GR3s, and some Pucarás at Goose Green were hit by *Ardent*'s gunfire. These losses represented a severe depletion of the Argentinian air strength. British air losses had been two Gazelle helicopters and one Harrier GR3 shot down, all by ground fire.

The British were forced to change their plans in the aftermath of the air attacks. A decision was taken to get as many of the landing ships out of the beach-head area that night. 42 Commando had remained on *Canberra* all day, ready to land on West Falkland if the Argentinian garrisons there had looked like stirring. Now this unit was hurriedly transferred to landing craft and put ashore at Port San Carlos. That night, *Canberra*, *Norland*, *Stromness* and *Europic Ferry* all sailed back to the carrier group with the damaged *Antrim* as escort. The departure of these ships removed the most valuable targets from the danger of subsequent air attack and also left room in the San Carlos anchorage for the warships previously forced to remain out in Falkland Sound.

The landing force settled down for the night. *Brilliant* steamed round to the far coast of West Falkland and quietly landed a party of S.B.S. men at King George's Bay, perhaps as an aircraft warning post. Captain Bill Canning of *Broadsword*, the senior of the two Sea Wolf ship captains, sent a signal to Admiral Woodward expressing the opinion that the ships of the amphibious force 'would be picked off one by one' unless changes were made in the use of the Sea Wolf ships. After an exchange of signals, it was agreed that Woodward would send *Coventry* forward to operate with *Broadsword* in a '42–22 combination missile trap' in the open sea area north of Falkland Sound.

Coventry's radar and medium-range Sea Dart system could then catch some of the Argentinian aircraft attacking the landing area; *Broadsword*'s Sea Wolf would protect *Coventry* and herself if Argentinian aircraft penetrated through to attack the combination. This was a major gamble, with Woodward risking in an exposed forward position *Broadsword* and the last remaining undamaged Type 42, although *Exeter* was due to join as a replacement 42 within hours.

It was only incomplete news of the landings which reached Brigadier-General Menéndez at Stanley. His first assessment was that only two British Battalions were ashore at San Carlos and that this was a diversion. He believed the main landings would later take place closer to Stanley. However, he attempted several moves to attack the beach-head with land forces. The first of these was an attempt to take two 105-mm guns of the 4th Air Portable Group from Stanley to reinforce Goose Green, which had no artillery. It was hoped that the Argentinians at Goose Green could then mount a counter-attack on the beach-head or at least shell it. The coast-guard vessel *Río Iguazú* was loaded with the two dismantled guns at Stanley, intending to sail round to Goose Green that night, but the loading was slow and the ship was caught next morning by Sea Harriers of 800 Squadron and attacked with cannon fire. One Argentinian machine-gunner was killed and the ship was so badly damaged that it had to be run aground. The guns were transferred to the shore where they were later helicoptered to Goose Green, although one was in a damaged condition and would never come into action. Two more guns were later helicoptered to Goose Green, arriving just in time to come into action against the 2 Para attack there.

On the day after the Harriers attacked the coast-guard vessel, more Harriers from the same squadron frustrated another Argentinian move. Four helicopters – three Pumas and an Augusta – were sent from Stanley to Port Howard, to deliver ammunition and stores to 5 Regiment at Port Howard and to bring back a Commando party there which was to be landed near the San Carlos beach-head and make an attack on it. Two more Harriers caught the helicopters and destroyed three of them, leaving only the Puma to bring the surviving crews and a few commandos back to Stanley. Several small parties of commandos were landed near the beach-head, with orders to infiltrate the landing area, but these proved ineffective and 40 Commando later captured some of the Argentinians, who were suffering from extreme exposure. Menéndez was having serious trouble with his helicopters. He dared not use them by day, because of the danger from Harriers, and he could not use them by night because his pilots had no passive night goggles. Dawn and dusk were the normal Argentinian helicopter operating times.

As the beach-head was steadily built up, General Galtieri pressed Menéndez hard to do something, saying that 'the army is falling behind the other services in the struggle', and quoting the Air Force attacks which were now taking place and the Navy's loss of the *Belgrano*. But Menéndez had based all his plans – previously agreed with Galtieri – on the basis of defending Stanley; he just could not get troops forward for lack of transportation. He said after the war:

It was well known what I could do and what I could not do. I had informed them of all my helicopter losses. I had told them that my main policy was to oppose a British landing at Puerto Argentino [Stanley] and they couldn't tell me that I had failed in this. And now they wanted me to do something else completely different and impossible.*

One move that was considered was the transfer from the mainland to Goose Green of the Córdoba Air Transportable Brigade (actually a reinforced regiment) which was earmarked as the strategic reserve for the Falklands. The second-in-command of the brigade had visited Menéndez in April and looked at suitable landing zones. Menéndez now asked if the brigade could be moved into Goose Green airfield so that it could attack the beach-head. But Galtieri refused; he was not prepared to accept the risk of the transport planes being caught by Sea Harriers. Instead, he offered to send the brigade in small parties by overnight transport plane to Stanley and let Menéndez helicopter them forward to Goose Green, and liaison officers from the Córdoba Brigade actually arrived to arrange this. But Menéndez declined the offer; it would take ten days to get the

* *Malvinas, Testimonio de su Gobernador*, Carlos M. Turolo, Editorial Sudamericana, 1983, p. 200.

entire brigade across to Stanley from the mainland and there
was still no way of getting them forward to Goose Green.

The Argentinians had to content themselves with a succession
of intense air attacks on the beach-head. Such attacks took place
nearly every day; there were at least twelve separate raids in the
six-day period 22 to 27 May, with possibly 130 sorties being
dispatched from the mainland. Some of the Argentinian tactics
improved. Airborne tanker aircraft were used more often. This
gave a proportion of the Argentinian attack aircraft a greater
endurance, enabling their pilots to avoid the Sea Harriers and
approach the anchorage from more favourable directions. The
Argentinians on the Falklands started using their radar sets
more effectively, plotting the flights of the Harrier patrols and
warning the mainland aircraft accordingly. An Argentinian
observation post on Chata Hill, although twenty miles from the
anchorage, was able to give some information about what was
happening at the beach-head. The Argentinians started sending
fast civil jets, Learjets, to probe the British defences and to
deceive the Sea Harriers and draw them out to action while the
Learjets turned away to safety. Admiral Woodward responded
by moving in his carriers another fifty dangerous miles on some
days. The Sea Harriers were moderately successful during this
period; they managed to intercept three out of twelve raids,
destroying four aircraft, all Daggers.

The main opposition to the Argentinian air attacks was
provided by the defences of the landing force. A vast amount of
ordnance was flung into the sky above the San Carlos area by
ships' missiles – Sea Cat and the old Sea Slug (the Sea Wolf ships
had been withdrawn; the enclosed anchorage was not suitable
for them), by guns of all kinds, from the 4·5-inch naval guns
down to infantrymen's machine-guns and rifles, and by the land-
based Rapier and Blowpipe missiles. Marine Steve Oyitch, of 45
Commando, describes the experience well.

You could set your watch by the raids. You'd say, right, get your wet
on – a cup of tea, sit down in your trench and watch them coming in and
getting blown out of the sky. They always came in two at a time – it was
just extraordinary. You'd see the Sea Cat missiles coming from the
frigates, the Rapiers coming up from the hillsides and tracer just erupt
from the whole valley of the San Carlos area. It seemed as though 5,000
men were all firing at once. I took my turn in the gun-trench on a

couple of occasions with our GPMG. It gave you a great feeling to be able to shoot back at them. You had no more than a couple of seconds but you could get a belt of fifty rounds off. I don't think I hit anything; they were moving too fast. I think they let us open fire, to give us the feeling that we were fighting back, instead of sitting there helpless.

When the first one went down, right in front of us, straight into the water off Ajax Bay, it seemed as though the whole Commando Brigade stood up and cheered and clapped and jumped for joy. The whole place erupted.

The main army anti-aircraft defence weapon was the Rapier missile system. Sergeant Bill Boyd was in charge of Rapier Post 33 Able, also at Ajax Bay. He describes an engagement on 25 May.

They came over the hill. I got the operator on to the second-to-lead aircraft; it was the best for us. I watched the missile go and everything seemed to go dead quiet. All around was hell breaking loose, with everyone else firing, but when you are concentrating on your own missile it all goes in slow motion. The missile hit the aircraft where the wings join the body. It was a grey aircraft. It carried on flying, doing what looked to me like a victory roll. I thought to myself, 'The cheeky bastard', because I had seen the bombs fall near the *Lancelot* and *Fearless*, so I ordered the operator to fire again and he did so. But, whilst it was flipping over – unknown to me – the pilot was ejecting the other side and then the wings fell off the aircraft, so I ordered, 'Slew down. Cancel,' because it might have hit one of the ships. The missile flew into a bank of earth 500 metres or so in front of us, nearly hitting two marines on a Blowpipe detachment. I spoke to them later; they were a bit surprised at something coming that fast from behind them – making a hell of a noise at Mach 2 then exploding.

There was a mass of claim and counter-claim for success by the different units and weapon systems at San Carlos. Commodore Clapp's operational staff attempted to refine these so that the Ministry of Defence in London could announce details of Argentinian losses. Fifteen Argentinian aircraft were claimed destroyed in and around the San Carlos area during the six days from 22 to 27 May (other successes outside the landing area are not being discussed here). But post-war figures based on private research* show that only five Argentinian losses can be credited to the San Carlos defences during this time plus a sixth shot down by its own

* See Ethell and Price, *Air War South Atlantic*.

side's anti-aircraft gunners at Goose Green; that last aircraft might have been damaged earlier over San Carlos.

The Argentinian pilots achieved only limited success during these days. Most of the attacks were against the ships in the anchorage, the Argentinian pilots coming in low over the hills, dropping down over the water, weaving through the ships and bombing and strafing whichever ship came into their sights. But so hectic was the ground fire that only three ships were hit by bombs and none of these bombs exploded, because the Argentinian aircraft were too low. The landing ships *Sir Galahad* and *Sir Lancelot* were hit on 24 May. One of *Galahad*'s bombs glanced off the ship's side and sank; a second passed through the officers' cabins and came to rest in a small compartment near the troops' cafeteria. The master, Captain Roberts, decided to evacuate the ship. Experts came on board and declared that the bomb was safe and part of the crew returned to carry on normal work while the bomb was removed during the next two days; it was then lowered into a Gemini craft where it was cushioned among packets of Corn Flakes and finally dumped in deep water. *Sir Lancelot* had an almost identical experience, although its bomb took five days to remove.

But the disposal of the bomb in the third ship hit during this period went all wrong. The Type 21 frigate *Antelope* was hit on 23 May. Although *Antelope* was an original member of the task force, she had only recently arrived in the Falklands, having been kept back to guard the amphibious ships at Ascension. Her period of duty in the Falklands was destined to be short and violent. She joined the battle group on 22 May and was immediately ordered to proceed to San Carlos as a replacement for her sister ship *Ardent*, lost the previous day. *Antelope* was allocated a position in the most open part of the bay, to intercept Argentinian aircraft flying through the main anchorage. Her crew were anxious to do well. Chief Petty Officer Alan Baker says:

> We were under no illusions; we knew exactly what we were in for. We knew as soon as we got there that we were in the most vulnerable position, but morale was still high.
>
> We thought that this was our moment to put all our training into practice. We knew that ships had been hit and men killed on the 21st – particularly our sister ship *Ardent* – and now was our chance to redress

the balance, as it were. Also, we felt we had been lucky to have missed the 21st when all the ships had been hit and felt it would carry on that way – we always felt it would be the other fellows, not us.

Antelope spent just four hours at San Carlos before being attacked. The action which followed was fast, violent and confused. Four Skyhawks came up Falkland Sound from the south, flew past the entrance to the anchorage and then disappeared behind Fanning Head. Two then reappeared over a hill to the east, making for *Antelope* and *Broadsword*, which was near by, but these Skyhawks turned away after drawing the attention of the ships' weapons; this move was probably a deliberate feint. The second pair then came straight over a much nearer hill to the north and 'came screaming straight at *Antelope*'. It is at this stage that it becomes difficult to disentangle different versions of the action. One of the two planes rose, swept over *Antelope* and struck the ship's rear mast with its drop tank; the top fifteen feet of the mast was bent right over but the Skyhawk survived. So low had been the approach of this plane that Sub-Lieutenant Mark Batten, on the emergency conning position near by, had been firing *down* on to the Skyhawk with a rifle. The second aircraft in the pair met a violent end. A seaman on *Antelope* was awarded a medal for shooting it down with a 20-mm Oerlikon; *Broadsword* was firing at it and also put in a claim and yet a third claimant was Sergeant G. J. 'Taff' Morgan's Rapier post – 33 Charlie – at Port San Carlos. Commander Nick Tobin of *Antelope* saw the Skyhawk 'disintegrate into a sheet of flame, an engine going one way and the pilot another in a position which would not support life'.

Antelope appeared to have survived unscathed but one bomb had skipped off the sea, entered the starboard side of the ship just below the bridge, and come to rest in a compartment above the join of the two engine rooms. The bomb had not exploded. A further Skyhawk, probably one of the first pair returning, now made a sudden attack which caught most weapons unprepared and put another bomb into *Antelope*'s port side. This too failed to explode but, according to Commander Tobin, 'rattled round the front of the ship', passing through the petty officers' mess and coming to rest in a cabin. One man was killed and two were injured.

Antelope joined in repelling another air raid and then Commander Tobin received permission to move into a more sheltered part of the anchorage for repairs and to await the arrival of a bomb-disposal team. Two Royal Engineers soon arrived from *Argonaut* and declared that the bombs could be defuzed. Most of *Antelope*'s crew were ordered to the upper deck. The bomb above the engine rooms was tackled first. All surrounding doors were closed, except for those above which were left open to provide a 'blast exit route' upwards if an explosion occurred. The two army men, assisted by two of *Antelope*'s engineers, started work. The bomb was a British made 1,000-pounder fitted with a Type 78 Mark 2 tail pistol fuze. Three unsuccessful attempts were made to remove this using a rocket wrench in which small 'slugs' were fired by remote control in an attempt to force the fuze out. It was then decided to use another device which fired a different type of charge, in an effort to shift the fuze. The device was fitted, the four men took cover, the device was fired, the expected small explosion was heard and, after a short wait, the party broke cover to inspect the results of their effort. The two army men were in front. There was a huge explosion and a nearby steel door was wrenched off its hinges – the door being nearly doubled over by the force of the explosion – and was flung towards the approaching men. It is probable that the small charge of the disposal device had penetrated the end of the fuze pistol and driven the striker forward. The fuze must have been fitted with a thirty-second delay, which is why the disposal party were caught in the open and struck by the door. Staff Sergeant Jim Prescott was killed; Warrant Officer Phillips had an arm torn off; the two naval men suffered less severe injuries.

Chief Petty Officer Robert Shadbolt describes the effect of the explosion.

I was stood just inside the hangar door when there was an almighty thud and bang, the loudest bang I have ever heard and I seemed to be lifted off the deck. It only took a split second to realize what had happened. The hangar and the upper deck filled with thick black acrid smoke which made it difficult to see and breathe. By looking over the starboard side of the ship, you could see the extent of the damage where the bomb had exploded. It was as though someone had gone up the side of the ship with a tin-opener. The most horrendous fire was raging; large electrical cables were flapping and sparking; the heat was intense. People were bravely trying to organize fire-fighting teams but the explosion had cut the hydrant ring main so that, when you turned the water on, nothing came out of the hoses.

It was dark now. The fiercely burning ship was watched by the whole anchorage. Marine Steve Oyitch was one of the men of 45 Commando on the high ground above Ajax Bay who watched in horror.

We had just got into our sleeping bags. We could see the crew in their day-glow survival suits on the flight deck. We could hear what we thought was screaming and we felt absolutely helpless, wishing to God they could get them off. Then the helicopters came in – all rules broken, all lights on and straight into the smoke. Then the landing craft from *Fearless* and *Intrepid* came in.

We watched them all gradually taken off – we were watching for about two hours. They just left it to burn in the end. Next day, our provo' corporal came up and told us that they hadn't been screaming, but it was the ship's sods' opera singing, 'Always Look on the Bright Side of Life' from the Monty Python film *Life of Brian* – a typical Jack Tar sense of humour, that was.

Commander Tobin's decision to abandon ship was well timed. No further member of *Antelope*'s crew was hurt. A few minutes after the last man left, the fire reached the small ready-use Sea Cat magazine, which exploded, and then the main Sea Cat magazine went up in a violent shower of sparks which broke *Antelope* into two parts. The dramatic moment was caught by Press Association photographer Martin Cleaver to become one of the war's most vivid images. Next morning, the bow and stern were still visible but these soon sank.

The Argentinian air effort slackened and there was only one more bad incident in the landing area, a sharp raid on the evening of 27 May which caught many men by surprise after nearly two days of quiet. Four Skyhawks came over the hills and swept, in pairs, along each side of San Carlos Water. For the first time, a deliberate attack was made on the units ashore. The first pair roared over San Carlos Settlement. A young marine (who does not wish to be named) describes the attack:

I was watching some local men excavating trenches for us with a tractor. My mate was telling the first one that he didn't expect to see any more bombing – the Argentinian air force had almost had it – when two planes came along our side of the bay, hugging the hillside we were on. There hadn't been any alert at San Carlos, though there might have been somewhere else. There was tracer going up after them but the planes were too fast and, after they had disappeared over the hill, a missile fired by one of the ships came over and nearly blew up 40 Commando; it landed in their lines.

I watched the bombs released; it was getting commonplace by now and you don't flap any more. They looked as though they were going miles over, but then the parachutes opened. That surprised us; we had never seen that before. The first ones landed on the beach near the jetty and the others worked their way up. They didn't all explode but some of them did – five, I think. The planes made a lot of noise, then the bombs exploded – that really was a terrific amount of noise. It was the same with every attack, the noise. Then we saw the smoke and a lot of guys rushing about. I was surprised to hear later that only two men were killed there.

The bombs had fallen just short of 3 Brigade Headquarters. One man, Sapper P. K. Ghandi, 'a nice, quiet Pakistani chap', was killed and Marine S. G. McAndrews, manning a machine-gun, would soon die of his wounds. Four other men were injured.

The second pair of Skyhawks dropped their bombs – also parachute-retarded – at Ajax Bay, on the opposite side of San Carlos Water. Here, in and around the abandoned refrigeration plant, was located the Brigade Maintenance Area with its piles of stores and ammunition, the medical station and an artillery battery. The medical station could not be marked with a Red Cross because of the presence of other units located in the same place; there were no other buildings available. Four bombs were dropped. One exploded just short of the refrigeration plant, killing five men outright and injuring twenty-two more of whom one would die later. Most of the casualties were from 45 Commando's B Echelon. Blast and fire damage wrecked one wing of the refrigeration plant but the medical unit had a near-miraculous escape from three bombs which scored direct hits but did not explode. One of the Skyhawks was damaged by fire from the San Carlos defences and crashed on West Falkland after its pilot had baled out.

This bombing raid had been remarkably accurate and well directed on to two important shore positions. It is possible that it was linked to the capture earlier that day of an Argentinian officer who had been found by 40 Commando's B Company hiding in the hills overlooking San Carlos. Beyond stating that he was a lieutenant-commander in the marines, this brave fellow would give no further information, but it was assumed that he had been observing the landing area and passing back information about where the British headquarters and most important stores were located. No radio was found but it would not have been difficult for him to hide this.

The landing force emerged from what became known as the 'Bomb Alley' period battered but with its beach-head intact. More than 10,000 tons of supplies had been landed, though there had been some setbacks and the desired scale of supplies would remain 'in arrears' until the end of the war. To be present in the beach-head during those days of air attack was an experience which none would forget. A naval officer remembers this contrast: 'One fine day, just after a hectic air attack, watching a farmer on a tractor taking fodder to his animals in that lovely pastoral setting.'

CHAPTER FIFTEEN

May the 25th

In an attempt to help with the intense air battle, Admiral Woodward had posted two of his ships, *Coventry* and *Broadsword*, as a '42–22 combination missile trap' in an exposed and vulnerable forward position outside the north end of Falkland Sound to engage Argentinian aircraft approaching and leaving the landing area. The first day on which the combination operated, 22 May, was quiet and the two ships were withdrawn that night. But, in their absence on the 23rd, the landing area was heavily attacked again and *Antelope* was hit. So, that night, *Coventry* and *Broadsword* went back in again, ready for action the following day. All went well on the 24th. The ships did not get the chance to use their missiles but two Sea Harriers of 800 Squadron were vectored on to four Argentinian Daggers approaching from the mainland and shot three of them down. Then came 25 May, Argentina's National Day, when a major enemy effort was expected. The Argentinians must have realized what these two ships were doing and how exposed they were. The two British captains wondered whether they would be recalled to the carrier group but Woodward told them to stay out for another day; he would move the carrier group nearer so that the Sea Harriers could have more time on patrol. The two captains – Bill Canning in *Broadsword* and David Hart-Dyke in *Coventry* – conferred on the radio about their exact dispositions for the coming day. Hart-Dyke asked to move further north, away from the Argentinian radars on East Falkland and into more open sea where his Sea Darts could operate more effectively. Canning wanted to stay closer to the land, where he could maintain better voice communications to the landing area and thus give them more effective early warning of approaching air attacks. The views of both captains were operationally valid but Captain Canning was the senior and his view prevailed. The two ships settled down in a position less than fifteen nautical miles away from the northern coast of West Falkland.

The Argentinians were indeed planning a big day but the early actions were inconclusive for the missile-trap ships. Soon after dawn, a Learjet made a high-level flight over the landing area and photographed the ships in the anchorage. *Coventry* picked up this plane on its radar, believing it to be one of the Boeing 707s so frequently encountered before. Captain Hart-Dyke describes events.

> He was about thirty-five miles away and I ordered two missiles to be launched. But the flash doors up to the launcher were jammed by dried salt. I had to send a sailor out with a bloody great hammer; there was only a small delay, but the Boeing's surveillance systems must have sniffed us and he was gone.
>
> It was very sad; there was a lot of gloom back at the task force and even at Fleet Headquarters. I had sent one signal and they were expecting a second – '*Coventry* splashed one Boeing'. Instead, I had to send a signal saying – 'Regret . . .' etc. I think I finished up with the words 'Woe is me' or some such wet comment.

The Argentinians then sent a series of raids to the landing area. *Coventry* twice fired Sea Dart missiles at Argentinian aircraft returning from these raids and believed that she had scored two successes. *Broadsword* thought her Sea Wolfs had got a Pucará or an Aeromacchi over the land. But these were all mistaken claims. The two captains again conferred by radio. There was a tremendous temptation to withdraw; the Spanish-speaking officer on *Coventry* had heard the Argentinian pilots reporting the position of the two ships. David Hart-Dyke says:

> On the one hand there was this feeling that we've just shot down three aeroplanes; let's get some more. Also, I knew that if we had turned back, the ships in the landing area would have got it. So, we stood our ground; I would love someone to tell me whether we made the right decision. We decided to stay out till the evening.
>
> We were actually discussing the wording of a signal, saying that we wanted to get out the next day. It was predictable that the Argentinians would be coming back for us; we actually heard their radio saying they were coming again and that they were going to get the Type 42 north of the Falklands.

The Argentinians now prepared a carefully coordinated

attack. Six Skyhawks would fly along the northern coast of West Falkland, as though to attack the landing area again, but would then turn and come out fast from the land to attack *Coventry* and *Broadsword*. Two of these aircraft had to turn back with technical problems but the remainder flew on. Timed to follow hard on the heels of this attack was another Exocet mission carried out by two Super Étendards refuelled from a tanker aircraft. The Étendards would approach the main carrier group from the north. The more forward position of Woodward's main group on this day had probably been detected.

The four Skyhawks found *Coventry* and *Broadsword* easily and commenced their attack, two aircraft making for each ship. It was now that several tactical shortcomings were highlighted. The position of the ships so close to the land prevented *Coventry* from making an early Sea Dart interception. The buoyant and determinedly aggressive mood of this ship and her captain then affected events further. *Coventry* had been in the exposed outer screen of the carrier for long periods, had seen her sister ship *Sheffield* lost and *Glasgow* damaged, but had never herself been attacked. She had done sterling work directing Sea Harriers and engaging targets with Sea Dart, though with less actual success with Sea Dart than she believed at the time. And, all the time, the press and radio were highlighting *Hermes* and *Invincible*, back at the relative safety of the main group. One *Coventry* survivor says, 'You should call this chapter *The Ship That Never Was*; we are still very bitter at our lack of recognition.' *Coventry* was undoubtedly hungry for further success.

When the Skyhawks made their approach, two Sea Harriers were available, orbiting just west of the two ships. *Coventry* initially directed these on to the oncoming raid, a perfect vector which brought the Harrier pilots into visual contact and rapidly closing within Sidewinder range. But, then, *Coventry* ordered the Harriers away, allocating her own Sea Darts to the engagement. The Harriers had no course but to obey, their pilots 'very miffed to have been taken off near sure kills'. This move is widely believed to have been a mistake. Sidewinders were more reliable than the shipborne missiles. *Coventry*'s intention was that the Sea Harriers would be placed in a position where they could engage any surviving Skyhawks departing after the raid. But that never happened; the Harriers were low on fuel by then.

Broadsword was attacked first. *Coventry* fired a Sea Dart and missed; the target aircraft was probably too close. Captain Canning describes *Broadsword*'s efforts to defend herself with one of the Navy's most modern close-range missile systems.

The first pair came straight for us. I was quite relaxed. I thought this was a perfect Sea Wolf situation. The aircraft were very low, so low that some of our look-outs could see the wake of their jet streams on the water – but Sea Wolf can get down there. The system alerted itself as it does, automatically; both systems were fully functional.

But the system did not engage at its best range – five to six kilometres – possibly because the two targets – wingtip to wingtip – confused the system. There was a frantic change to an alternative mode of operation. We took about three seconds; but that was two seconds too long. There was no way the human finger could have worked the keyboard in the time. By the time we had done it, they were over the top of us.

There were four bombs. Three missed us; one hit. The one that hit bounced off the sea, came in through the starboard side of the ship aft, five feet above the waterline, and, as it was climbing upwards after bouncing, it came up through the flight deck, hitting the Lynx as it passed through, and carried on into the sea the other side. The other three bombs fell – one ahead and two over the top of us, all very close. One of the look-outs said that, if he had stood up, he could have touched one of them.

The second pair of Skyhawks had swung round to the south-east and made for *Coventry*. The exact sequence of events here is not clear but this close-range attack was again a situation which *Broadsword*'s Sea Wolf should have taken care of; this ship was in the combination solely for the close-range protection duty. Her Sea Wolf system was now operating in the secondary mode and had successfully 'acquired' the Skyhawks but, then, *Coventry* started a turn to avoid the air attack. The turn placed *Coventry* between *Broadsword* and the incoming Skyhawks. No Sea Wolfs were fired.

Another grudge held by the *Coventry* survivors is towards the B.B.C., or the unknown person who leaked the recently broadcast item that a high proportion of Argentinian bombs dropped in the San Carlos area were not exploding. It is believed that the Skyhawks dropped 400-kg French SAMP bombs fitted with a direct-action nose fuze, rather than the tail-fuzed bombs which had performed so poorly in recent days. The nose-fuzed bombs armed themselves on release, were activated upon striking something substantial and exploded a few milliseconds later. The bomb which hit *Broadsword* had gone in and out without hitting anything substantial, but three bombs hit *Coventry*, which was unfortunately end on to the Skyhawks because of her recent turn, and the bombs plunged deep into and along the ship until they exploded. One burst near the Computer Room; the other two were in the centre of the ship near the forward engine room. Severe internal damage was caused and holes were blown outwards in the ship's port side. *Coventry* heeled over and started to sink. Nineteen men were killed; the figure could have been so much greater.

The rescue operation took nearly two hours. *Broadsword*, still shaken by her own bomb hit, watched for further air attack and rescued some of the men in *Coventry*'s life-rafts. A call for help was picked up by *Hermes* and at San Carlos; both launched a stream of helicopters in a well-executed combined operation. Alf Tupper was a crewman on one of the helicopters sent from San Carlos.

Coventry had her hull upturned and was on fire. I was horrified. I had been in the Navy eighteen years and had never seen anything like that before. We could see quite a lot of life-rafts; they all looked very full and we could see heads poking out from under the canopies. We did a full orbit round the flotilla of life-rafts to see if there was anyone in the water needing help first. We only found a Chinese but he was dead.

Then we went to the first life-raft. As I was wet anyway, from going down to the Chinese, I indicated to the crewman that I would stay down there. I unhooked and stayed in the life-raft. There were at least twenty men there. One or two were very badly shocked and I think some were burned, though I couldn't see this properly because they had their once-only suits on. They were all wet and cold. They behaved quite well; they could hear the helicopter overhead but knew it was friendly because this English-speaking chap had been thrust among them. I got one guy to sit on the end of the life-raft to help me and we started to get them out, two or three at a time, on to the roof – partly to get them ready for rescue, partly to collapse the canopy so that it didn't act as a combination of a sail and a parachute and cause it to drift away or blow over, which would have been disastrous.

Tupper stayed on the life-rafts and helped to load four or five helicopters, clearing three life-rafts of more than sixty survivors;

he was decorated for this work. *Broadsword* stayed until after dark, ensuring that there were no more survivors. *Coventry* remained afloat, upside down, until she sank the following day.

The tragedy of this action was that four Argentinian aircraft, making a straightforward attack in clear weather conditions, had penetrated four layers of defence – Sea Harrier, Sea Dart, Sea Wolf and close weapons – to sink a ship which had served the task force so well in recent weeks.

Only twelve minutes after the news reached *Hermes* that *Coventry* had capsized, the carrier group was itself under attack. The two Super Étendards carrying Exoccts had been successfully refuelled by a Hercules tanker aircraft and had approached the British ships from the north. It was a similar attack to the one which had hit *Sheffield* three weeks earlier, though with different Argentinian pilots. The two Étendards picked up the radar echoes of the carrier group and released their Exocets. They were briefly spotted on the radar of one of the British escorts but swung away to make a safe return to the mainland. The same escort – the recently arrived frigate *Ambuscade* – then detected that missiles had been released and all the ships in the carrier group with the necessary equipment fired off their chaff rockets. *Ambuscade* was the nearest ship to the attack. One Exocet was seen to be heading towards her but her cloud of chaff caused the missile to veer away to continue on, its radar searching for a further target among the larger ships forming the main group. Some accounts have credited a specially equipped Lynx helicopter with decoying this Exocet away from the two aircraft-carriers but this was not so; these special helicopters were not yet in action. It was the defensive layout of the group, with a line of supply ships shielding the two carriers from Exocet attack, which saved *Hermes* and *Invincible* but cost a valuable ship. The Exocet picked up the ship at the northern end of the supply-ship screen and veered straight on to that new target. The ship was *Atlantic Conveyor* and she was not equipped with chaff rockets.

The *Atlantic Conveyor* (a Cunard roll-on, roll-off ship of just under 15,000 tons, commanded by Captain Ian North, with approximately 160 men on board) had been with the carrier group for six days. Fourteen Harriers brought down from

Ascension on her upper deck had been flown off but the constant air attacks on the landing area had prevented this large and valuable ship from being sent into San Carlos and discharging the rest of her cargo. One Chinook and one Wessex had taken off that day for their first proving flights after the long voyage and these were being used to move stores around the carrier group when the Exocet attack took place. But *Atlantic Conveyor*'s upper deck was still loaded with three of the huge Chinooks and five Wessexes – all intended for helicopter work with the landing forces. Below decks were thousands of tons of valuable stores, also earmarked for the landing forces. *Atlantic Conveyor* was preparing to go into the landing area for the first time that night and would have left the carrier group about an hour after the Exocet attack.

Atlantic Conveyor tried to turn away from the oncoming missile, a standard response to present a smaller radar profile and to place her large vehicle loading ramp on the stern, stowed in its normal vertical sea-going position, as a shield against a possible missile hit. But the ship did not complete its turn and the first Exocet caught her on the port side, well aft. Initially, the missile performed as programmed; it entered the ship, turned to run along the length of the ship, and passed through the engine control room and under the bridge area before coming to rest. The warhead of the Exocet which had hit *Sheffield* three weeks earlier failed to explode but there is less certainty about this one. The missile buried itself deep in the stores stacked in what would normally have been the vehicle and container decks; the actual terminal point was near the junction of Numbers 2 and 3 decks. No one ever got near to that area. The missile may have exploded, its effect muffled by the stores, but the general consensus is that it did not, and the subsequent fire was the result – as in *Sheffield* – of the unspent fuel igniting. What happened to the second Exocet? No other ship was hit and no one reported seeing it at any time. It may have failed soon after being launched, but there is a theory that it also hit the *Atlantic Conveyor*, its radar having locked on to the first missile and following that first missile into the ship. But only one, large jagged hole was all that could be seen in the ship's side.

Roger Green was a Fleet Air Arm man on the upper deck, in charge of a party testing a Wessex helicopter.

There was a great big, vibrating thud and, then, almost instantaneously, the explosion happened inside the ship. It really rocked the ship. I saw Robby running, thrown off his feet. The rest of us hit the deck and just lay there for thirty seconds at most. I saw one of our lads coming round the corner of the superstructure – the front of his shirt was all open, his face was black, tears were streaming down his face. All he kept saying, over and over again, was, 'Fucking great ball of fire.' I got hold of him and asked him if he knew who I was but he kept on saying the same thing. Then I noticed small bits of metal in his chest and some cuts on his face and realized he was wounded. I grabbed hold of him and took him into the officers' dining room, which was the emergency sick bay, and was just handing him over to our naval medical orderly when there was another explosion which shook us, because it was directly below us, but not as bad as the first one. I found out later that it was nitrogen bottles exploding.

I waited again until I saw someone else moving after that explosion and went outside; the whole ship was now covered in a mass of thick black smoke billowing out from every little crevice down below.

Nigel Stronach was another Fleet Air Arm man, in charge of a party of men taking an early dinner and enjoying a bottle of Moselle on their last night before going ashore at San Carlos.

There was a thump, a big thump, you could feel it through your feet. They piped that we had been hit and we were to hit the deck. I thought they had got it the wrong way round; it made it difficult later on because we didn't know whether we were still under attack. I tried to get in touch with the bridge but the telephone had gone out. A message was piped to send eight blokes to act as a stretcher party. I sent them off under an R.A.F. sergeant but they came back a few seconds later because of the smoke. Their eyes were streaming and even their respirators couldn't help. There was no way they could make it. I detailed off one of the lads to go to the bridge and tell them we couldn't get to the back end of the ship.

Then one of the petty officers of my squadron came in, looking just like a black-and-white minstrel show. He had been down below and had actually been blown over by the missile. He was black all over, except for two white rings around the eyes where his glasses had been. I must admit, it seemed very funny at the time. Another man came in soon, from the same place. He was going on about his life jacket being blown off and he had got a couple of small burns on his back. He makes quite a thing about them now, showing the girls his war wounds.

The explosions continued; they seemed to be coming closer. One of the things that I remember was the faces of some of the younger lads. I remember one of them, sitting crouched up. He looked scared. I gave him a wink to try and cheer him up; he tried ever so hard to smile but it was only half-hearted. We had been a training squadron before we were sent down south and we had a lot of young eighteen- and nineteen-year-old naval airmen. I found out later that a lot of our young lads at the front of the ship were fighting the fires and doing remarkably well.

Like *Sheffield*, *Atlantic Conveyor* was destroyed by fire. Many tributes are paid to the mixed crew of Merchant Navy, Royal Navy and R.A.F. men aboard who were conscious of the ship's value to the Falklands operation and tried so hard to save her. There was no water pressure in the hoses for the first fifteen minutes. *Alacrity* and *Brilliant* came dangerously close in to this large burning ship, which was rolling on a strong swell, and tried to fight the fire. But all the efforts were defeated by the density of the smoke and the deep-seatedness of the fire, which was fuelled by a huge mass of inflammable stores. Night started to fall, making the rescue task more difficult. There were 200 bombs on board which might explode when the fire reached them. The order to abandon ship was given ninety minutes after the Exocet hit.

Atlantic Conveyor, with a regular crew of thirty men, was not well fitted to evacuate more than five times that number – in the growing darkness and with a cold and heavy swell running. A helicopter took a few men off the bow but most men queued patiently to go down a pair of pilots' ladders at the stern. A cluster of life-rafts waited at the bottom. There was 'confusion but not panic'. The rafts were eventually pulled across to *Alacrity* who picked up seventy-four survivors. *Brilliant* took on another thirty-two. Twelve men died, three in the original Exocet strike and nine during the fire-fighting and abandon-ship phases. One of those lost was Captain Ian North, the *Atlantic Conveyor*'s fifty-seven-year-old master, a veteran of the Second World War and a much admired skipper. He was the last to leave his ship, could not find a place in the first life-raft and disappeared while trying to reach another one. Petty Officer Len Cobbett describes how he nearly joined the list of dead. He was a naval photographer, transferred from *Hermes* only two hours earlier so that he could be landed ashore that night. Unfortunately his cameras were stored below decks and could not be reached because of the fire, otherwise he might have produced some of the war's best photographs.

I jumped off the ladder forty feet up and inflated my life-jacket while under the water by pulling the CO_2 knob. I came up and saw the life-rafts but was hit by something, I don't know what. I didn't remember anything then – not for some time. The next I knew was that I was stuck between *Atlantic Conveyor* and a life-raft – the raft on my chest and my back against the ship. The sea was quite heavy. I managed to push the raft off me and get round to its entrance – a very laboured job. I didn't know at the time, but my once-only suit was badly ripped and the water had started to come into contact with my body and the cold was affecting me.

There was a little set of steps to the life-raft but I couldn't feel my legs – there was nothing there. I tried to pull myself up with my arms but saw that it was full. I couldn't hold on any longer and fell back into the sea. It was all quite hazy by now – like coming out of a deep sleep is the nearest I can put it.

I was just drifting off – very tired and finding it quite difficult to breathe – when someone grabbed me and manhandled me into another life-raft that had more room in it. I remember lying face down in the floor but I couldn't move or breathe until they moved me to the side and made more space. Someone said I had broken my back and the next I knew was that I was being strapped into a Neil Robinson stretcher – the bamboo stretcher they use for moving people in awkward conditions. I got so that I could see up through the entrance of the life-raft and see the sky and then the life-raft seemed to drop away from me; a swell had come up at the same time as they had taken the strain on the rope and the next thing I saw was the side of *Alacrity* smashing me in the face – a helluva bang, but I was so cold I didn't feel it.

Len Cobbett was not seriously hurt and was evacuated two days later in a helicopter, the co-pilot of which was Prince Andrew, to a tanker which took him to Ascension.

Atlantic Conveyor was later taken in tow by the tug *Irishman*, but she sank three days after being hit. The Argentinian pilots had destroyed a valuable target. The landing force was denied the use of enough tentage and living equipment for 10,000 troops, a mile of portable steel runway, many vehicles, helicopter and aircraft spares, bombs and nine helicopters. The loss of the three R.A.F. Chinooks was a particularly hard blow. All were new or nearly new, each costing £3·5 million; one only had three hours' flying time. The land campaign would be severely handicapped by the loss of their load-carrying capacity. One Chinook survived, being away from *Atlantic Conveyor* when the ship was hit. This helicopter, the famous 'Bravo November', landed on *Hermes*, where it was nearly pushed over the side because its bulk was hindering normal carrier operations; it later flew ashore to San Carlos where it made a heroic and legendary contribution to the land campaign. Two Chinook crews were sent ashore to operate this helicopter; the remainder were sent back to Britain.

The Argentinians soon learnt what ship they had hit, claiming it as 'a Harrier-carrying container ship'. A total claims list up to this day, 25 May, was published in their Falklands forces newspaper and probably on the mainland. Claimed as sunk were five warships, the *Canberra*, the *Atlantic Conveyor* and a 'troop transporter'; *Hermes*, seven other warships and three landing ships were 'seriously damaged'. (*Hermes* was regularly reported as being damaged in this period, an elaborate story being added about the President of Curaçao (Suriname) refusing to provide repair facilities.) Fourteen Sea Harriers and twelve British helicopters shot down were also claimed. Though these claims were grossly inflated the Argentinians had none the less inflicted severe losses on the British naval forces: two frigates, two Type 42 destroyers and the *Atlantic Conveyor* were all sunk or about to sink and several other ships had been damaged. But the British aircraft-carriers were still untouched and the Harrier force, recently reinforced by *Atlantic Conveyor*, was stronger than ever, while the Argentinian air strength was a rapidly wasting asset. The 25th of May was a great boost to Argentina on her National Day but it marked the end of a phase and the whole nature of the Falklands war was about to change.

CHAPTER SIXTEEN

Goose Green

The units ashore at San Carlos had made no major move during the period in which the Argentinian air force had been sinking the British ships. For the men on the British ships being attacked and for the commanders and politicians watching from London, it had seemed an interminably long period of waiting and there was much frustration at the lack of military movement. In fact, it was only five clear days since the day of the landings and Brigadier Thompson had basically been following his original orders: *To establish and secure a beach-head and to await the arrival of Major-General Moore and of 5 Brigade.* The exact texts of further orders issued at this time have not been released but, according to Brigadier Thompson, his further instructions during this waiting period were: *When firmly established and not before, to patrol aggressively and mount operations to sap the enemy's morale and will to fight.* The beach-head had been secured but the only serious effort so far to mount operations to sap the Argentinian 'morale and will to fight' had come to naught. This was a proposed raid by 2 Para on the Argentinian garrison at Darwin–Goose Green, fifteen miles south of the beach-head area. This plan had been shelved, for reasons which will be described later.

Because the Argentinians made no attempt to attack the beach-head, there had thus been an absence of significant land action. For most of the infantry units it was a time of digging, patrolling, watching and waiting, not unlike a prolonged exercise, though with the added display of the daily Argentinian air attacks. The guns of 29 Commando Regiment had been called upon to shell suspected Argentinian patrols or helicopters seen in the distance. The guns opened fire, on super charge at near-maximum ranges, but most or all of these targets were later classified as 'phoney'; they were usually wandering sheep. The most serious incident occurred during the day of 23 May, when a patrol from A Company, 3 Para, went out into the hills north of Port San Carlos to outflank a suspected Argentinian patrol. The neighbouring company, C Company, were notified of the patrol and its route but then saw men moving in another area. They opened fire with machine-guns and called for artillery support. But the target was the A Company patrol, which was later accused of being a thousand metres away from its advised route. Seven paras and a sapper were injured, two of the paras suffering severe head wounds which were feared might be fatal, but they recovered. A Sea King helicopter sent to give medical aid then landed heavily and damaged its tail rotor.

Most of the paras and marines were anxious for this phase to end and keen to get at their enemy, but it had been necessary to wait for the logistic build-up. There was no shortage of stores and ammunition; the original task-force ships had sailed with food, stores and ammunition for sixty to seventy days of operations, and radar, electronic and weapons spares for ninety days. There was, however, some anxiety over the rate of usage of some kinds of ammunition – Sidewinder missiles, naval shells and chaff rockets were being consumed quickly, but none of the items needed by the units ashore were in short supply. The problem was of getting these stores out of the ships and on to the land. Some disruption had been caused by the air attacks, but by 27 May it was estimated that thirty days of war stocks were ashore, although the ship carrying the main ammunition reserves, *Elk*, had not yet been unloaded.

The process of settling into the beach-head and waiting for 5 Brigade was interrupted by new orders. The news of ship after ship being lost – but no movement on land – was too much for London and there was domestic political pressure for an early move. Admiral Lewin says that there was also 'much international anxiety and pressure to cease fire and negotiate in order to save further casualties'. A cease-fire at this time would have left the British forces ashore at a huge disadvantage, out in the open, with winter coming on fast, but with the Argentinians safely in the shelter of Stanley and the other settlements. Admiral Fieldhouse, the overall commander, was anxious to get on for other reasons. Sandy Woodward made this entry in his diary after *Coventry* and *Atlantic Conveyor* were lost: 'The battle is high risk at sea and in the air. It must now go high risk on land.'

Woodward made no direct complaint to Fieldhouse but Fieldhouse must have been aware of the feelings in the British ships that the naval part of the task force was seemingly being asked to bear the brunt of the Argentinian air attacks while the troops ashore did nothing. Fieldhouse had already given new orders to Brigadier Thompson: *3 Commando Brigade was not now to wait for 5 Brigade but was to start moving out from the beach-head with the ultimate intention of investing the defences of Stanley*. This message actually arrived just before *Atlantic Conveyor* was sunk, probably during the day of the 25th. Thompson's staff prepared a plan for a mass movement forward by most of his units, using the helicopters expected that night from the *Atlantic Conveyor*. The helicopter lift would take these units right up to the ring of hills around Stanley; he could thus 'invest' the Stanley defences in one bound. The plan was just being completed when news arrived that *Atlantic Conveyor*, nearly all of her helicopters and all of her stores were lost. There was much dismay; Major John Chester, brigade major at 3 Brigade Headquarters, says, 'We regarded *Atlantic Conveyor* as the fulfilment of all our hopes and the basis of all our plans.' Fieldhouse now spoke again to Thompson, with the news of the loss of *Coventry* and *Atlantic Conveyor* fresh in London spurring him on. '*Some action is required*', Thompson was told. Thompson decided to resurrect the plan for 2 Para to raid the Darwin–Goose Green area and to start pushing other units forward *on foot*, overland in the direction of Stanley. 45 Commando and 3 Para were the units chosen to start this overland trek to the east. 42 Commando would move later, possibly by helicopter. 40 Commando would have to remain behind to defend the base at San Carlos, it being intended to relieve 40 Commando of that duty with 2 Para after the Darwin–Goose Green operation (but that exchange never took place).

These orders committed one of the British units, 2 Para, to the first major set-piece engagement of the war, the Battle of Goose Green. (The objectives were really Goose Green and Darwin, but Darwin is little more than the residence of the manager of the Falkland Island Company; Goose Green was a major settlement, the largest place outside Stanley. The battle was to be fought on a small part of the Goose Green property which, at 400,000 acres, was the largest in the Falklands, almost as large as Anglesey, the Isle of Man and the Isle of Wight combined.) It was only by chance that 2 Para was selected for the raid; this battalion happened to be holding the beach-head sector nearest to Goose Green. The word 'raid' is a common one in the Royal Marines but not in the Army. Brigadier Thompson says the intention was *'to inflict as much damage as possible, to destroy defences and then withdraw'*. There was no real enthusiasm for the operation at 3 Commando Brigade Headquarters; Thompson would have preferred to bypass and mask Goose Green rather than commit one of his five infantry units to a major engagement on a flank at this stage. 2 Para had started to move off on this operation on 24 May but had then been recalled. Various reasons have been suggested for the change of plan; the exact reason is not important. Lieutenant-Colonel H. Jones of 2 Para (always called 'H' because he disliked the Herbert for which it stood) had some unkind remarks to make about the Royal Marines and their green berets when the original operation was cancelled, after his patience in waiting twenty years for a chance of action.

Now, however, Brigadier Thompson had received new orders from London and the old plan to raid Goose Green was resurrected and implemented unchanged, except that 2 Para was not to return after the operation; 'raid and return' became 'capture and stay'. The hallmarks of the Goose Green operation were that it was run on a shoestring of resources and was rashly optimistic in its assessment of Argentinian opposition. It should not have succeeded. 2 Para were to march more than ten miles from their position on Sussex Mountain, first to a holding position, then to the Start Line for their attack, with no guarantee that they would not meet any Argentinian troops on the way or be shelled or be attacked by the Pucarás stationed at the nearby airfield. They were then to attack and capture an isthmus of land five miles long but averaging little over a mile wide and thus without the ability to make any major outflanking move. The whole of this ground was of a completely open nature in which the location and strength of the Argentinian defences had not yet been accurately plotted. The presence of Falkland civilians at the far end of the objective further complicated the situation.

The intelligence assessment of the Argentinian strength was

vague. A four-man S.A.S. patrol, commanded by Corporal Trevor Brookes, had been in the Goose Green area for two weeks, attempting to assess the Argentinian strength. There had been an interesting episode during this time when the commander of the Argentinian artillery at Goose Green decided his guns needed some target practice and prepared to fire on a wooden hulk on the other side of Darwin Harbour. Eric Goss, the settlement manager, protested, saying that this wreck was part of the Falklands heritage and the Argentinians fired elsewhere. Mr Goss found out after the war that the hulk had been one of the S.A.S. 'hides' at that time. The S.A.S. had not been able to send back much useful information. The best estimate that 3 Commando Brigade could give was that there were two, possibly three, companies of infantry, with limited artillery; there were also many air-force personnel. What was even more open to doubt were the fighting qualities and state of morale of the Argentinians; the general opinion was that they would prove to be a poor enemy.

There was actually the equivalent of a reinforced battalion at Goose Green. When the Argentinian air force established a Pucará base at the airfield in mid-April, C Company of the 25th regiment moved in; these were good-quality troops from the symbolic national unit sent early to the Falklands. Lieutenant Stephan, described by Mr Goss at Goose Green as 'a true soldier', was probably their commander. He is believed to have been killed in the fighting. But then, when a new brigade arrived in the Falklands in May, most of the 12th Regiment, commanded by Lieutenant-Colonel Ítalo Piaggi, appeared at Goose Green and Piaggi became the senior army officer. Piaggi's men were described as being 'a poor lot of conscripts', looked down upon by the other local Argentinians. The Goose Green garrison also received four field guns – only three were usable – just before 2 Para's attack. Alongside the army units at Goose Green was the 'Condor Air Base' commanded by Air Commodore Wilson D. Pedroza. 2 Para would be outnumbered nearly three to one by the total number of Argentinians at Goose Green.

The command set-up was a typical Argentinian muddle. Air Commodore Pedroza was the senior officer – rejoicing in the title of Commander of Task Force Mercedes – but Lieutenant-Colonel Piaggi would have to fight the ground battle. Two days before this battle commenced, however, Brigadier-General Menéndez ordered the commander of III Brigade – Brigadier-General Omar Parada – to move his headquarters from Stanley to Goose Green and take command, not only of the immediate area, but of the whole area of operations around the British landing zone on both sides of the Falkland Sound. But Parada delayed the move, remained in the house at Stanley which he had taken over, and would attempt to fight the Battle of Goose Green by radio.

The overall plan for the Goose Green operation had been prepared by the 3 Commando Brigade Headquarters staff. There was no way the battalion could be parachuted into action. The ideal solution would have been a helicopter lift to a location south of Goose Green, where the Argentinian defences were almost non-existent. This was considered but dismissed, because a move by day would be too risky without air superiority – Pucará versus helicopter would have been no contest – and a night move had to be ruled out because so few helicopter crews were trained to use passive night goggles. The helicopter option disappeared completely when *Atlantic Conveyor* was lost. A move by landing craft was considered next. The full battalion could easily be carried in one lift but Major Southby-Tailyour, the landing-craft and local-waters expert, studied the approach and declared that kelp (thick Falklands seaweed beds) and submerged rocks would only give a fifty-fifty chance of success. Colonel Jones considered this and stated that he believed the operation only had a 75 per cent chance of success anyway and that to halve that figure meant failure. He preferred the only remaining option, to march to the battle. It became known later that a landing-craft assault would have met mined and defended beaches.

Next came the question of supporting arms. The artillery batteries in the landing area were out of range of Goose Green, unless perhaps they fired on super charge, a prolonged period of which would ruin the guns for the rest of the campaign. It was decided to lift half of 8 Battery – three 105 mm guns – to an intermediate position, and the frigate *Arrow* was allocated to give further gunfire support, the rapid if less accurate fire of her one 4·5-inch gun being approximately equal to that of six of the

Royal Artillery's guns. (*Arrow* would use the mast of her sister ship *Ardent*, which was sticking out the water where *Ardent* sank, as a datum marker in the coming operation.) GR3 Harriers from *Hermes* would also be available, weather permitting, to give close support. The tanks of the Blues and Royals were considered, but there was still reluctance to use these away from firm tracks and they were not included in the operation. Later experience showed that they could have operated without difficulty and they missed their best opportunity of the war of supporting an infantry attack over suitable tank country. Similarly, none of the marines' Volvo 'over snow' vehicles were included and the paras had to leave most of their heavy mortars and much other equipment behind.

Colonel Jones was whisked back from his battalion position on Sussex Mountain to Brigade Headquarters on the afternoon of 26 May and enthusiastically received orders that the operation cancelled two days previously was now on again. There were few preparations to be made; one officer says that Goose Green was 'a come-as-you-are party'.

That evening, the paras handed over their old positions on Sussex Mountain to L Company of 42 Commando and set off for the eight-mile trek to their intermediate position at Camilla Creek House, an empty building sometimes used for outstation work on the Goose Green property. There was some desultory Argentinian shelling well to the left of the line of march; Major Tony Rice, the artillery officer attached to 2 Para, could tell that three 105-mm guns were firing from Goose Green, weapons which had not been reported by intelligence.

Most of the battalion stayed at Camilla Creek House during the daylight hours of the 27th but two small reconnaissance parties went forward and placed themselves where they could see down either side of the isthmus, though neither party had a completely clear view. The locations of three new groups of Argentinian trenches were plotted. One of these groups of trenches, on a small rise later known as Darwin Hill, was to have important consequences. One of the observation parties had a forward air controller and a Harrier GR3 raid took place during the day. But the controller and the Harrier pilots never made direct contact, possibly because the Harriers were coming in too low; all communications with the aircraft were through an air liaison officer at brigade headquarters. A first raid took place on gun and infantry positions and then a second raid was ordered – always a risky prospect, going back to the same target so soon – but the Harrier pilots thought that 2 Para must be in serious difficulties. They did not know that these were only preliminary softening-up missions. The second sortie was flown by Wing Commander Pete Squire and Squadron Leader Bob Iveson; Iveson had also flown the first sortie. The Harriers made two passes and then Bob Iveson decided, on his own initiative, to make a third run to attack some trenches, ironically the trenches on Darwin Hill that were later to cause so much trouble to 2 Para.

It was a silly move, a very poor tactical move, and it cost me because it was on that pass that I was hit. I was coming off target, heading roughly west, and was hit by two fairly large-calibre shells – probably 30 or 35 mills – a hell of a thump which rattled me round in the cockpit, banging my head on the side of the canopy – it moved the aircraft sideways that much. Things went very wrong, very quickly, after that. The controls stopped working; there was smoke in the cockpit and finally fire, at which point I pulled the handle. I was in a good attitude to eject but I was a bit fast and a bit low – about 450ish knots and about 100 feet.

The 'chute opened and I found myself going right towards the fireball of my aircraft in front of me. I couldn't tell whether it had struck the ground or not. I think now it had blown up in the air. I was only on the 'chute for about ten to fifteen seconds before I hit the ground, in a bit of a heap and very winded.

The Harrier's speed had taken it well away from Goose Green. Iveson suffered a compression fracture of his back and wind-blast damage to his eyes. Immediately after landing, he ran away from a line of dots seen on the horizon – sheep – and later hid in Paragon House, another of the outlying Goose Green property shepherds' houses, but on the wrong side of a bay from British forces. Mr Goss had earlier stocked this house with a trailer load of food in case the Goose Green civilians were forced to flee the settlement. Iveson only took one tin of baked beans; he had his own emergency rations. He made contact with friendly aircraft on his survival radio and was picked up by a naval helicopter two days later. He recovered from his injuries and returned to the Falklands for a second tour after the war, meeting Mr Goss and thanking him for the baked beans. His flying helmet was recovered by one of Goss's shepherds.

The paras' two reconnaissance parties were later spotted by the Argentinians and engaged by long-range machine-gun fire. They withdrew, capturing on the way a civilian Land-Rover carrying Argentinian reconnaissance troops, who are reputed to have revealed the true strength of the Argentinian forces at Goose Green. This news was not well received by Colonel Jones, who would soon learn that the B.B.C. had announced that lunchtime that '2 Para were moving towards Darwin'. This news item was the result of a premature release of information by the Ministry of Defence. That afternoon Mrs Thatcher informed the House of Commons that British forces were moving out of the beach-head. These were incredible errors of disclosure and an angry Colonel Jones threatened to sue the B.B.C. and everyone else involved for manslaughter on his return to England. A study of the Argentinian moves shows that no further reinforcements were sent to Goose Green until after 2 Para attacked, but the news items probably alerted the existing garrison.

That evening, the half battery of guns was brought up in a shuttle of eighteen helicopter lifts – one lift for the twenty-eight gunners, three for the guns, and fourteen for the 840 rounds of ammunition, which was loaded loose in nets to save the weight of boxes and pallets. The guns set up their positions 800 metres north of Camilla Creek House, four kilometres from the infantry Start Line and eleven from the final objective.

The reconnaissance parties reported in. Captain Allen Coulson, the intelligence officer, plotted the latest details of Argentinian trenches on his map. He was satisfied that he knew the location of nearly every Argentinian position with the exception of those around the airfield at the far end of the objective. Colonel Jones had already prepared his plan on the basis of the earlier radio reports from the reconnaissance parties. Because of the open nature of the defended area and its approaches, there would have to be a night attack, an operation frequently exercised by all Nato armies because of Russian air strength. The advance to contact would be in silence – with no artillery fire – but, as soon as contact was made, the attack would become 'noisy' – 'shell them hard, then move in fast'. Jones hoped that the main Argentinian positions would be defeated by dawn, leaving the clearing of their base units in and around the civilian settlement to be done in daylight. 'H' Jones was now ready for the culmination of his career as a regular soldier, the issuing of orders to his officers for a full-scale battalion attack. Before coming to the Falklands, he could never have dreamt that this moment would come in the shelter of a gorse hedge outside an empty shepherd's house nearer to the Antarctic than to the Iron Curtain. He outlined the main plan. There would be six phases, attacking six separate defended locations. Only one company should be in action at a time; if it stuck, another company could hook round the opposition. Next to speak was Lieutenant John Thurman, a Royal Marines officer who had served in the Falklands before and who gave details of the terrain. Captain Coulson had to wait for Thurman to finish before he could give details of the enemy forces and the locations of their positions. Colonel Jones intervened to urge that the process be hastened. It was getting dark; the company commanders were anxious to be finished so that they could give out their own orders. Some details given out about the enemy trenches were missed.

The battalion moved off well after dark to cover the final four and a half miles to the Start Line. A and B Companies then deployed themselves either side of Burntside Pond in the middle of the head of the isthmus. A Company moved at Zero Hour, 3.30 a.m. local time, the first objective being a suspected Argentinian platoon position at another shepherd's house, Burntside House; it was assumed that there would be no civilians here. The house was reached five minutes later and attacked with typical para vigour. Second Lieutenant Guy Wallis's 3 Platoon went to ground and put down heavy fire on the house; 1 and 2 Platoons moved around, shooting as they came. Anti-tank rockets were fired at doors and windows; grenades were thrown. Thirty or so Argentinian soldiers had been in the nearby shed for several days but the only occupants of the house itself were four civilians: Gerald Morrison, the shepherd in charge of the Goose Green pedigree breeding flock, his wife, his mother and a friend. Mr Morrison states:

We knew the paras were on the way; we had heard it on the B.B.C. World Service. An Argentinian soldier came at about 8.0 o'clock from the lambing shed where they had a small party. They told us to turn the generator off because it was making too much noise. They used to come down two at a time, pinch the food and then bugger off. We never knew how many there were. We had run out of candles and had no paraffin for the tilly lamps, so we were blacked out. We all went to bed on the floor.

There had been a lot of shelling or mortaring earlier in the night about 200 yards away down the valley. I couldn't tell what it was; the most I had heard before in my life was a 0·22. We were all lying in the corners; we would have got under the carpet tiles if we could. It was a wooden bungalow which gave no protection. If they were going to kill us, we wanted them to do it quickly and not go through hell first. We were terrified, but the shelling eased off.

Then the house was hit, bullet fire first, all over the place – mother had a row of eleven bullet holes just above her head. A grenade came through the window of one bedroom but luckily there was no one in it. It lifted the roof of the bedroom about five inches and all the furniture caught fire. We thought we were caught in cross-fire but then we heard them shouting orders in English and a bit of swearing. We started shouting that we were civilians. It seemed a lifetime before they heard us, but I don't suppose it was long. One fellow answered but I don't think you would want to write down exactly what he said. He wanted us to surrender. Jim and I went out and they made us stand with our hands against the house. They were full of apologies then – but it was a bit late.

We counted 130 holes in the house afterwards and then gave up counting.

The Argentinian troops are believed to have withdrawn; they were probably an outpost and were able to alert the main garrison. The paras reported some 'return fire' but this was later believed to be their own bullets ricocheting off rocks. The only casualty in Burntside House was a dog which was hit in the mouth by a bullet and had to be destroyed three months later.

B Company, on the right, crossed its Start Line half an hour later. D Company followed, intending to pass through B when the first objectives in this area were secured. But both companies were soon in action, B Company having unknowingly bypassed some Argentinian positions in the dark. Scattered fighting now commenced. The artillery of both sides was called into action. It was still very dark and cold and it started to rain. But the paras made steady progress in a series of actions. Major Philip Neame, commander of D Company, describes a typical engagement.

It was a platoon position on top of a little hill – a small knoll; forward of it was a series of machine-gun positions in a defile. They had earlier fired across at A Company and the Colonel on their trek south. 'H' came back and told me to deal with it. It took about thirty seconds to prepare my attack. We were already in basic one-up and two-back formation, and gunfire was already directed on to the position – or the area of it – no one could see what was really going on. I just decided on a night advance to contact. I said on the radio: 'Same formation. Situation – enemy on hill top. Axis – such and such a bearing. Move now.'

We advanced up the hill. They must have seen us and we attracted sporadic small-arms fire but it was very wild and ineffective. We got on to their position and started to clear through it. The sections broke down into four-man fire units and started to clear through with bullets and grenades. The Argentinians adopted what I call the ostrich complex; most of them had their heads buried in their trenches or their sleeping bags and appeared to be hoping we weren't there.

We had started the clearing process when 10 Platoon, my right rear platoon, came under heavy and effective automatic fire from these machine-guns which were situated between myself and B Company; that was the first we knew about them. 10 Platoon assaulted one of these positions, and took it, but were then basically pinned down. They lost two men killed. After that, confusion was fairly paramount. By this time, one of 12 Platoon's sections had wandered off and got separated. I lost comms with 11 Platoon intermittently and, more by their platoon commander's initiative than my action, they came across and put in a left flanking attack on the machine-guns 10 Platoon were facing, with 10 Platoon putting down covering fire. I was quite impressed by the degree of initiative and control this young lad showed.

In the course of his attack he lost two men. That was really the end of all the opposition. We took three prisoners; we didn't bother to count the number of dead; our main priority was to reorganize in the completely featureless dark, with our dead and wounded having to be found.

To sum up, it had been complete chaos and confusion. We had no proper Start Line, no firm idea of where the enemy were, no firm idea of where B Company were but, despite that, I think we came out of it quite well because of initiative at every level – and aggression. The prerequisite had been to keep the momentum going. That was H's abiding instruction and it had been fought in that vein.

The young platoon commander who did so well here was Second Lieutenant Chris Waddington. 10 Platoon's two dead – Lance-Corporal Tony Cork and Private Mark Fletcher – were 2 Para's first fatal casualties in the battle. Cork was hit first. Fletcher's body was found sprawled across Cork's; he had been applying a wound dressing to Cork when hit himself. One of the two men killed in 11 Platoon, Lance-Corporal Gary Bingley, and an uninjured colleague, Private Barry Grayling, were awarded Military Medals for attacking a four-man Argentinian machine-gun post. The infantry battles of the Falklands would be characterized by some superb work by the corporals and lance-corporals of all the British units.

The first Argentinian prisoners started coming back to Battalion Main Headquarters, where they were given a preliminary interrogation by Captain Coulson, the intelligence officer, and Captain Rod Bell, the Royal Marine interpreter. Coulson remembers:

The first lot in appeared to be rear-echelon troops – cooks and such like. There were no officers; they had gone. We also found some prisoners ourselves, hiding in sleeping bags in the bottom of trenches in the foetal position – not unusual in warfare, we were told later. We got absolutely nothing from them. They were just like little children.

Meanwhile, A Company was making good progress on the left flank. The company moved tactically against a suspected Argentinian company position on Coronation Point but this proved to be empty. The company commander, Major Dair Farrar-Hockley, left one of his platoons here, to provide a firm fire-support base for the next bound, and carried on down a track skirting a small bay and intending to move to encircle the few houses in Darwin Settlement less than a mile away. Colonel Jones and his Tactical Headquarters had been following hard on A Company's heels and Jones now came up and consulted Farrar-Hockley. The two officers verified their position and Jones urged the company to push on towards Darwin; he believed that the main Argentinian positions had now been penetrated. Ahead of them was a small rise, so small that it had no name even on large-scale maps; it is now known as Darwin Hill. This was one of the locations on which the reconnaissance parties out the previous day thought they had spotted Argentinian positions, but these were not marked on the company commander's map. In fact, there was a line of good trenches across this hill and another small rise to the west (to the right of A Company's line of advance).

The Argentinian position on and around Darwin Hill had not been long occupied. When 2 Para's attack had commenced around Burntside House two hours earlier, the Argentinians had sent a company out from Goose Green and this unit had only just moved into these positions. An Argentinian officer, Lieutenant Roberto Estévez, is credited with moving this company up and rallying the remnants falling back from trenches further forward. He died later in the day. The Darwin Hill line was really the main defence position; the trenches further forward had been manned by lower-quality troops to absorb the first shock.

Major Farrar-Hockley was unaware of all this. He intended to deploy his own two platoons on the hill before dawn, which was now breaking, and then sweep down to the left to capture Darwin Settlement. Second Lieutenant Mark Coe set off with the leading platoon.

As we approached the hill, we saw three men on the top of the hill about a hundred metres away. None of them appeared to be armed. They waved to us. We thought perhaps they were civilians; one man even thought they were walking their dogs perhaps. No one wanted to fire in case they were civilians. Later, we thought perhaps they were trying to surrender. Just as we did identify them as Argentinians, fire was opened on us from a position to the right. The tracer came straight over the top of us. I didn't have to give many orders. The leading section and myself went straight into a gorse gulley on the left to gain cover but 3 Section under Corporal Adams rushed into the small fold in front of him where he ran into fire. He was hit and badly wounded in the shoulder but extricated himself and a machine-gunner, who had two bullets in the head and one in the leg.

It is probable that the three Argentinians seen waving had thought the oncoming paras were Argentinian survivors falling back from the forward positions.

The two platoons and company headquarters had all come under fire. Several men were hit; the remainder rushed forward to take cover in the dead ground at the foot of the low hill or in

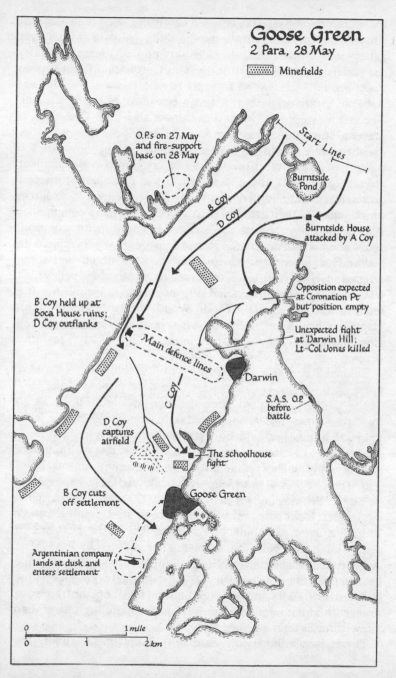

Goose Green
2 Para, 28 May

░░░░ Minefields

Start Lines

O.P.s on 27 May
and fire-support
base on 28 May

Burntside
Pond

B Coy

D Coy

Burntside House
attacked by A Coy

Opposition expected
at Coronation Pt
but position empty

B Coy held up at
Boca House ruins;
D Coy outflanks

Main defence lines

Unexpected fight
at Darwin Hill;
Lt-Col Jones killed

Darwin

S.A.S. O.P.
before
battle

C Coy

D Coy
captures
airfield

The schoolhouse
fight

B Coy cuts
off settlement

Goose Green

Argentinian company
lands at dusk and
enters settlement

0   1 mile
0   1   2 km

the gorse gulley on the left. A Company was in a difficult situation and one that would require skill to recover from. It was getting light. Most of the company were pinned down right under the enemy's nose. The platoon left on Coronation Point was on lower ground – well-sited to support an attack on Darwin, but not able to fire on the Argentinian positions on top of the hill. The enemy were too close to call down shell fire. One tank, if present, could easily have broken the deadlock. Colonel Jones caught up, A Company's sergeant-major asking him not too politely to keep out of the way while his company commander attempted to sort out the situation. It was a conflict of interest between the battalion commander who wanted rapid progress for the sake of the battalion plan and the company commander who needed time to sort out a complex and difficult situation. Jones and his staff continued to move up and they joined up with Farrar-Hockley's men. Various moves were urged and some attempted. The men were reluctant to move; they believed it was hopeless. The officers led off, intending to establish a fire team on the hill. Captain David Wood, the battalion adjutant, Captain Chris Dent, A Company's second-in-command, and Corporal David Hardman were immediately killed. The move petered out.

Colonel Jones now decided he would try to lead a flanking move on the right. He crept round the dead ground at the base of the hillock and called upon the men around him, 'Let's go forward.' Jones ran around the right-hand side of the rise, followed by his personal bodyguard, Sergeant Barry Norman. He was seen to run upwards to attack the nearest of the line of Argentinian trenches on the hill, firing his Sterling sub-machine-gun. Then he was seen to leap over backwards. He appeared to be hit but this was not so; he was merely changing the magazine on his gun. Jones rose and started a second attack on the trench. Thirty yards or so behind him was the first of the next line of Argentinian trenches running inland, not yet seen by Jones. The Argentinians manning that trench had an easy shot against the lone British officer across the dip between the two hillocks. One bullet struck Jones in the back and he fell. Sergeant Norman dragged him back into shelter but his commanding officer died a few minutes later without speaking.

There is a comparison here with Captain Giachino, the

commander of the Argentinian commandos who attacked Government House at Stanley on 2 April. Both officers were severely wounded at the head of their men in the first attacking moves made by their respective sides in the Falklands, and both bled to death. Both became national heroes. Lieutenant-Colonel H. Jones was awarded a posthumous Victoria Cross.

Major Chris Keeble, the battalion second-in-command, came up from the rear to take over command of the battalion after the death of Colonel Jones. Keeble inherited a difficult situation. A Company was still fighting in the Darwin Hill area. B Company, on the other side of the isthmus, was held up by Argentinian defences in the Boca area. Boca House was merely the scattered stones of an old building; a 0·5-inch machine-gun post there was proving to be a particularly difficult obstacle. It was now full daylight and the paras had a long way still to go, over ground devoid of cover. Only half of the ground between the Start Line and the main objective, Goose Green Settlement, had so far been taken. The Argentinian artillery was firing steadily; the exact positions of the enemy guns had not yet been located. 2 Para was losing the initiative and its supporting firepower, never strong, was diminishing. The three British guns continued to give support but the gusty wind blowing over the battlefield prevented really close support being given to the forward para companies. 2 Para had only brought two of its six 81-mm mortars and the battle had now put these out of range. They were being carried forward and more ammunition urgently requested by helicopter. H.M.S. *Arrow* had been two hours late in opening fire with her gun because of a fault. She had then performed well but only by putting a man in a supposedly automatic turret to hold a broken microswitch in place with his finger! But *Arrow* had now departed, ordered back to the protection of the San Carlos air defences; she had, however, fired off most of her ammunition. The bad weather was preventing Harrier ground support being given.

Officers of several ranks and from several units speak with admiration of Chris Keeble's performance on taking over 2 Para in these difficult circumstances. Keeble believed that he needed 'a point of main effort where I could concentrate a large amount of violence in a small sector'. He decided not to reinforce A

Company at Darwin Hill on the left; he believed that Major Farrar-Hockley could sort out the situation there with his own resources. D Company, which had been clearing bypassed trenches on the right, was ordered to come up on the extreme right and attempt to find a way round the Boca House position which was holding up B Company. Keeble sent forward as many supporting weapons as he could to this right flank, including the heavy Milan anti-tank launchers which he had always believed would make good 'bunker busters'. His intention was to push his main weight through on the right and capture the relatively high ground from which he could dominate the airfield and Goose Green Settlement and then, perhaps, encircle the entire Argentinian position.

Unknown to Major Keeble the Argentinians were reinforced at this time. Brigadier-General Parada, who was trying to control the battle from Stanley, had scraped together about eighty men of 12 Regiment. Half of them were the men under the Lieutenant Esteban who had been in Port San Carlos on the day of the initial landings and had shot down two British helicopters there. Now Esteban and his men were helicoptered to Goose Green and thrown into the battle. Preparations were being made to fly in another company of Argentinian troops later in the day. Parada ordered one of the companies already at Goose Green to move forward but its commander replied that he could not do so because he would have to move over ground on to which an Argentinian helicopter had scattered plastic anti-personnel mines as part of the Goose Green defences. What the Argentinians were able to do now that daylight had come was to use the 20- and 35-mm anti-aircraft guns at the airfield in the ground-defence role – fearful weapons with a high rate of fire using explosive cannon shells; many paras were injured by these.

Menéndez had asked the mainland for air support at Goose Green but no help was forthcoming from that source, the weather was too bad. Menéndez then asked the local naval and air commanders to make a maximum effort with their units at Stanley airfield. Three Pucarás and two Aeromacchis took off and attacked the British artillery positions and some of the paras, but without much effect. But one Pucará caught a couple of Scout helicopters of 3 Brigade Air Squadron's B Flight which were bringing in ammunition and taking out casualties. The action took place near Camilla Creek House. One helicopter evaded the attack but Lieutenant Dick Nunn's helicopter was hit. Nunn was killed at once; his crewman, Sergeant Belcher, was thrown clear when the helicopter crashed, but lost a leg.

Out on Darwin Hill, A Company's battle turned slowly in favour of the paras. 2 Platoon on the left started to make some slow progress in destroying the Argentinian trenches on the hill itself but the real progress was being made on the right, in the area where Colonel Jones had been hit. In response to Jones's move, parts of 3 Platoon and Company Headquarters gradually edged forward and found their way into better positions. The platoon commander was injured. While 2 Platoon fought on the left and distracted the Argentinian attention, two N.C.O.s of 3 Platoon were able to engage the Argentinians on the second hillock. Sergeant Terry Barrett was at last able to bring machine-gun fire on to the trenches but the real damage was done by Corporal David Abols. Abols was with the company sergeant-major, who had a 66-mm light anti-tank rocket launcher. The sergeant-major fired a rocket at the first trench on the second hillock and missed. Abols tried and scored a direct hit, the charge causing devastation in the same covered trench from which Colonel Jones had been shot. The removal of this key trench enabled the positions on Darwin Hill itself to be taken one by one from the extreme left. The Argentinians had made the mistake of constructing their trenches in a line and not in a staggered pattern and were not able to give each other mutual protection. The paras on Darwin Hill then covered 3 Platoon, who similarly knocked out trench after trench on the second feature, Corporal Abols doing further good work with his rocket launcher. The Argentinians surrendered twenty minutes later. Corporal Abols, a twenty-five-year-old married man from Liverpool, received the Distinguished Conduct Medal, and Sergeant Barrett received the Military Medal.

Good infantry work and handling of weapons, without much outside support, had defeated a well-defended position. It was now full daylight, still raining, and smoke from a burning gorse hedge blew across the tattered battlefield. There were eighteen Argentinian dead and seventy-four prisoners, thirty-seven of them wounded.

The rest of Major Keeble's plan was also bearing fruit. One officer says that the plan 'worked like clockwork'. B Company withdrew a little from its exposed position in front of Boca House, the company suffering several losses including one man killed. D Company had found a narrow path at the edge of the sea, out of sight of the Argentinians at Boca House. (In fact, this way round the Argentinian flank was discovered and reported to Colonel Jones just before his tragic involvement in the Darwin Hill action.) Major Neame led D Company in single file along the path. His men then deployed along the sea edge and were ready to fire into the flank of the Argentinian positions. Major Neame describes the success which followed.

B Company were blasting the enemy out of the bunkers with the Milans and we opened fire with all our weapons. Every time someone moved, we put down a hail of fire. This combination resulted in an enormous number of white flags being shown. I suggested to Battalion that we go forward and take the surrender. We ensured that all weapons were ready but we held our fire. It took half an hour to arrange everything; the radio links were getting poor. The tide was coming in by then and I told my 2 i/c, who was handling the comms, to get a move on; we were getting our feet wet. There was a faint air of the ridiculous about all this. Eventually, everything was laid on and we moved up and took the surrender. I think we took about twenty prisoners and I believe there were a dozen dead. The rest had run off.

The hard fight at Darwin Hill and the skilful outflanking of the Boca House position broke the back of the Argentinian defences. There would still be some tough action but the paras now had the initiative again and had room to manoeuvre. Major Keeble ordered three companies to advance. (See the map on page 263.) On the right, D Company was to move directly from Boca House to the airfield and B Company was to move further down the west coast of the isthmus before turning inland to cut off the Goose Green Settlement from the south. C Company was to advance from Darwin Hill and complete the encirclement of the settlement from the north.

C Company's advance down the forward slope of Darwin Hill was seen and came under fire. Artillery fire put company headquarters out of action, with one man killed and most of the remainder wounded. The platoons were pinned down for a while.

They were in the classic infantry soldiers' dilemma – to fulfil their duty they must stand up in broad daylight and move forward into aimed fire. Captains Paul Farrar and Colin Connor, the commanders of Patrol and Reconnaissance Platoons respectively, conferred and dashed forward together. Their platoons were not seriously hit and advanced steadily.

D Company struck off well from Boca House but was diverted from the airfield by a minefield. The company finished up overlapping C Company's approach to the settlement and both units became involved in a fight for some outlying buildings of the settlement – the schoolhouse and some store buildings – which were quite clearly defended by Argentinians and empty of civilians. Lance-Corporal Michael Robbins was a machine-gunner in 11 Platoon.

We came under fire when we started down the side of the airfield. There was small-arms fire and the anti-aircraft guns were also firing in spells – I can remember the 'crack, crack, crack'; it was a slow-firing weapon. But, by that time, the adrenalin was pumping and, unless you were hit or pinned down, you wouldn't know whether it was close to you or not. We went to ground while we tried to discover where the small-arms fire was coming from. Then we started to take the first of the buildings. I followed my section commander. I think someone threw a phosphorus grenade into the first building because it burnt really fiercely. I don't know whether anyone was inside – I didn't hear any screams. The whole platoon was around these buildings – only tin shacks, outhouses and stores; there must have been some ammunition in them; you would hear an explosion now and again. Some shells or mortars started falling – artillery defensive fire, I think. They even dropped a round right in the schoolhouse which their own men occupied. I don't think they could pinpoint us but Stan Dixon was hit in the back.

Private Hanley, my mucker, and I were firing with the machine-gun from the flank and Corporal Harley ran forward with another man and each put a phosphorus grenade inside the nearest building. I don't know whether there was anyone in there but there wouldn't be anyone after that machine-gun fire and the grenades. We used a fair bit of ammunition; I think we put 300 rounds into it. We went on, clearing the buildings one by one, until we reached the school.

While the schoolhouse fight was taking place, the paras suffered a setback through a misunderstanding. D Company's 12 Platoon commander, Lieutenant Jim Barry, 'a forceful, determined para officer', reported that white flags were flying at an

Argentinian position on the airfield. The company commander, Major Neame, was mindful of the recent success at Boca House where the Argentinians had surrendered a strong position.

Jim and I had been discussing that while moving up. We decided that getting them to surrender was the best way of making progress, provided everything was safely tied up with our own people near by.

It was a crucial position. If Jim could get up there, he would take the pressure off the other platoons. My immediate reaction was to wait until I had tidied it up with Battalion but I don't know whether that message got through to Jim or not. I didn't speak to him myself but I did pass it on on the company net.

The next message I got was that, in the process of taking the surrender, Jim and the section with him had all been shot.

Lieutenant Barry and two of his N.C.O.s – Corporal Paul Sullivan and Lance-Corporal Nigel Smith – were killed and several more men were wounded in this incident. It appears that, as Barry and his men approached the Argentinian position, a distant British machine-gunner fired by mistake on the Argentinians, possibly thinking that he was giving covering fire for a British attack. At this, another Argentinian position (not the one attempting to surrender) opened fire on Lieutenant Barry and his men. No surrender took place; fighting resumed; the Argentinian position was destroyed thirty minutes later.

It was at this time that aircraft of both sides appeared. Argentinian Aeromacchis and Pucarás made a well coordinated attack. Major Neame describes what happened.

We were trying to reorganize after the schoolhouse battle and after losing Jim Barry. The first one came over firing cannon – quite frightening. We reckoned the minefield was the lesser of two evils and dispersed into it. We should have been an easy target but no one was hit.

While we were doing this, the Pucarás attacked; the second one came right over the company aid post and dropped a small napalm bomb; there was a fireball on the ground but it missed everyone. It seemed to be a totally inappropriate weapon for the circumstances. Everyone opened up on that guy and his aircraft was hit. The pilot baled out and the aircraft crashed. He was captured by our company, an officer in his late twenties, looking a bit pale but not injured.

An Aeromacchi was also shot down. Three GR3 Harriers then

appeared in response to urgent requests for help and attacked the Argentinian artillery positions. Lance-Corporal Robbins was watching.

They told us an air strike was coming in. Three Harriers came over; two dropped cluster bombs and one fired rockets at some guns only 200 metres away. That was a big morale booster; they came right over our heads and we thought it was great. We had been lying on our stomachs but we all jumped up and cheered.

Many British soldiers say that Argentinian resistance crumbled fast after that 'devastatingly and terrifyingly accurate cluster-bomb attack'.

Evening was now approaching. The fighting had lasted for more than fifteen hours. The paras had pressed the Argentinians back until they were hemmed in on the southern edge of the airfield and in the settlement. The civilians were still locked up in the Community Hall. The Harrier strike had silenced most of the Argentinian guns. There had only been one development which posed a further threat to the British. 12 Regiment, the main Argentinian unit at Goose Green, still had its B Company, ninety men strong, in the Mount Kent area just west of Stanley, which had never been sent forward to Goose Green. The Argentinian helicopters which had earlier in the day brought in one batch of reinforcements had been pressed repeatedly to pick up this company and get it up to Goose Green; the helicopter commander had delayed, saying that the weather was so bad that ice would form on the helicopter blades. Menéndez insisted that the attempt be made and a Chinook and four small helicopters successfully landed this fresh company a mile south of Goose Green in the late afternoon. The Argentinian intention was that the company would make contact with the main garrison and help it to withdraw to the south and disengage.

But the main garrison was now cut off, 2 Para's own B Company having completed the encirclement from the south. Two Argentinians found their way through the para lines, using a gulley, and made contact with the newly arrived Argentinian company. The company commander decided that he would rather move into Goose Green and fight on from there with his own regiment. The whole company then passed through the paras' positions in the darkness and joined the already congested and confused Argentinian garrison in the settlement.

Brigadier Thompson was also sending reinforcements forward. Major Mike Norman's J Company of 42 Commando was helicoptered to the northern end of the isthmus, but would be too late to fight at Goose Green. More British guns and mortars and ammunition were also flown in. Two Argentinian prisoners of war were sent into the settlement to open negotiations which would last for most of the night. The Argentinians were in a near hopeless position. They still had plenty of ammunition and a superior force of men, but they had been defeated at every stage of the action so far and could see no prospect of success the next day. Their best defences were gone; they had no room to manoeuvre. The British pointed out that the Argentinians were surrounded by 'the best unit in the British Army', but the limited strength of that unit was not disclosed in case the Argentinians would feel ashamed to surrender to an inferior force. A major demonstration of firepower and more Harrier attacks were promised for the next morning. No further help could be expected from Stanley and the nearest Argentinian troops were thirty-five miles away. The two senior Argentinian officers at Goose Green asked Stanley for permission to surrender. Brigadier-Generals Menéndez and Parada replied that Goose Green should make its own decision. The Argentinians paraded their men the next morning and surrendered. One para says, 'My main memory of that ceremony was the surprise on their faces when they saw how few we were.'

Goose Green was an outstanding British victory, attributable almost entirely to the fighting qualities of 2 Para. There were conflicting reports of Argentinian casualties in the confusion following the battle. About 1,500 Argentinian prisoners were taken. (*Norland* later took 1,536 prisoners from San Carlos to Uruguay; most of these were from Goose Green.) There were wild estimates that 250 had been killed, but this is several times greater than the true number; the official British figure is forty-five Argentinian deaths. The Argentinian dead were collected from the battlefield and buried in a mass grave north of Darwin Hill. Mr Goss says that there were thirty-nine bodies here, including that of an Aeromacchi pilot. The British military-graves service later removed thirty-seven bodies from this grave. Menéndez, in an interview after the war, said that 100 men of the 12th Regiment and thirty-one from the 25th Regiment were 'killed or wounded'. He objected to the British claim that 500 paras defeated 1,500 Argentinians and points out that many of the Argentinians were administrative men or air-force technicians who could not be classed as combat troops. He also claims that the British had brought Scorpion tanks into the battle and, 'at their last gasp', had been forced to call in a company of marines as reinforcements; the claim about the tanks is not valid.

The British suffered seventeen fatal casualties, fifteen from 2 Para, one from the Royal Engineers and a Royal Marine pilot. The dead included the commanding officer of 2 Para, four other officers and seven junior N.C.O.s. D Company, with eight dead, suffered the heaviest casualties. Between thirty and forty paras were injured.

There would be five further set-piece battalion actions in the Falklands but this one at Goose Green had several significant features. It was the first real conflict on land between the two sides. 2 Para's objectives were – in geographical terms – at least four times larger than those in any other battalion action and the Argentinian forces faced by 2 Para were also greater than in any other action. It was the only action which spilled over into daylight hours, with all the resulting problems of fighting on the open Falklands terrain. It was the only one where civilians were present; not one of them was hurt. (After the war, seventeen-year-old Kevin Browning from Goose Green came to Britain, joined the Parachute Regiment and was posted to 2 Para after his training.)

CHAPTER SEVENTEEN

The Investment of Stanley

While the Battle of Goose Green was being fought, there had been a major British advance to the north, less dramatic than the action at Goose Green but just as significant. In response to his orders to move out of the beach-head and 'invest the Stanley defences', Brigadier Thompson had intended to move the main strength of his brigade across country north of the ridge of high ground which ran the length of East Falkland Island. (See the map on page 280.) If *Atlantic Conveyor* had not been sunk, the move along that northern axis would have been by helicopter, but now most of Thompson's men would have to walk. Their initial objective was Teal Inlet, twenty-five miles from San Carlos but still twenty miles short of the Argentinian defences around Stanley. No major opposition was expected. Teams of the S.B.S. and the Mountain and Arctic Warfare Cadre had been inserted into the Teal area by small boat and helicopter and reported it clear of Argentinians. The main danger was Pucará air attack on the line of march – and the weather.

The original plan was for 45 Commando to lead off and march to Teal Inlet via Douglas Settlement; this route would make use of a track for three quarters of the way. 3 Para were to follow the marines as far as Douglas, then pass through and lead on into Teal. There was, of course, much rivalry between the red and green berets and, on this occasion, the paras literally stole a march on the marines. Lieutenant-Colonel Hew Pike of 3 Para discussed the route with Alan Miller* at Port San Carlos Settlement. Mr Miller advised that a more direct route to Teal – missing out the dog-leg to Douglas – was feasible, despite the absence of a marked track, and Mr Miller offered to lend his son

---

* *Author's Note*. Sad news of the untimely death of Alan Miller in January 1984 reached me in a letter from his family just half an hour before I wrote this part of the book! His son Philip received the British Empire Medal for his work with the paras.

Philip and two tractors to carry the battalion's mortars. This suggestion was accepted by Brigadier Thompson and 3 Para thus led the march all the way.

The two units moved out of the beach-head on 27 May. The marines were away first, crossing from Ajax Bay to Port San Carlos in landing craft. Marine Steve Oyitch was there.

We were pleased to be on the move; we hadn't liked just sitting about and waiting. We just wanted to push them into the sea and get home, although we still thought they would back off without a fight at that stage.

We passed right through the position of the paras – 3 Para, I think – but we didn't speak to paras; we never did. We call them the cherryberries and they call us cabbage heads, because of our green berets, and we never speak to each other – scrapping sometimes – but they usually try to keep us apart.

45 Commando's first 'yomp' was a thirteen-hour stint of fourteen miles during which fifteen men fell out, to be picked up by one of the unit's Volvo cross-country tracked vehicles or the Scout helicopter which accompanied the column. This is Captain Ian Gardiner's account of the march.

The walk from Port San Carlos to New House, some twenty kilometres, was the worst of my life. The weather was not too bad but the ground was boggy. Where it was not boggy, there were strong lumps and tufts of grass which, however one stands on them, even in daylight, one stands a good chance of turning one's ankle. In places it was pretty steep but all faded into insignificance compared to the cursed weight we were carrying – much of which I knew to be wholly unnecessary. I probably made things worse for myself by allowing my bitterness to burn up energy – but the marines were magnificent. We lost the first man after 200 yards – a man known to be the Company skate – and about six more over the next few hours. They were mostly the weaker-spirited men who, although they possibly did have something wrong with them, would probably have found some pretext or other to roll around in agony in any event. The rest went on with the greatest of stoicism and good humour all day and through until 2 o'clock the following morning. I was immensely proud of them. If possible, marching in darkness was worse than daylight, and, for those at the tail end of a queue of 600 men bumping and stumbling through the black night, life must have been hell. I was fairly preoccupied by trying to keep people together and perhaps didn't notice so much, but by the time we leaguered up, I was near my wit's end.

At 2 a.m. I gave the order to bed down without erecting bivouacs. Our bivvies were simply a waterproof groundsheet supported by a small stick and some rubber bungies. This was a bad mistake. It rained during the early morning, and the plastic bags in which our sleeping bags were stretched did not keep the water out. Our sleeping bags were soaked. The last citadel of a man's morale is his sleeping bag. The comfort and resource it offers is amazing. On subsequent occasions, when one was being shelled or heard bombing close by, it was an instinctive automatic reaction to wriggle deeper into the 'green slug'. When all else failed, when the world crumbled around us, even if the sleeping bags had failed to turn up in the evening, one always was vaguely comforted by the prospect of climbing in to a dry bag eventually. When my citadel was drenched by rain, my morale was at its lowest . . . I was subsequently to spend more uncomfortable and more bitter nights and it was not the last time my citadel was breached – but one hardened. The men who survived that march basically stuck it right through to the end.

45 Commando dumped their heavy Bergen haversacks that second morning and marched to Douglas Settlement to be welcomed by the twenty-two civilians who had earlier received the unwelcome attention of Lieutenant Esteban and the Argentinians who had withdrawn from Port San Carlos on the day of the landings (and who were that day fighting at Goose Green). The marines then moved on to Teal Inlet.

The experiences of 3 Para – 460 men strong on the march – were similar to those of 45 Commando, except that they carried lighter loads from the outset and travelled a shorter distance directly to Teal. The paras' word for a long march is 'tabbing' as opposed to the marines' 'yomping'. 3 Para reached Teal in only two stages, covering the twenty miles of trackless terrain in thirty-three hours and spending a bitterly cold night in the open without sleeping bags. Fourteen exposure cases – all private soldiers – had to be evacuated and one man shot himself in the shoulder with a 'negligent discharge'. Both units were accompanied by a tank troop of the Blues and Royals, their drivers proving all the sceptics wrong by moving successfully over trackless ground. It had been forecast that the weight of the tanks would penetrate the turf and cause the tank to stick in the peat below. The tank drivers quickly learned to look ahead for the green moss which covered the softest patches, select the best line through it and then move fast without having to 'stick', that is pull one of the sticks which slowed the speed of a track to alter direction – any change of direction would have caused a track to tear the fragile turf and ruin the route through for the next tank. Peat diggings near the settlements, ten feet deep sometimes, had to be avoided at all costs.

3 Para approached Teal Inlet tactically during the night of 28 May but only one Argentinian, either sick or wounded, was found in the settlement. 45 Commando came in two nights later, disappointed to find that their rivals had arrived first and had the best accommodation. Some of the press reporters give Teal Inlet a bad name for lack of hospitality but 3 Para and 45 Commando both speak well of their reception; the marines found it 'a delightful place, a few houses amongst a hillside with green meadows leading down to the flat calm inlet'.

An even more audacious move forward took place the night after 3 Para reached Teal Inlet. Mount Kent, 1,500 feet high, was the largest of the hills around Stanley; it was only twelve miles west of the town and was sixteen miles further forward than Teal Inlet. The S.A.S. reported that Mount Kent was only thinly held by the Argentinians. Now that the Goose Green action was over, Brigadier Thompson felt that he could spare both the helicopters and the troops to attempt a night assault on Mount Kent. Sea Kings of 846 Squadron would carry up the infantry and the R.A.F. Chinook would follow with half a battery of guns. 42 Commando would provide the infantry, 7 Battery of 29 Commando Regiment the three guns. If this operation was successful, the British would have established themselves on a feature which overlooked all of the main Argentinian defences around Stanley, which were on a series of lower hills whose homely names would soon become part of British military history – Mount Harriet, Two Sisters, Mount Longdon, Tumbledown Mountain, Mount William, Wireless Ridge, Sapper Hill.

The operation commenced on the night of 29/30 May. Three Sea Kings picked up part of K Company, 42 Commando, and set off for the forty-mile flight to Mount Kent, but had to turn back because of snowstorms. The helicopters were more successful the following night. The three Sea Kings each made three flights and took forward 42 Commando's Tactical Headquarters, the whole of K Company, four heavy mortar teams and a Blowpipe detachment. The first landings took place just short of Mount Kent as the S.A.S. were having a brief firefight with an Argentinian patrol, but there were no British casualties. K Company then advanced on to the mountain and found it deserted. Menéndez later stated that he would have liked to have occupied Mount Kent in strength but, after he had sent the spare company of 12 Regiment from Mount Kent to reinforce its parent regiment during the Goose Green battle, he had insufficient troops in that area to occupy both Mount Kent and Two Sisters. He decided to keep a tight ring of defences in the Mount Longdon, Two Sisters and Mount Harriet arc of hills and leave Mount Kent with only the light screen which was dealt with by the S.A.S.

The artillery lift did not go so smoothly. The Chinook had the interesting task of moving forward three 105-mm guns, their ammunition and fifty or so gunners from a unit that had never worked with Chinooks before. The three guns – one internal and two underslung – and twenty-two gunners ('gun bunnies' to the Chinook crew) were successfully put down behind Mount Kent, but near-disaster struck on the return flight. Flight Lieutenant Andy Lawless, co-pilot in the Chinook, describes what happened.

We were turning tight, avoiding a known enemy position, when we were caught in a snow shower and lost sight of the ground. We hit a lake.

It was fairly alarming at the time. We were travelling at about 120 knots and we hit the water with the underside; we just belly-flopped. The engines partially wound down with the shock loadings – the jolt – and they ingested some water. Fortunately, the Chinook is designed for water operations and she had a flat underside, so we wallowed on the water for about fifteen seconds, reducing speed to a walking pace. Then the engines recovered and we were able to lift off. The gods were smiling on us. Squadron Leader Langworthy was flying it at the time and he had started to reduce speed in the snow shower and that had started to flare the aircraft; it brought the nose up. If we hadn't been in that attitude, we would probably have crashed; the forward undercarriage would have dug in and pulled the aircraft under the water.

The aerials for the radio were on the underside and the radio was u/s. We had to nurse her back to Carlos by a different route than the one planned and that route hadn't been cleared by the Rapier sites. But we got in by flashing our landing lamps and got down safely – quite glad to be back.

The damage to the Chinook prevented its flying again that night, leaving the three guns at Mount Kent without any ammunition. The three Sea King crews agreed to ignore the flying hours limitation drawn up for passive night-goggle flying and flew to the artillery site at San Carlos to start picking up the ammunition, finding that Battery Sergeant-Major J. Francis had already anticipated the change and repacked the ammunition into smaller loads.

(Chinook Bravo November – ZA 718 – was not seriously damaged and went on to do further sterling work during the Falklands campaign. Squadron Leader Dick Langworthy, the senior pilot with the helicopter, was awarded the Distinguished Flying Cross. This Chinook remained in the Falklands after the war, 'The Survivor' painted on its nose. Sadly, Squadron Leader Langworthy died of natural causes in early 1984 while doing his second Falklands tour at the Chinook base at Kelly's Garden, near San Carlos Settlement.)

This move to Mount Kent on a shoestring of helicopter resources was a major achievement, placing a unit in fresh condition on this important feature. More men of 42 Commando were brought in the next night and the Argentinians made no effort to dislodge the marines. The three guns, under Gun Position Officer Lieutenant Mark Williams, were soon in action, delighted to be so after the lack of targets at San Carlos. The first call was on to 'a mass of Argentinian defenders' in the Moody Brook area, ten miles away. All the Argentinian positions as far as the racecourse on the outskirts of Stanley were now within range if the guns used super charge.

Brigadier Thompson's next move was to allow 3 Para and 42 Commando to edge their way further towards Stanley but not to seek a major battle for which support and supplies were not yet available. These moves were preceded by four-man patrol groups and observation parties of the Mountain and Arctic Warfare Cadre. One of the patrols, scouting ahead of 3 Para, saw two Argentinian helicopters land seventeen men at Top Malo House, another isolated and empty shepherd's house. This was reported

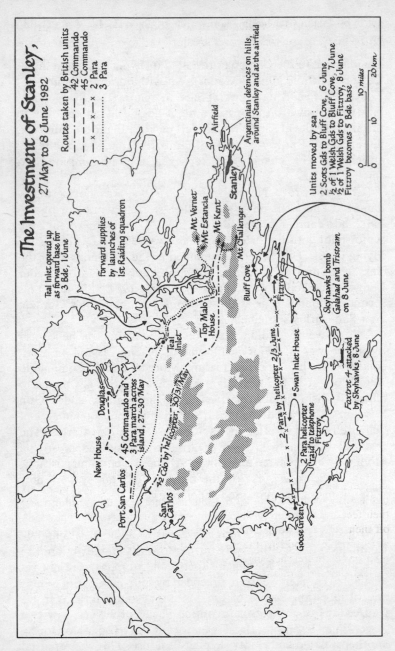

The Investment of Stanley,
27 May to 8 June 1982

Routes taken by British units
— · · — 42 Commando
— · — · — 45 Commando
— x — x — x 2 Para
· · · · · · · · 3 Para

Airfield

Argentinian defences on hills,
around Stanley and at the airfield

Stanley

Mt Vernet
Mt Estancia
Mt Kent
Mt Challenger

Units moved by sea :
2 Scots Gds to Bluff Cove, 6 June
½ of 1 Welsh Gds to Bluff Cove, 7 June
½ of 1 Welsh Gds to Fitzroy, 8 June
Fitzroy becomes 5 Bde base

10 miles
20 km
10
0

Teal Inlet opened up
as forward base for
3 Bde, 1 June

Forward supplies
by launches of
1st Raiding squadron

Bluff Cove

Fitzroy

Skyhawks bomb
Galahad and Tristram
on 8 June

Top Malo
House

Teal
Inlet

42 Cdo by helicopter, 30/31 May

2 Para by helicopter 2/3 June

Swan Inlet House

Foxtrot 4 attacked
by Skyhawks, 8 June

New House

Douglas

45 Commando and
3 Para, march across
island, 27–30 May

2 Para helicopter
'raid' to telephone
Fitzroy

Port San Carlos

San
Carlos

Goose Green

to Captain Rod Boswell, the cadre's commander at San Carlos. Top Malo House was beyond artillery range and it was too late in the day for Harriers to be dispatched for an air strike, so Brigadier Thompson said that Boswell's unit could tackle the Argentinian party. Boswell gathered together nineteen men and was taken into action by a Sea King the next morning, Boswell stating that it was 'a super forty-five kilometres approach at an average height of ten feet, the helicopter's wheels being inside the river beds – the best flight I have ever had'. The marines were landed near Top Malo House and, in a brisk and neat action, killed five Argentinians and took twelve prisoners, seven of whom were injured. The Argentinians fought well; their officer was hit four times before surrendering. (This was Anglo-Argentinian Second Lieutenant Luis Albert Brown; he survived his wounds.) Three marines were injured but they also survived. Fourteen other Argentinians who watched the action from their positions on Mount Simon and Malo Hill, came in and surrendered to 3 Para the next day.

3 Para continued its further march in worsening weather. They were being guided by Terry Peck, before the war one of the most outspoken of the Falklands local politicians opposed to any attempt to transfer sovereignty to the Argentinians, and now making himself useful in many ways to the British troops. The paras reached Estancia House, which was fortunately occupied only by friendly civilians. Richard Stevens, a civilian there, remembers the paras' arrival.

Terry Peck brought a patrol in for a cup of tea and a good feed. They were tired; some were exhausted. I noticed the mixture – one or two dragging their feet, others looking as though they were ready to start out again immediately. A lot of them had bad feet; they couldn't feel their feet and they were blistered – more than blisters really. They took off their socks, so we had drying socks all over the house. Their boots were terrible – sodden; the Argies had better boots. They rested an hour or two in the shearing shed and then left.

Next day most of the battalion came in. The medics were in the sitting-room and Roger Patton, the second-in-command, set up in one of the sheds outside. The Intelligence were in the dairy and others were in the meat-house. These were all rooms no bigger than this table. Some even slept in the hen-house and that was like sleeping on the floor of a telephone kiosk.

I have a high opinion of 3 Para; they were very hard men. Some of them said they had only joined because they could kill legally. They were all psyched up. Some said that's what they had joined for; there wouldn't be another war like this, so they would leave after this. Others put it differently; they felt the Argies had invaded a country that was British and they were determined to put them out. They took it as an affront to Britain. 'We'll put it right for you, mate,' one of them said, this hard para sergeant up in the hills.

3 Para found that Mounts Estancia and Vernet were unoccupied and took over those peaks, local civilians providing tractors and trailers to help with the move. These advances by 3 Para and 42 Commando left 3 Brigade occupying a seven-mile-long chain of peaks facing the main Argentinian positions, with a 'no man's land' of low ground varying in width between two miles in the south to five miles in the north. The S.A.S. and S.B.S. were also probing forward, attempting to establish the location of the Argentinian main positions and drive in their outposts.

The presence of both types of secretive Special Forces operating in the same area led to a tragic clash on 2 June, when the S.A.S. fired on an S.B.S. party and Sergeant I. N. ('Kiwi') Hunt of the Royal Marines was killed, the only fatal S.B.S. casualty of the war. As with all 'blue-on-blue' fatalities, the authorities attempted to cover up this event for as long as possible. It would surely be better if such incidents, inevitable in any war, were announced as soon after the event as possible. The obsession with service secrecy and the gradual discovering of the Falklands 'blue-on-blues', spread over several succeeding years, caused much anguish to the families concerned and almost as much worry among all the other families until it was known that the disclosures had run their course.

But Brigadier Thompson's units were now at the end of a long and tenuous line of communications. Two of the British units were positioned only thinly on high ground in worsening winter weather; the first snow fell on 1 June. Frantic efforts were made to turn Teal Inlet into a forward base but the flow of helicopter supplies was drastically reduced by other calls and by bad weather. Some fine Royal Marine initiatives cleared the way for supply ships to sail right up to Teal, which was situated twelve miles inland from the entrance of rambling Port Salvador. Rigid raiders of the 1st Raiding Squadron established S.B.S. parties on two inland islands and a Carl Gustav detachment of 45 Commando on Centre Island, which controlled the entrance to Port Salvador. On the night of 31 May/1 June, *Intrepid* sent in one of her small landing craft as a makeshift minesweeper. A large electric magnet was installed to detonate magnetic mines and a barrel, in which a Black & Decker drill was running, was towed to set off acoustic mines. The coxswain, Corporal Gordon Cook, was wrapped up in padded foam and flak jackets for protection. No mines were found and the entrance was declared clear. *Sir Percivale* made the first supply run from San Carlos the next night, bringing a Royal Corps of Transport Mexeflote raft for landing supplies. This was the start of a regular shuttle by *Sir Percivale* and *Sir Geraint* and the building up of stores at Teal for the coming battles for the hills around Stanley. These ships would be in serious danger from air attack until Rapier defences could be moved forward.

The establishment of Teal as a forward base – to which Brigadier Thompson also moved his headquarters – marked the end of a most favourable phase for the British land forces. After the slow build-up and the air attacks on the beach-head shipping, there had been the major victory by 2 Para at Goose Green and three battalions had been moved forward by one means or other more than three quarters of the way across East Falkland Island, the leading units being right up to the main Argentinian defences. In particular, the ability of the paras and marines to cross the Falkland terrain without major support from vehicles or helicopters was a major achievement. All of this had been achieved within seven days of Brigadier Thompson being ordered to get his troops on the move out of the beach-head.

CHAPTER EIGHTEEN

Distant Action

Naval and air action was now playing an increasingly subordinate role to the land war. Rear-Admiral Woodward's task group remained on station east of the Falklands, providing constant air patrols, close air and gunfire support, receiving new supply ships and escorting them into the landing area. Seven new warships had made timely arrivals which more than replaced the losses suffered earlier. The new ships were *Active*, *Andromeda*, *Avenger*, *Bristol*, *Cardiff*, *Minerva* and *Penelope*. The Argentinian navy failed to put in any appearance during this period and, as so often, it was air action which predominated. The Harrier force suffered three further losses in four days. On 29 May, a Sea Harrier of 801 Squadron slid across the wet deck of *Invincible*, as the ship rolled in heavy weather, and the Harrier tipped over the side and went nose first into the sea. Its pilot, Lieutenant-Commander Mike Broadwater, had been sitting fully strapped in on deck alert and was able to escape in his ejection seat, to be picked up unhurt from the sea by a helicopter.

The next Harrier to be lost, on the following day, was a GR3 on a ground-attack mission. Squadron Leader Jerry Pook and Flight Lieutenant John Rochfort took off, their aircraft armed with 2-inch rocket pods, seventy-two rockets per aircraft. The cluster bombs which the GR3 would normally have carried were in short supply; most of the replacement supply had gone down in the *Atlantic Conveyor*. The two Harriers were tasked to attack an Argentinian helicopter landing area ten miles outside Stanley, but found no targets there. Squadron Leader Pook takes up the story.

At that point, we flew across a road and I felt a small but significant and meaningful clunk down at the back end of the aircraft. John said, 'Hey! You're leaking fuel,' and he also said, 'Did you see the vehicles on the road?' Apparently, they had been firing at me.

I had a quick look round the cockpit and there was no indication on the instruments yet that there was anything else wrong. I turned off left, hard, from the empty target, about 90 degrees, and rolled out. I looked across at John and could see that he was lined up exactly on an artillery position we knew was somewhere in this area. It stood out quite well; it was poorly camouflaged – a whole lot of gun positions and stacked ammunition and their vehicles. We had been looking for targets like that for a long time; this was too good to miss. It was only luck that we caught them just like that.

I watched John fire his rockets; they hit right in the middle. Splendid! I turned back hard and went in myself from the opposite direction. I fired – a great roar as the rockets all left the aircraft; it's a tremendous weapon. When they hit the ground they looked like a lot of flash bulbs going off. That must have lowered their morale.

I wanted to get home then; I was leaking fuel badly. We climbed to gain height and use as little fuel as possible. The radio had failed again and I had lost John. The Harrier has had a known history of poor radio communications throughout its R.A.F. service life. I levelled off at 25,000 feet and the gauge settled down. I realized the fuel was going down really fast. I'd had it; the aeroplane wasn't going to get me home. I flew as fast as possible, to get as far as possible before it ran out. John caught me up again. He told me later that it was easy to find me; he just had to follow this huge great trail of vaporized fuel.

It was a bit like a bad dream watching the fuel gauges. It all ran out and the engine just stopped. I couldn't talk to John but I gave him the visual signal. He had already called in and got the helicopters airborne and was keeping them informed. I felt that I had a fifty-fifty chance but I don't remember too clearly; it's still all a dream.

Three helicopters rushed to rescue the Harrier pilot; it was a Sea King from 826 Squadron which picked him up after only five minutes in the water. After one day's rest with a stiff neck, Jerry Pook was flying again. The R.A.F. Harriers on *Hermes* had now lost three of their original six aircraft on this hazardous ground-attack work but four replacement GR3s were flown down in one long hop from Ascension early in June – epic flights with much danger for the pilots if anything went wrong.

Another Harrier was lost on 1 June, this time a Sea Harrier of 801 Squadron from *Invincible*, the fourth loss by that squadron but the first through enemy action. Flight Lieutenant Ian Mortimer, one of the R.A.F. pilots on loan to the Fleet Air Arm, was on patrol high over the Falklands when he was shot at by a Roland missile. Mortimer believed that, at 13,000 feet, he was above the Roland's ceiling. He saw the missile coming up and climbed for more safety but his aircraft was hit. This time the incident can be described through the eyes of a Falkland Islander. Paul Howe was having tea in the kitchen at the Upland Goose Hotel.

I went out to the deep freeze in a shed and I heard the whoosh of a missile; it sounded as though it was from Moody Brook. I could hear the aircraft; you can normally tell the height but this one was much lower than normal. Then I heard this terrific crack. I ran back quickly to get a better view of what it was. It was a day of low cloud but I saw the aircraft spiralling and tumbling to the south-east about four to five miles away. I could tell by the shape of the wings that it was a Harrier. I jumped up to try and see a parachute but I couldn't. I was very concerned about the pilot. I got what I wanted from the deep freeze and went back inside and told the others that I had just seen a Harrier shot down. No one believed me and I got quite mad because I knew damn well I had seen it, but next day the B.B.C. announced that a Harrier had been lost over Port Stanley.

Later, after the conflict, many of the Sea Harrier pilots stayed in this hotel and I kept pursuing the matter and eventually I was able to meet the pilot. He was very interested in what I had seen, because he had this idea that he may have ejected prematurely; he was very relieved when I told him there were flames coming from the back of his plane.

Flight Lieutenant Mortimer was rescued from the sea that night by a Sea King. This was the last Harrier to be lost in the air. Three pilots had been killed in accidents and one by ground fire but every British pilot who attempted to eject survived. They had great faith in the Martin Baker Mark 9 ejection seat with which the Harriers were equipped; many of the Argentinian aircraft had a poorer seat and a poorer survival record.

The Sea Harriers on combat patrol achieved little success during this period, mainly because the Argentinian air units were only making brief appearances. There was, however, a spectacular success earlier in the day on which Flight Lieutenant Mortimer was shot down. A Hercules, returning to the mainland from an overnight supply run to Stanley, climbed a little from its normally low-level exit route and was spotted on radar by a British ship, the *Minerva*, which directed a Harrier on to it. Lieutenant-Commander Nigel 'Sharkey' Ward shot it down with a Sidewinder missile and cannon fire. The seven Argentinian crewmen were all killed.

*

The Argentinian unit, Grupo 1, with nine Hercules aircraft, was a resourceful unit. Its two tanker aircraft carried out many refuelling flights and the transport aircraft regularly ran the dangerous blockade-breaking flights to the Falklands. On 29 May, two days before Lieutenant-Commander Ward's success in shooting one of these aircraft down, a Hercules adapted to carry bombs made the first attack on the lines of supply to the British naval force off the Falklands. The Hercules found an unarmed British tanker about 800 miles north-east of the Falklands. This ship was the *British Wye*, a 16,000-ton British Petroleum ship which was one of a chain of tankers on the supply route topping up warships and supply ships sailing into the operational area. The Hercules flew low over the tanker and released eight bombs. The press reported at the time that these were rolled off the open rear ramp of the Hercules, but Second Officer P. P. Akers, on duty in the tanker's wheelhouse, says that the bombs were dropped from racks fitted to the wings of the Hercules. Four bombs fell wide and did not explode; three were nearer and exploded in the sea and one bomb hit the deck of the *British Wye* but failed to explode and bounced into the sea. Fleet Headquarters later asked if the ship could report any distinguishing features of the Argentinian aircraft. Admiral Fieldhouse is said not to have expected too much from this request, thinking that the Officer of the Watch would 'have had his hands full at the time'. But 'the most comprehensive reply imaginable' was received and Fieldhouse sent back a 'thank you' signal with the comment that 'the only thing missing was the colour of the pilot's socks'.

Sandy Woodward's naval task group spent a relatively peaceful four days after the attacks on *Coventry* and *Atlantic Conveyor* but Sunday, 30 May, brought another of the much-dreaded Exocet attacks. The Argentinians dispatched two Super Étendards and four Skyhawks. There was only one Exocet missile left of the five airborne versions held by the Argentinians at the opening of the war; the second Super Étondard was included in the raid so that it could help if the missile-carrying aircraft suffered a radar failure. The Skyhawks each carried two 500-lb bombs. The Argentinians took a great deal of care in the planning of this attack. They realized that previous Exocet attacks had not reached the British aircraft-carriers because the attacking aircraft had not penetrated far enough east. This time they provided two refuellings *en route*, to allow the attack planes to get behind the British task force and attack from the east. If they could do this, Woodward's aircraft-carriers would be the first large ships picked up by the Exocet's radar. In the past two weeks, Argentinian radar stations at Stanley (about which more later) had been plotting the appearance and disappearance of Harriers on their screens at the beginning and end of each Harrier mission and felt able to predict from these the exact whereabouts of the British carriers. In their turn, the task-force air operations officers had been ordering the Harrier pilots to use indirect routes and to fly at low level, below the Argentinian radar cover, when leaving and approaching the carriers.

The six Argentinian planes set off in the late morning and reached the operational area without serious difficulty. The centre of the British task group was 135 nautical miles (155 statute miles) east-south-east of Stanley – relatively close to the islands. But Woodward's ships were well spread out at this time, with much activity between the task group and the islands and between the task group and the area to the west known as the TRALA (Tug, Repair and Logistic Area), where ships under repair and supply ships waiting to unload were placed. No less than twelve of Woodward's escort vessels were away from the main group on various tasks. The Argentinian aircraft came in low from the south-east, the latest information from Stanley indicating to them that they were headed for the carriers. The two Super Étendards emerged briefly over the radar horizon, located the echo of ships, aimed their one Exocet at the largest echo and turned away. The four Skyhawks followed the smoke of the speeding Exocet, hoping this would lead them to a crippled aircraft-carrier.

But the Argentinians had not gone far enough east; they were about forty miles short of the carriers. The indirect routeings of the Harriers had probably misled the Argentinians at Stanley. The Exocet was making for the recently arrived frigate *Avenger*. On this day she had twenty-four S.B.S. men on board and was on her way back in towards the Falklands to try again to land these men at various places in the Volunteer Beach area, sixteen miles north of Stanley; the operation had been abandoned the previous night because of a fault with *Avenger*'s Lynx helicopter. At the time of the Exocet attack, *Avenger* was passing just south of the carrier group's Type 42 air-defence screen. *Exeter*, ten miles to the north, was the nearest ship in the screen.

It was *Exeter* who detected the attack first, but *Exeter*'s urgent warning was not broadcast clearly and, in *Avenger*'s Operations Room, the Exocet was reported as coming in from the north instead of the south. Most of her officers were watching a film in the Wardroom; everyone agreed that the next few minutes were more exciting than the film. *Avenger* turned to place her stern and her Sea Cat missile launcher towards the believed direction of the threat and worked up to full speed. But *Avenger*'s own warning devices then picked up the threat and it was realized that the ship was steaming hard *towards* the Exocet and that the chaff rockets which were being fired furiously were being left behind the ship and would be of little use. Captain Hugo White decided that he could not risk reversing course and presenting his ship as a broadside target, but he reduced speed to the minimum. *Avenger*'s radar then locked on to a small, fast echo approaching from dead ahead and the ship fired radar-aimed shells at what was believed to be one of two Exocets in flight. A fireball was seen at the place where *Avenger*'s shells were firing and the ship claimed one Exocet destroyed in flight. But *Ambuscade*, about ten miles nearer the aircraft-carriers, is believed to have visually sighted an Exocet passing harmlessly between herself and *Avenger*.

The Skyhawks came in next; in fact, one had already been destroyed. The fireball seen by *Avenger* was the Skyhawk formation leader's aircraft exploding. It had been hit by a Sea Dart missile fired by *Exeter* over or past *Avenger* – a good shot. The remaining three Skyhawks made straight for *Avenger*. Her earlier turn in the wrong direction left *Avenger* with the Sea Cat launcher on the aft of the ship not able to engage and she could only defend herself with her 4·5-inch gun and a few light weapons. The gun was switched to emergency control and fired a barrage of air-burst shells to explode in front of the Skyhawks, but the Argentinian pilots flew through this, although one Skyhawk may have been damaged. Master-at-Arms Bill Jarvis had earlier rushed to the bridge, thinking that this was the safe end of the ship in an Exocet attack. He describes the next, dramatic action. He still thought that an Exocet missile was on its way.

I looked at the weather and thought it was too nice a day to die. I wondered if anyone would find anything of me if the missile struck the bridge and consoled myself with the thought that I wouldn't know a thing about it if it did.

I watched the three Skyhawks as they approached, two fine on the port bow and one fine on the starboard bow, bobbing and weaving fast and low. I focused my binoculars on the nearest one on the port bow and thought how evil it looked, heading straight for us, for me. As the aircraft reached the bow, the First Lieutenant gave the order 'Take cover' and we all hit the deck, face down, hands clasped behind the neck. The First Lieutenant landed virtually on top of me. The two port aircraft flew down the port side below the level of the bridge windows and released their bombs harmlessly into the sea. The starboard aircraft released his bomb just ahead of the ship and, as the bomb hit the sea, he flipped over to port and cartwheeled along the surface of the sea down our starboard side. Suddenly all was peace and quiet again.

Wreckage from the splashed Skyhawk was strewn on the surface, so we launched our seaboat to recover and identify it. Floating around alongside us we could clearly see pieces of flying-clothing, a wallet, some banknotes and part of the pilot's torso with a bit of leg attached. The First Lieutenant wanted the swimmer of the watch to recover this grisly relic and bring it inboard, but he refused (or failed to understand) and instead picked up a liquid oxygen tank and a holed dinghy.

The whole attack, from the first detection of the Super Étendards to the end of the Skyhawk attack, lasted no more than four minutes. *Hermes* and *Invincible*, forty miles away, had picked up the Exocet warning and carried out their usual defensive moves. There had been a humorous incident on *Invincible* – humorous only in retrospect – when the ship turned towards the Exocet threat instead of away as expected and members of her crew all rushed in the wrong direction.

The Argentinians made great publicity of this attack, which was billed as 'The *Invincible* Raid'. The two returning Skyhawk pilots claimed to have flown over the *Invincible*, burning and crippled from the Exocet attack, and to have hit it with their bombs. This claim was persisted with long after the war and even after *Invincible* had returned to England undamaged.

The Argentinians now had no more airborne Exocets left.

*

Admiral Woodward's task group had devoted much effort to establishing which types of radar sets were being used by the Argentinians around Stanley. These radars were not only being used to establish the location of the British aircraft-carriers, by plotting the Harrier flight patterns, but were also plotting the approach of every Harrier sortie into the Stanley area, where most of the Argentinian forces were located and where the final land battles would soon be fought. The British established, from radar emissions, that the Argentinians definitely had a Westinghouse TPS 43 three-dimensional radar set and possibly a more advanced TPS 44 set as well. A secondary danger was the Skyguard radar sets, which controlled the light anti-aircraft guns that were such a menace to low-flying Harriers. At least one Skyguard set was known to be in the Stanley area, up on Sapper Hill; there may have been another set nearer the town. But all of these Argentinian radars were mobile and their locations were moved frequently. Their estimated positions were bombarded at night by British warships but, so far, the radar sets were always ready for action next time the Harriers came in.

Neither Sandy Woodward's ships nor his Harrier squadrons were equipped with anti-radar missiles but the R.A.F. in Britain had already started to rectify this gap in the British defences. The result was a new series of Vulcan–Victor Black Buck operations from Ascension. The R.A.F. had an anti-radar missile – the French-built Martel – and some of these were brought to Ascension slung under the wings of Vulcans in transit from Britain, but they were not considered suitable for action around Stanley because the Martel could be 'seduced' by too great a range of radars and might finish up in Stanley town and cause harm to civilians. Instead, the British hastily purchased some of the more advanced American Shrike missiles which were more selective in their detection of radar emissions.

Squadron Leader Neil McDougall of 50 Squadron and his crew had flown their Vulcan to Ascension and were ready to fly a purely anti-radar operation, carrying two Shrikes and no bombs. The absence of the heavy bomb load allowed the Vulcan to carry an extra 16,000 pounds of fuel in bomb-bay tanks, resulting in fewer Victor tankers being required and also allowing the Vulcan time to cruise around the Stanley area and select the correct radar emission to attack.

Black Buck Four took off from Ascension on the evening of 28 May, Squadron Leader McDougall flying Vulcan XM 597 which would carry out all the anti-radar flights. All went well until just before the penultimate refuelling point on the flight south, where two Victors failed to pass fuel to each other because of a failure in the mechanism of the hose-drum of the donor aircraft. The Victor pilot acting as 'tanker lead' at that stage of the flight realized that the operation would have to be aborted and calmly broke radio silence and ordered the Vulcan back. McDougall tried to persuade the second tanker to refuel his Vulcan with just enough fuel to fly to Stanley and then divert to Brazil on the return flight, although McDougall realized that this was 'clutching at straws'. But the Victors were not interested in such makeshift ideas and the tanker lead gave an emphatic 'no'. Everyone flew back to Ascension, McDougall's crew 'disappointed to miss our first operation but this tinged with relief', the Victor crews much dismayed to have let down the Vulcan.

Two nights later, on 30/31 May, McDougall and his crew tried again. This time all went well except that the last fuel transfer on the outward flight coincided with a particularly severe bout of 'inter-tropical-zone' turbulence in which trying to insert the Vulcan's probe into the Victor's hose basket was, according to one of the Vulcan crew, 'like trying to thread a needle on a Wurlitzer' or, more crudely, 'like trying to stuff wet spaghetti up a wild cat's bum'. The Vulcan was still taking on fuel when the Victor had to start turning back to Ascension and the last few pounds of fuel were taken on just in time for the Vulcan to turn south again and carry on to Stanley.

The Vulcan's arrival over Stanley was timed to coincide with a pre-dawn flight by one of the task force's Harriers which would, it was hoped, force the Argentinians to start using their radars. A forty-five minute 'cat-and-mouse game' then followed. The Vulcan's two navigators and the air electronics officer had to sort out which radars were operating and where the sets were located. The crew was under strict instructions not to endanger the civilians of Stanley. The primary target was the Westinghouse TPS 43 and the Vulcan crew had been given four possible locations for this. The Westinghouse set was detected but its operators were being very careful, only using it for short periods. But the Vulcan crew's patience was eventually rewarded and the two Shrikes were launched. The radar emissions ceased at the same time as the first missile was seen to explode and it was hoped that the set had been destroyed. In fact, only part of the 'wave guide' circuitry had been damaged and the set was back in action thirty-six hours later, after a replacement part was flown in from the mainland.

A further mission, Black Buck Six, was flown three nights later, on 2/3 June, Squadron Leader McDougall and his crew again being the primary crew, flying their third long mission in six nights. This time the Vulcan carried four Shrikes, two targeted for attack on the Westinghouse radar and two on the Skyguards. The Argentinians were more careful on this occasion. A Harrier flight from the task force was cancelled because of fog and the Argentinian radars were not being used. McDougall and his crew cruised around over the Stanley area, trying everything they knew to induce the Argentinians to open up their radars. As the end of its allotted period over Stanley approached, the Vulcan was preparing to launch its missiles blindly into the airfield area when the emissions of a Skyguard radar were detected. Two missiles were fired and the Argentinian forces newspaper later admitted the destruction of a Skyguard set on Sapper Hill and the deaths of a lieutenant, a sergeant and two soldiers of a unit described as 'GADA 601'.

On the return flight, the Vulcan's fuel probe inexplicably broke off while refuelling from a Victor and Squadron Leader McDougall had no option but to put down at Rio de Janeiro, having first thrown overboard as many secret documents as possible. His 'Mayday' calls were picked up from Rio and the air-traffic controllers gave the Vulcan priority. It landed with five minutes of fuel remaining. The Nimrod from Ascension which acted as shepherd on all of the Vulcan return flights followed the Vulcan in and listened to the radio conversations, right up to the taxi instructions, so that it could report that the Vulcan was safely down. The Vulcan crew was treated decently and correctly and the aircraft and crew were released seven days later, but the unexpended missiles had to be left behind.

CHAPTER NINETEEN

5 Brigade Comes Up

On Sunday, 30 May, the command ship H.M.S. *Fearless* sailed into San Carlos after having left that anchorage three days earlier to collect Major-General Jeremy Moore and Brigadier Tony Wilson from *Antrim*, which in turn had sailed from the task force to transfer those officers from the *QE2* at South Georgia. At San Carlos, Jeremy Moore assumed the duties of Commander Land Forces Falkland Islands; Brigadier Wilson was waiting for his 5 Brigade to arrive. The War Cabinet had not been prepared to risk the *QE2* – half as large again as *Canberra* – in Falklands waters and the 67,000-ton liner had been sent instead to South Georgia where the units had been transferred to *Canberra* and *Norland* for the voyage to San Carlos.

The arrival of Major-General Moore immediately released Brigadier Thompson from his role of land forces commander and also from the tiresome chores of looking after the base area at San Carlos and keeping in touch with the Task Force commander back at Northwood over policy decisions. Thompson quickly handed over to Moore and moved 3 Commando Brigade's Headquarters to Teal Inlet to be up with his troops. Jeremy Moore inherited a most favourable situation. A secure base had been established at San Carlos; most of 3 Commando Brigade was up to the Stanley defences; 2 Para had won its victory at Goose Green, 'a marvellous gift for a new commander to have', says Moore. There was some rearrangement of units. 2 Para and 29 Battery (of 4 Field Regiment), which had been detached from 5 Brigade in the early days of the Falklands crisis to sail on Operation *Corporate* with 3 Commando Brigade, now reverted to 5 Brigade command. 40 Commando was ordered to remain at San Carlos to defend the base, a decision governed more by inter Army–Royal Marines diplomacy than by the relative value of units.

The future role of 5 Brigade had never been clearly defined.

At the time of its dispatch from England, three weeks earlier, no one knew what situation the brigade would face on its landing. Perhaps the Argentinians would have surrendered; then, 5 Brigade would become the Falklands garrison, allowing 3 Commando Brigade an early return home. A more realistic scenario was that 3 Commando Brigade would be held up in the hills surrounding Stanley, unable to bombard the Argentinians because of the presence of the civilian community and with winter approaching. Sir Terence Lewin, Chief of the Defence Staff, certainly saw 5 Brigade's possible role as that of a relieving force in a prolonged siege of Stanley. But the process of landing and moving out from the beach-head had not progressed as fast as London expected and here was 5 Brigade, with an energetic commander whose brigade had been messed about ever since the Falklands crisis erupted, looking for a worthwhile task. Major-General Moore and Brigadier Wilson had travelled together in *QE2* and Wilson had persuaded Moore that 5 Brigade would not be relegated to the role of base defence or kept back as a reserve. Moore, perhaps rashly as it turned out, had agreed that 5 Brigade could come up into the front line on the southern flank facing the Stanley defences *and that 5 Brigade could have parity in supporting resources with 3 Brigade already up on the northern sector*.

The Gurkhas were the first of the 5 Brigade units to be given a task. Most of the battalion was moved from San Carlos to Goose Green by the Chinook, this move being marked by much impetuosity on the part of the Gurkhas, with their companies mostly being landed in the wrong places at Goose Green and with much marching and counter-marching ensuing. The Gurkhas took over responsibility for the Goose Green area from 2 Para and commenced a vigorous round of patrolling, sometimes going out on civilian tractors and trailers – 'a fun way of going to war' – and sometimes with 656 Squadron's helicopters. Seven Argentinians were captured in a joint helicopter and Gurkha operation at Egg Harbour House, eighteen miles from Goose Green and overlooking the Falkland Sound. On being ordered to lie down to be searched by the Gurkhas, all the prisoners did so except the officer in charge. A Gurkha rifleman drew his *kukri* knife and the officer complied at once. Three more Argentinians walked in later and were also taken prisoner. Ten anti-aircraft missiles were also found and the Argentinians said that five other similarly equipped teams were at other isolated locations. This led to a further search by Gurkha helicopter patrols all round the large sub-island of Lafonia, but no further Argentinians were found.

The Gurkhas believed themselves fortunate that they were the first unit of 5 Brigade to be given a task but this led to them being the last unit in the brigade to reach the Stanley area and to missing a major battle at the end of the campaign. The *kukri* drawn at Egg Harbour House was the only one to be so drawn in anger during the Falklands war.

Major-General Moore's intention that he would treat his two brigadiers equally in the matter of resources could not immediately be carried out. By the time Moore took over, 3 Commando Brigade had moved three of its battalions forward on the northern flank in response to the direct order from London. With the loss of the three Chinooks on *Atlantic Conveyor*, Moore just did not have sufficient helicopter resources to move 5 Brigade up on the southern flank at the same time as keeping 3 Brigade's forward units supplied and fed. Brigadier Wilson then decided that he would march his brigade along the track south of the central mountains. There were several drawbacks to this proposal, the main one being that there were no settlements or decent harbours where the marching units could have rested and received succour on the thirty-five mile stretch from Goose Green to Fitzroy Settlement. There were also doubts that the two Guards battalions, in particular, would have been able to stand the privations of a long trek.

It was 2 Para who cut through this dilemma. On the afternoon of 2 June, Major Keeble suggested that a small party be flown forward by helicopter to Swan Inlet House – half-way between Goose Green and Fitzroy; it was believed that the house was deserted and that the telephone line from there to Fitzroy might still be intact. 5 Brigade just happened to have five Scout helicopters of 656 Squadron at Goose Green; they had been trying to mount a raid with 2 Para on a possible Argentinian observation post on Mount Usborne but were prevented from doing so by thick mist. 5 Brigade also happened to have been given the use of the Chinook helicopter for the day, to move stores around San Carlos and to Goose Green. A rapid plan was made, a plan which Brigadier Wilson, who just happened to be at Goose Green that day, approved of enthusiastically. The Scout helicopters would take a small party of paras to Swan Inlet House and attempt to telephone Fitzroy to see if there were any Argentinians at that settlement. If Fitzroy was clear, 2 Para could start moving up, using the Chinook and the Scouts. Major-General Moore's headquarters was not informed of the operation.

All went superbly. This is a composite account by some of the Scout helicopter crewmen involved.

The mist was not too bad at low level. We went along the coast, over the water, flying below fifty feet. Two missile and machine-gun-armed helicopters went in first, up Swan Inlet only six feet up; they couldn't see over the banks. Our orders were that, if there was any hostile fire, we were to return. It all happened very quickly; there was no time to think but it was potentially quite hazardous. It should have been done to the same music as the *Dam Busters* film.

We fired four missiles, aimed just to miss the buildings, with the intention of keeping their heads down; two misfired, one had the wire break and only one exploded properly. The troop-carrying helicopters – with four paras in each – were 800 metres to the south, listening to the count-down, 'Firing now – five, four, three, two, one, hit!' and then the three Scouts went in with their twelve paras.

There were no Argentinians at Swan Inlet House. Major John Crosland of the paras cranked the telephone and twelve-year-old Michelle Binney, daughter of Fitzroy's manager Reg Binney, answered the telephone. Mr Binney soon took over and told Crosland that the few Argentinians who had been at Fitzroy and at nearby Bluff Cove Settlement had all departed. The paras and the helicopters returned to Goose Green; this part of the operation had been completed in less than an hour.

A furious loading and flying of helicopters now commenced. The Chinook was taken off its load-carrying work; it could carry more than fifty paras. The Scout pilots removed their machine-guns; all of them could now carry four paras. Major Farrar-Hockley's A Company, with a mortar detachment, carried out the first lift. One officer says, 'It was just like a cavalry charge; it would make a lovely film.' The flight nearly ended in disaster. Eight miles away, behind Mount Kent, was 3 Commando

Brigade's 7 Battery and its 105-mm guns. An observation post reported a Chinook landing men at Bluff Cove and the battery prepared to open fire. Fortunately, the observation post realized that the Chinook was British and the order to fire was cancelled. The helicopters flew several more sorties, having to refuel from drums of fuel left by the Argentinians. Most of two companies and 2 Para's Headquarters were flown in before dark. Bluff Cove Settlement, three miles away, was also occupied. (2 Para received a new commanding officer to replace 'H' Jones on that day, Lieutenant-Colonel David Chaundler having been flown out from England and parachuted into the sea near H.M.S. *Penelope*.)

There were mixed reactions when news of Brigadier Wilson's leap forward along the coast reached Jeremy Moore's head-quarters. There was some dismay that Wilson had committed part of his brigade so far forward without the means of supporting them or supplying them. If the Argentinians discov-ered 2 Para and turned upon them in strength, the paras – still weary and below strength after Goose Green – might suffer severely. There was no artillery, no air defence, just one weak and isolated battalion. But Moore buckled down to make the best of the situation, although it was to bring both problems and loss, and it led to the disfavour in which Brigadier Wilson found himself at the end of the war. Some Sea Kings were found the next day and these lifted the rest of 2 Para forward. Helicopters could not, however, be provided for any further move of a major unit and it was decided that the rest of 5 Brigade and its vital air defences should be brought up by a series of night lifts by sea. There were sufficient ships available; the only question was whether they could be risked so near to Stanley and so far away from the air defences of the San Carlos anchorage. The larger of the amphibious ships, *Fearless* and *Intrepid*, could easily have brought the remainder of 5 Brigade and its air defences round in one night's lift, but these 12,000-ton ships would have been exposed in the Fitzroy–Bluff Cove area at dawn, before the Rapiers had been put ashore. It was decided that these ships could not be risked, and a compromise was reached. *Fearless* and *Intrepid* would be dispatched – but on separate nights – half-way to Bluff Cove and then trans-ship the units they were carrying to landing craft for the remainder of the voyage, themselves returning to the protection of the San Carlos defences before dawn. The first lift did not take place until the night of 5/6 June, three nights after 2 Para's move forward. The paras were, fortunately, not attacked in that interval. *Intrepid* set off at dusk with the Scots Guards. The guardsmen had spent three wet and cold days at San Carlos, in harsh contrast to their recent voyage on the *QE2* and their more distant memories of ceremonial duty at Buckingham Palace. They were delighted to board *Intrepid* to be greeted with a hot meal and have the chance to sleep for a few hours in dry conditions. *Intrepid* steamed round to Lively Island and there the guardsmen transferred to the four landing craft being carried inside *Intrepid*, for a miserable, seven-hour, rain and spray-soaked passage to Bluff Cove.

Intrepid returned safely to San Carlos. *Fearless* set off the next night with the Welsh Guards but with only two landing craft; the four at Fitzroy were intended to come out and meet *Fearless*. It was at this point that things started to go wrong, and the consequences of the *ad hoc* nature of 5 Brigade's move forward started to be compounded, leading to the *Sir Galahad* disaster. 5 Brigade's signal system was not as modern as that being used by 3 Commando Brigade. Neither Moore's headquarters on *Fearless* nor 5 Brigade's headquarters at Goose Green could pass signals directly to the units at Fitzroy and Bluff Cove. A relay station had been inserted by helicopter half-way up Mount Wickham but this had not managed to fill the communications gap.(A Gazelle helicopter carrying 5 Brigade's Signals Officer had been lost while flying to the relay station the previous night and four men were killed. This was one of the unfortunate 'blue-on-blues' which was hushed up for as long as possible, the helicopter having been shot down in error by a British missile. The Type 42 destroyer, H.M.S. *Cardiff*, was stationed off the coast with *Yarmouth* near the helicopter's route, standing by to give gunfire support on to Argentinian positions if required. 5 Brigade Headquarters had no naval liaison officer attached, and did not realize that the proposed flight should have been signalled to Moore's headquarters so that the Navy could be informed of the flight: the brigade believed that it had an autonomous right to fly helicopters in its own area. *Cardiff*'s radar picked up a slow-moving contact, flying from west to east, and it was assumed that this was an Argentinian supply aircraft coming into Stanley airfield. The Gazelle's I.F.F. (Identification, Friend or Foe) was not switched on because it had been found that its use was disturbing other electronic equipment. *Cardiff* fired two Sea Darts; the helicopter was destroyed and its crew and two Royal Signals passengers were killed.

The story was covered up immediately. One family was told that the helicopter had 'crashed in bad weather', although it was a clear, calm night. Other reports stated that the Gazelle had been shot down by an Argentinian missile. An enquiry at Farnborough indicated that the missile damage found in the helicopter wreckage had not been caused by a Sea Dart missile, although a Sea Dart casing had been found near the crash site. Those of us who knew the true story soon after the war were persuaded not to publish it 'for the sake of the families'. Captain Michael Harris, of the *Cardiff*, answered my questions in July 1983 in a letter thus:

We were keeping very alert for enemy aircraft which we knew were re-supplying Port Stanley from the mainland; also there were reports of Pucarra movements over the islands at night. Thus, when our radar detected an aircraft to the West at about 0400Z, apparently heading for Port Stanley, I shot it down. As to whether I could have told the difference between slowish enemy aircraft and friendly army helicopters on my radar: no, I could not, and I was not expecting any friendly air movements in my area that night . . .

It was the first time that *Cardiff* had hit an aircraft, and we believed it to be an enemy one. When, the next day, it seemed probable that it was a friendly helicopter that had been hit, I elected not to tell my Ship's Company because I felt that the news would have done harm to their confidence at a time when it needed to be high . . .

I wrote to General Moore as soon as it had become apparent what had happened, expressing my great sorrow. Beyond that, I have had to live with it.

When pressed by a national newspaper in May 1986, the Ministry of Defence informed the families of the four men what had happened and then reluctantly admitted the matter publicly.) The relay station did come into use later, although its two-man crew suffered from exposure. So, when *Fearless* sailed with the Welsh Guards on the night of 6/7 June, there were misunderstandings over the arrangements for the transfer of the Welsh Guards from the midway point on to Bluff Cove. Major Southby-Tailyour, in charge of the four landing craft now at Fitzroy, did not feel able to sail to meet *Fearless* because of the battering his craft and their crews had suffered the previous night while carrying the Scots Guards. This decision did not reach *Fearless*, which waited as long as she could at the rendezvous and then put half of the Welsh Guards into the only two landing craft she was carrying and returned to San Carlos. These landing craft reached Bluff Cove safely but half of the Welsh Guards were taken back to San Carlos and thought was given to a better way of moving troops up than this unsatisfactory landing-craft shuttle.

During the day of 7 June, any final thoughts of using *Intrepid* or *Fearless* again were dismissed when a categorical order arrived from Admiral Fieldhouse that these large ships were not to be risked in this way. But a simple solution appeared to be available. Commodore Clapp had six of the slower and smaller 'Sirs'. Two of these were already being used forward at Teal Inlet and one, the *Sir Tristram*, was at Fitzroy unloading 1,500 tons of artillery ammunition. Teal could not be observed from the Argentinian main positions on the hills around Stanley but, in clear weather, Fitzroy could. The *Sir Galahad* was being prepared at San Carlos to sail that night, loaded with four Rapier launchers and the vehicles and men of 16 Field Ambulance, 5 Brigade's medical unit. The destination was Fitzroy, where 5 Brigade's base area was being established. It seemed a simple thing to transfer the two companies of Welsh Guards still remaining on *Fearless* to *Sir Galahad* and, if the ship sailed promptly at dusk, the Welsh Guards could be delivered to their ultimate destination, Bluff Cove, in the early hours and *Galahad* could return to Fitzroy before dawn with the Rapiers and the Field Ambulance. *Majors Guy Sayle and Charles Bremner, the two Welsh Guards company commanders, were told that their destination was Bluff Cove.*

The Rapiers and the Welsh Guards were quickly loaded on to the *Galahad* but, because of communications difficulties at San Carlos and lack of landing craft and helicopters, 16 Field Ambulance took six hours to load and the ship was not ready for sailing until about five hours after dusk. Captain Philip Roberts of the *Galahad* believed that it was now too late to set sail. He told the Welsh Guards to get off to sleep and sent a signal to Commodore Clapp's staff saying that he intended to stay at San

Carlos and sail at dusk the following day. A return signal came, however, ordering him to sail at once – *but to go only to Fitzroy, which could be reached before dawn.* The Rapiers and the Field Ambulance were to be unloaded at Fitzroy but precisely what was to happen to the Welsh Guards was left in the air. *No signal reached 5 Brigade Headquarters, which was now at Fitzroy; no extra landing craft or helicopters were made available at Fitzroy for the unloading; the Welsh Guards officers now asleep on the* Galahad *were not told of the change.*

Sir Galahad sailed round to Fitzroy, fifty miles by land but more than three times that distance by sea. It was a clear night, with a full moon, and there were fears that Argentinian observation posts on the west side of Falkland Sound might have seen the ship. If *Sir Galahad*'s passage was seen, however, that was not the cause of her later misfortune. She sailed into Port Pleasant, Fitzroy's harbour, soon after 8 a.m. local time and anchored about a mile from the nearby settlement. There had recently been a long spell of misty weather, free from any air attack, and there was now a certain amount of complacency about the capability of the Argentinians to mount further air attacks. The large number of aircraft claimed shot down by all types of British units led to the assumption that the main Argentinian air strength had been broken. The morning that *Galahad* sailed into Port Pleasant was clear and bright. *Sir Tristram* had been here all the previous day and had not been attacked. The scene looked peaceful enough. There had been six landing craft working in the anchorage the previous day, more than enough to unload the *Galahad* quickly and take the Welsh Guards on to Bluff Cove. But five of these landing craft had gone, four to return to their parent ship *Intrepid* and the fifth to Goose Green to fetch up some of 5 Brigade's much-needed signals vehicles. Only one landing craft remained – Colour-Sergeant Michael Francis's *Foxtrot One* – and a Mexeflote pontoon raft (of 17 Port Regiment, Royal Corps of Transport). These two vessels were alongside *Tristram*, partly loaded with ammunition and waiting for the tide to turn.

There were two kinds of people at Fitzroy on that day, 8 June: those who had experienced the Argentinian air attacks in the San Carlos area soon after the 21 May landings and those who had not. All of the 5 Brigade men had been in *QE2* at that time.

Majors Tony Todd (Royal Corps of Transport) and Ewan Southby-Tailyour (Royal Marines), were handling the unloading of ships at Fitzroy. These two officers had both been at San Carlos throughout the earlier raids. They came out post-haste to *Galahad* to solve what for them was the appalling problem of getting the 350 Welsh Guardsmen off the ship as quickly as possible. An immediate solution was offered; the loaded Mexeflote would be towed to *Galahad*, the guardsmen could perch themselves on top of the stacked ammunition and be put ashore – all within an hour. They could then march to Bluff Cove. The Welsh Guards company commanders were not keen. They had been told that they were to be delivered by ship to Bluff Cove. They were unwilling to weary their men by what they saw as an unnecessary overland march, although it was only five miles to Bluff Cove by way of a wooden bridge across an intervening inlet which had been repaired after having been damaged by the Argentinians. Admittedly, it was very poor terrain between the two places and the Welsh Guards had not coped well with an earlier march in the San Carlos area. There was another factor. The march overland solution would require the Welsh Guards to leave all their heavy equipment in *Galahad*, hoping it would find its way round to Bluff Cove later, but there was obviously the risk that this would not happen in the reigning confusion (there was also a good deal of stealing of equipment by some units at this time). The two Guards majors declared that they were not prepared to be separated from their heavy equipment on the eve of reaching the front line.

The next plan suggested was for the one landing craft present to be unloaded of its cargo of ammunition so as to embark as many guardsmen and their equipment as possible and run them round to Bluff Cove. Local tractors and trailers and a 20-foot cutter requisitioned by the Royal Engineers at Fitzroy could also be used. It was agreed that this would be done. All during this period, a solitary Sea King (of 846 Squadron) which had come round from San Carlos on the *Galahad* was unloading the Rapier units, but this required eighteen helicopter lifts. The landing craft came alongside *Galahad* at noon and announced that it was ready to start embarking the Welsh Guards but Lieutenant-Colonel John Roberts, commander of 16 Field Ambulance and the senior ranking officer present, now stated that it was essential that the leading echelon of his unit should have priority use of the landing craft, the Welsh Guards having turned down one offer of leaving the ship. Twelve men and nine vehicles of the medical unit were then transported ashore, an operation which took up another hour. Again the landing craft returned and should at last have started to embark the waiting guardsmen, but then another snag developed. The landing craft had damaged its loading ramp on the last journey and this could not be lowered to allow the loading of the guards' heavy weapons and equipment at *Galahad*'s stern doors. The troops could have quickly climbed down into the landing craft, but it was decided to leave them in *Galahad* until their equipment was loaded by crane – a slow process. *Sir Galahad* had been five hours at Fitzroy; *Sir Tristram*, now empty of stores, was still present.

The Argentinian troops on Mount Harriet, ten miles away, could see the two ships at Fitzroy across the intervening low ground. The sighting report had reached the mainland. Two Argentinian air force units were ordered to take off, Grupo 5 with eight Skyhawks from Río Gallegos and Grupo 6 with six Daggers from Río Grande. The British at San Carlos were given a little warning of a possible attack – probably from the submarine *Valiant* which was patrolling off Río Grande – *but this warning did not reach the Galahad*. Three of the Skyhawks and one Dagger returned early because of technical trouble but the remaining aircraft of the two formations reached the Falklands, intending to fly either side of Lafonia to reach Fitzroy. But the Daggers, flying up Falkland Sound before turning to attack Fitzroy from the west, spotted the British frigate *Plymouth* and decided to attack her. *Plymouth* was on her way to carry out a bombardment of a suspected Argentinian observation post on Mount Rosalie, which overlooked the landing area from the other side of Falkland Sound. The ship was hit by four bombs, but once again the Argentinian pilots were too low and none of the bombs exploded. There was, however, much superficial damage and a depth charge near the flight deck blew up and started a fire. Four men were injured and one was overcome by smoke. *Plymouth* claimed two Daggers destroyed but only one of the Argentinian aircraft was slightly damaged.

*

Forty-five miles away at Fitzroy, there were no effective defences against air attack. No warships had been sent to the harbour. Two Sea Harriers had been on routine patrol to the south; they were now in hot pursuit of the Daggers which had just bombed *Plymouth*. Both *Galahad* and *Tristram* had Bofors guns and Blowpipe teams aboard but these were not weapons that could stop a determined attack by five jets. All depended on the four Rapier missile launchers of Lieutenant Adrian Waddell's H Troop which had arrived that morning on the *Galahad* and which were now all ashore. A good Rapier team could set up its launcher only fifteen minutes after being landed by helicopter but the Rapiers were disposed according to a plan whereby they were to protect the supply base and 5 Brigade Headquarters being set up around Fitzroy Settlement – *not the anchorage*. However, two of the Rapier launchers were posted to the south of Fitzroy and their fields of fire did cover the anchorage; Sergeant Bob Pearson's post covered any approach from the east, Bombardier McMartin's from the west. By yet one more of those cruel strokes of misfortune which led up to the *Galahad* tragedy, one of these posts was suffering from a mechanical defect and was not yet ready for action. Sergeant Pearson's launcher had been severely jarred while on *Galahad* and a replacement tracker optic was required. The spare part was back at San Carlos. The Sea King flew back there, picked up a technician and the spare part, and had just landed alongside Sergeant Pearson's launcher when the Argentinian attack came in – *from the east*.

The Argentinian pilots had been given Port Fitzroy as the location of the British ships but Port Fitzroy was a stretch of water north of the settlement, while *Galahad* and *Tristram* were in Port Pleasant to the south. The five Skyhawks flew in from the west and along Port Fitzroy, finding no ships but being fired at ineffectively by British troops there. The Argentinian formation leader, Lieutenant Cachón, then ordered his fellow pilots to sweep out south over the sea and return home; their fuel was limited. While turning, one of the other pilots glanced up Port Pleasant and saw the two ships. The Skyhawks simply tightened their turn and came straight into the attack. Sergeant Pearson's Rapier team could do nothing except watch the jets fly by.

We watched it all from our post. We were in a good position. If they had come in ten minutes later, I believe we could have got at least one and then we would be talking dead Argies. If we had got the leader, the others would probably have pulled away or at least 'gone tactical' and started jinking around on their bomb runs.

The attack also caught the defences of *Galahad* and *Tristram* unprepared. An air-raid warning had been received at Fitzroy twelve minutes earlier *but there was no way of passing the warning to the two ships*. Royal Marine gunners on *Galahad* opened fire and one Blowpipe missile was fired from the stern of *Tristram* but the explosion of a bomb in the ship underneath the operator caused him to lose control of the missile guidance. Three of the Skyhawks attacked the *Galahad* and two the *Tristram*. All five aircraft returned safely to their base and their pilots all survived the war.

The *Sir Tristram*, with fewer than a hundred men on board – mostly Hong Kong Chinese crewmen – and with most of her cargo discharged, was hit by two bombs on the starboard side towards the stern. One bomb passed straight through the ship without exploding; the other burst inside a small compartment and buckled the steel wall of the nearby tank and vehicle deck. Two Chinese crewmen were killed, one of them sixty years old.

The effect upon the *Galahad* was vastly more serious. The Argentinians attacking *Galahad* came in at a higher altitude and their bombs took a more plunging trajectory. One came through a hatch and exploded in the tank deck; another exploded in the engine-room and galley area, killing a Chinese; the third bomb burst in the officers' quarters, killing an engineering officer. (These bombs did not explode in the conventional sense; they 'deflagrated', with the casings bursting open and then the contents burning fiercely rather than exploding inside the casing.)

Lance-Corporal Barry Gilbert was a Royal Marine of the 1st Raiding Squadron, sailing in the *Galahad*, hoping eventually to rejoin his unit at Teal Inlet.

It was a lovely day, flat calm; the sun was out and I was feeling in remarkably good spirits. I caught a glimpse of them, out of the corner of my eye, these two aircraft coming in at one hell of a rate. The speed they came in at! No sooner you'd seen them coming than there was the unforgettable sound of a jet passing right over your head. It was horrendous and I know I'll never go to another air show. They were so close and so low that you could see the pilots. The planes were whitish, with red and blue markings.

We all reacted and dived to the floor, literally as the ship was hit. The deck lifted from under us. Then we were hit again. It's like a car accident; everything seems to go in slow motion, seems to take an awful long time. Then, when everything starts to come back to normal again, a few seconds later, it all speeds up. It goes from one extreme to the other. It was terrifying. I had been in so many attacks – I had been on *Lancelot* when she was hit at San Carlos – and every time it happened there was a kind of surge inside you. You want to react but there is nothing you can do; you are in God's hands literally I must admit there was an instant fear of death.

Lance-Corporal Bill Skinner, of 9 Parachute Squadron, Royal Engineers, had come out to *Galahad* from the shore in the civilian cutter which had been requisitioned, waiting for the long-delayed move of the Welsh Guards to Bluff Cove.

When the bombs went off, it was just like the scoring of a goal at a football match; everyone went 'WOOOH' – a surprise shout. I was knocked over by a large blast of hot air which propelled me about twelve feet through some panelling which was also smashed down by the blast, and I finished up lying on top of some stairs – ship's panelling and fittings all around me – knee-deep in places.

I stood up and there was a bloke in front of me who was injured in the back, bits of fitting stuck into his back. Two blokes were helping him out. I started to make my way out of the door on to the deck. I was in a queue of six people on my side when someone shouted 'Stand still!' It was a daft thing to say; there was no panic. A few seconds later, someone else said, 'No. Let's go; the smoke's bloody killing us.' It was thick, acrid, multi-coloured, very pungent, of the electrical sort and it had bits in it like the big bits of soot which float off when an acetylene torch burns. I decided to go back and started making my way down the stairs. Half-way down I found this bloke; he was sliding slowly down, head first – bump, bump, bump. He had obviously been knocked over by the blast. He wasn't badly hurt, just bashed about by the blast. I picked him up by the webbing straps on his back and just dragged him down the last few steps on his knees. He stood up then, brushed himself off, and we both made our way out of the door on to the middle deck. There were people moving about there – mainly injured men and others looking after them. One of them had a head wound and it must have been bad because blood was soaking through the dressing which another man was putting on him. But it was mainly blokes with severe burns. You couldn't tell which unit they came from; we had all lost our berets by that time.

But the worst scenes were in the completely open tank and vehicle deck, where most of the troops were concentrated. A bomb exploded here among the Welsh Guards, who suffered terribly. There were also twenty tons of ammunition here: mortar shells, small arms and grenades, and a large quantity of petrol. This became an inferno. The men of the 16th Field Ambulance and some S.A.S. men who were aboard were further forward and not so badly affected. I made no request to interview the Welsh Guards and have no personal accounts from what must have been a nightmare episode with the effects of the bomb explosion, the fire and dense smoke from burning ammunition, where dazed and wounded men tried to make some sense out of chaos. At least forty-three men died in that open tank and vehicle deck and up to 150 were burned or injured, many of them very seriously.

The order to abandon ship was given. Detailed descriptions for the rescue phase are not necessary either. There can be few people who will ever forget the dramatic television filming of the helicopters flying into the smoke around the *Galahad*, to the background of exploding ammunition in the ship's hold, of small boats bringing ashore wounded guardsmen. The scene of a burly Welshman, holding the stump of his leg aloft as he was carried ashore, is particularly vivid. Many of the less seriously injured and unhurt survivors escaped by ladders into Colour-Sergeant Francis's landing craft, which had fortunately been on the port side and was not damaged by the bombs, and in other boats. Some got away on the Mexeflote pontoon, whose operator was at first reluctant to come too close to the side of the *Galahad* for fear of a major ammunition explosion, 'until someone threatened to shoot him and that got him to move in closer'. The 846 Squadron Sea King, which had been lifting the Rapiers, was the first helicopter to start taking injured men off the *Galahad*'s deck. Two more Sea Kings, of 825 Squadron, at least one Wessex (probably from 847 Squadron) and an army Gazelle of 656 Squadron all joined in. One helicopter crewman says that, 'no one controlled the rescue but it all went like clockwork'. The helicopters delivered the wounded to the Fitzroy landing site, only a mile away, where an unnamed marine corporal in a red fluorescent helicopter marshalling jacket was observed to do sterling work, guiding in the helicopters, sorting out the worst casualties, 'running around all over the place with sweat pouring from him'. First aid was given by the advance party of 16th Field Ambulance, which had come off the *Galahad* earlier that morning, and by the medics of 2 Para, and there was soon a great shuttle of helicopters taking injured men to the better facilities at Ajax Bay and, eventually, out to the hospital ship *Uganda*, which had its busiest day of the war, receiving 159 casualties.

Captain Philip Roberts, *Galahad*'s master, was the last man off his ship, about three quarters of an hour after the Skyhawk attack. The *Galahad* was left to burn herself out. She was towed out to sea later in June and sunk as a war grave for the dead men left inside her in this, the most serious setback of the war for the British forces. The *Tristram* returned to Britain to be repaired. Forty-eight men died on the *Sir Galahad*: thirty-two Welsh Guardsmen (mostly members of the Mortar Platoon), four Army Catering Corps men, three members of the R.A.M.C., two each from 9 Para Squadron R.E. and from the R.E.M.E., three British ship's officers and two Chinese crew members. All of the dead were of junior ranks except for an R.A.M.C. major and the three ship's officers. There was naturally a great public interest in the reasons why the two ships were caught in an exposed position and why the *Galahad* still had so many men aboard when she was bombed. It was felt that someone had blundered and that someone should be blamed. Most of the steps leading up to the tragedy were the results of unforeseen operational circumstances: the loss of *Atlantic Conveyor*'s helicopters, the recent clearing of the weather, the lack of a reliable signals link between the British main base and Fitzroy, the mechanical failure in a landing craft's ramp, the failure in a Rapier's tracker optic. It is true that there was some complacency over the danger of air attack and some lack of liaison between the parties involved, but this was against a background of an intense effort to get the British forces forward as fast as possible and get the war finished before the winter weather set in, hardly a cause for serious blame.

*

Argentinian aircraft returned to Fitzroy less than two hours after the attack on the *Galahad* and the *Tristram*. Four Skyhawks of Grupo 4 followed the approach path made by the earlier raid and achieved a small, initial, bloodless success while approaching Fitzroy from the west. Sergeant Dick Kalinski, pilot of an army Scout helicopter of 656 Squadron, saw the Skyhawks and was manoeuvring violently to take shelter behind a hillock when the drive shaft to his machine's tail rotor broke and the helicopter spun rapidly round its axis and plunged into a pond. Kalinski and his crewman were able to wade ashore and the helicopter was later salved. The Skyhawks flew on, over Bluff Cove where they were greeted by a mass of small-arms fire from the well-organized Scots Guards, who had received warning of the air attack and had just heard of the tragedy suffered by their colleagues on the *Sir Galahad*. The battalion quartermaster later issued 18,500 rounds of ammunition to replace the amount fired off in about forty-five seconds at the Skyhawks. The battalion claimed 'two, possibly three' Skyhawks shot down but these claims are not valid.

The Skyhawks banked and approached Fitzroy from the north-east, intending to attack the ships in Port Pleasant again. But this time the Rapier posts were ready and the north-easterly rather than the easterly flight path allowed all four posts to open fire. Missiles zoomed and exploded all over the place; the Argentinian pilots counted at least seven missiles in flight. The Rapiers certainly frustrated this attack and no bombs were dropped. Sergeant Bob Pearson's post, which had not been able to fire when the earlier attack came in, was credited with one Skyhawk shot down, 'crashed out of sight and confirmed later by ground troops', but this was another over-optimistic claim. The Argentinian aircraft all reached the mainland, although some were badly damaged by the various ordnance flung at them by the Scots Guards and the Rapier posts.

The Argentinian air effort was still not finished. A formation of four more Skyhawks from Grupo 5 – the same unit which had earlier attacked the *Galahad* and the *Tristram* – again probed the southern part of East Falkland, perhaps intending to attack Fitzroy again. Unusually, they were operating in conjunction with some missile-armed Mirages which were flying as high cover. Steaming peacefully to the south was one of *Fearless*'s landing craft – Colour-Sergeant Brian Johnston's *Foxtrot Four*, with a crew of eight men and carrying 5 Brigade's signals Land-Rovers and trailers and other vehicles. The Skyhawks caught *Foxtrot Four* half-way between Goose Green and Fitzroy. The crew opened fire with two machine-guns, but a devastating bomb and rocket attack blew the wheelhouse away and the port engine right out of the craft. Johnston and five of his crew – three more Royal Marines and two naval ratings – were killed. Colour-Sergeant Johnston had been one of the heroes of the *Antelope* bombing, having brought *Foxtrot Four* alongside the frigate to rescue most of her crew when there was an unexploded bomb aboard the frigate. He would be awarded a posthumous Queen's Gallantry Medal. One of *Foxtrot Four*'s passengers was injured. This was Air Trooper Mark Price, an Australian who was the driver of 656 Squadron's commanding officer's Land-Rover. Price was partially blinded but recovered, although not to the standard required for helicopter aircrew training which had been his ambition.

Retribution came swiftly. Sea Harriers had been trying to maintain a constant air patrol since the first attack on Fitzroy. Flight Lieutenant David Morgan and Lieutenant David Smith of 800 Squadron spotted the Skyhawks while they were attacking the landing craft and promptly shot three of them down with Sidewinders. All three Argentinian pilots were killed. These were the last Sea Harrier air successes of the war.

The Argentinians at Stanley picked up British signals about the Fitzroy disaster and, that night, Brigadier-General Menéndez discussed with his superiors on the mainland whether a unit from the defences around Stanley should be dispatched to attack the Bluff Cove–Fitzroy area while it was still disorganized. But the operation never got off the ground; Menéndez decided that the British were too well established and that the task was beyond the capabilities of his units.

There was one other major event on this day. Argentinian Hercules aircraft were still attempting to bomb ships on the British lines of communications down to the Falklands. A Hercules spotted a large tanker 450 miles off the Argentinian mainland. The Hercules attacked without establishing the identity of the tanker or why it was so far west of the route from

Ascension to the Falklands. The ship was hit by at least two bombs. One blew a huge hole in an internal tank; another penetrated two more tanks but did not explode. No one was hurt. Ironically, this ship was also named *Hercules*. She was owned by the Liberian subsidiary of a New York company, the Maritime Overseas Corporation, and was sailing in ballast round Cape Horn from the Virgin Islands to the Alaskan oilfields. The owners and the United States Government made remarkably little public fuss over this attack on an unarmed neutral ship. The owners considered disarming the bomb but decided to scuttle her instead and this was done on 21 July. The ship was insured for a war risk of $20 million but, with a world surplus of super-tankers, the owners settled for $9 million on the basis of the original bomb damage. The *Hercules*, at 220,000 tons, was the Argentinians' biggest 'success' of the war; her tonnage was more than four times greater than the combined tonnage of all the British and Argentinian ships sunk in the war!

Thus ended another dramatic day in the Falklands war. Fifty-six British and Chinese had died, making a total of 192 members of the Task Force who had so far lost their lives in Operation *Corporate*. Only three Argentinians were killed, the three Skyhawk pilots shot down by Sea Harriers. Argentinina losses so far in the war were approximately 560.

CHAPTER TWENTY

Final Preparations

Recent moves and actions had brought the two sides to the verge of the climax of the war, the final confrontation battle in front of Stanley. Before describing that battle – really a series of battles – it is appropriate to look at the condition of the Argentinian defenders and of the civilians in Stanley.

The Argentinians still had more than 11,000 well-armed troops in the Falklands and still outnumbered the British forces. But nearly 2,000 of Menéndez's men were of no use to him at all. These were the two regiments and their supporting troops who formed the Port Howard and Fox Bay garrisons on West Falkland, which were now like stranded whales, consuming food and supplies but not able to move and influence the coming battle around Stanley, although several times urged to do so. The approximate strength of the units in the Stanley area was between 8,500 and 9,000 men, of whom fewer than 5,500 were infantry, mostly conscripts but in good positions and still outnumbering the British infantry. Menéndez's original planning had been based on two beliefs: firstly, that a diplomatic solution might be reached before any serious fighting took place; secondly, that if it did come to a battle, the British would land near Stanley. The Argentinians assumed that a landing at or near the main objective supported by gunfire and carrier-borne aircraft, even against a defended beach, was preferable to a long march over trackless terrain from an indirect landing. That was how the Argentinians had captured Stanley on 2 April; that was how Menéndez believed the British would return and he had thus kept his best troops and his main strength in the Stanley area. The British suspected that this would be the Argentinian thinking and, as has been described earlier, carried out a deception campaign to bolster that belief.

When the British landed at San Carlos, Menéndez had found himself unable to move effectively against the beach-head. As late as 8 June, nineteen days after the landings, he sent his chief of staff,

Brigadier-General Daher, and two staff colonels back to the mainland on an aircraft to beg Galtieri to make some sort of move from
the mainland against the British rear at San Carlos. But Galtieri
refused this plea. He was not prepared to risk further loss by
mounting such a hazardous operation. Menéndez then urged Galtieri to accept United Nations Resolution 502, passed on 2 April,
which had called for the Argentinian withdrawal from the islands.
Galtieri refused. He reminded Menéndez of the large force he had
at Stanley and urged him to be more 'mobile and aggressive'.

Much work had been done on the Stanley defences. Trenches
and gun positions had been dug; minefields had been laid in great
profusion. But most of these defences were sited, and the troops
disposed, to face a landing, either on the airport peninsula east of
Stanley or on the beaches to the south. (The map on page 327
shows the layout of the Stanley defences.) Now the British were
approaching through the mountains to the west. Menéndez
believed that the British were still capable of carrying out another
landing in the Stanley area and he made only one small adjustment
in response to the advance through the mountains. The 4th Regiment had occupied three heights – Mount Challenger, Wall Mountain and Mount Harriet – with their main positions facing south.
This regiment was now turned north, abandoning the positions on
Challenger and Wall but occupying Two Sisters. (That was the
reason why 42 Commando, occupying Mount Challenger, found
empty trenches and abandoned equipment.) More minefields were
hurriedly laid in the path of the approaching British, many of them
unmarked on minefield maps, some laid indiscriminately by helicopters. The Argentinians were never short of mines; the world
arms trade had been awash with surplus mines and Argentina had
easily been able to stockpile these before the war. The British later
identified mines made in five different countries.

The Argentinian supply position was patchy. The initial British
blockade – by submarine bluff – had not worked. The Argentinians had rushed a mass of supplies in by ship between the date of
their original occupation early in April and the arrival of the task
force at the beginning of May. There was no shortage of clothes,
nor of most foodstuffs, nor of many luxury items like wine and
cigarettes and a large surplus of such items was left in containers
dumped on Stanley's waterfront. The American firms which
owned most of the containers asked for their return after the war

but the British refused, saying that they regarded the containers as spoils of war and suggested that the Argentinians be asked for compensation. (The Falklands civilians became beneficiaries in a big way from this Argentinian stockpile. At an interview in one Stanley house eighteen months after the war, I was given an Argentinian sweet by the daughter of the household, told that the sugar and tea in the cup I was holding were Argentinian, was given as souvenirs a pair of Argentinian socks – one of fifty-two pairs 'liberated' from the containers by that family – and three bars of chocolate from a stock in the house of more than a hundred bars. The chocolate had been sent by patriotic civilians in Argentina as comforts for the soldiers; many of them had written messages of encouragement. One says, 'Good luck to you all and we hope the English don't get you.' Another, from a schoolgirl, has a pair of lips blowing a kiss and a drawing of a soldier waving an Argentinian flag. And the bar sent by 'Jessica' of Buenos Aires did not reach an Argentinian soldier; it is in my collection of souvenirs.)

A large part of this mass of supplies never reached the ordinary Argentinian soldiers. It is probable that Menéndez was preparing for a long siege, but there were certainly grave deficiencies in his distribution system. Most of the Argentinian regiments were moved out to the allocated defensive areas as soon as they arrived. A mass of vehicles was landed but these were only able to move out of Stanley for short distances along the local tracks, not into the open ground where most of the infantry was deployed. This, together with the Argentinian shortage of helicopters, led to some units being seriously short of rations. Argentinian helicopter losses were so heavy, mainly through Harrier action, that the last two serviceable Chinooks were withdrawn to the mainland on 8 June. Some units had been rushed in from the mainland without their cooking equipment and had to depend on the goodwill of units near by for hot food. Basic rations – different packs for officers and men – reached most units, but the conscripts in the poorer units frequently complained that officers and N.C.O.s stole their rations. There was a severe shortage of fresh meat and bread, items of which the Argentinians were very fond. Every single sheep in the whole of the Stanley defended area had now been killed and eaten, and a start was being made on the local dairy herd. The disabling by Harriers of the *Bahía Buen Suceso* at Fox Bay in mid-May had been a severe setback; she was loaded with flour from the mainland. Small boats were able to bring a little of this flour round to Stanley and a local bakery was persuaded to bake some bread. The nightly air bridge kept going until the end and performed wonders, but the average nightly lift of supplies was only about two tons; most of the aircraft used were Fokker Fellowships, with C-130 Hercules sometimes bringing in larger loads. Most of the Argentinian wounded were evacuated in these aircraft so there was no great medical problem at Stanley. There was plenty of small-arms ammunition but ammunition for the artillery was always in short supply.

The general level of Argentinian morale was low in this cold and alien land. They could see that the Falkland Islanders were not anxious to accept the Argentinian 'liberation'. Rotation with mainland units had been promised but had not materialized. The Argentinian soldiers were beginning to realize that much of the propaganda fed to them about British setbacks was lies. The British had not so far lost a battle and were just outside Stanley. British artillery fire and the hated Harriers seemed to dominate the battlefield.

The Argentinian soldiers often came into Stanley from the positions up on the mountains or out near the airport. Mrs Jill Harris describes how Argentinian soldiers tried to obtain food at the civilian West Store.

It was an everyday occurrence; we used to call it 'running the gauntlet'. They would hang about and wait for a woman on her own. There were always three or four Argies and they would approach a woman on her own or an old person. They tried to get us to buy the food for them, but most of us tried to avoid doing it. Sometimes they offered large quantities of money.

There was a notice at the store, in the window, saying in Spanish, 'No military personnel in uniform allowed in the store.' Menéndez went in one day, in uniform. Isabelle Castle was on the check-out and told him he wasn't allowed in. Menéndez said, 'I know. I wrote the notice and signed it.' She said, 'You should know better, then; you'd better leave,' but she was shattered afterwards when she realized what she had done.

There were also regular begging or stealing missions to the houses in Stanley. Les Harris describes a typical incident.

One morning I found that the ducks had gone. I automatically thought they had been stolen in the night, during the curfew. Children being children, Jane and Ralph said they were going up to have a look. Jane came racing back and said, 'Come quick, Dad. They're still there.' I went and found two very young, very wet Argentinian soldiers hiding behind the hen-house in some bushes. I speak Spanish and I gave them five minutes to get off our property. They said they were from the hills and that they were starving. I still gave them five minutes and told them I would fetch the military police.

I waited exactly five minutes and then went to the Argentinian marines who were in the Social Club. The man in charge asked me whether they were army or marines. I said army. He told his men to get their weapons and then set off to catch them. Half-way up the street, they cocked their weapons. They looked very efficient; their marines were always clean and smart. My wife, meanwhile, had found the two children watching over the soldiers; they weren't going to let them get away. My wife was very frightened.

The two soldiers had been round the back of the hen-house in the meantime and left the inevitable Argentinian 'visiting card' – they had relieved themselves. They did that wherever they went, specially in houses they broke into, and they seemed to do an awful lot of it, too. I think they were using a lot of bad meat and vegetables in their rations.

The marines arrested the two soldiers without any trouble and there were our five ducks – all dead – and some vegetables. We all went down to the gymnasium where the military police were based. An officer came out and said, 'What, you two! Not again!' and gave them quite a bollicking. He told me I would not be troubled by those particular two again. He also said he would keep the ducks as evidence and would return them tomorrow – but tomorrow never came.

The civilian leaders who emerged in Stanley had already organized several measures to protect and help the more vulnerable members of their community. One problem was the confinement by the curfew of people in houses of wooden or corrugated-iron construction which gave no protection against shelling; the curfew also isolated solitary people with medical problems. Michael Bleaney was chosen as civilian spokesman and, after a meeting at his house at which an Argentinian civilian administrator was present, the safe-house system was instituted. Six 'section leaders' identified the best stone or brick built houses in their areas and these were marked with a sign to identify them for 'when the British forces came'; the Argentinian insisted on 'if'. The houses were then further strengthened with sandbags and stocked up with mattresses, first-aid supplies and reserves of food. Only one householder refused to help. Now, with British naval shells falling around Stanley nearly every night, the safe-house system was in regular use. Mrs Nidge Buckett, an English housewife from Harrogate, describes life in the safe houses and morale at this time, with the British not far away.

Relations were mostly very amicable. The only few cases of dissent in the safe houses were over such things as failure to contribute food or friction between young people who wanted the video on late and older people who wanted to go to sleep. The feeling within the town, as far as people getting on with one another, was the best we have ever known. I am only a short-term islander but most of the people I spoke to agreed that the community spirit had never been better. There was a lot of good conversation in the safe houses, especially from the older people who had spent their early life in the Camp. There were lots and lots of sweaters knitted. There was always pre-meal drinks – hot rum toddies for most of the men, martini or sherry for the ladies. There were rosters for the preparation of meals, washing up and the evening brew. Most people dispersed after the early morning tea or coffee and went home for their mid-morning breakfast – an old Falkland custom.

We wanted the British to come in and push the Argies out of Stanley and there was no doubt in anyone's mind that it would soon happen; although they came in sooner than we thought. Most of the people were very worried and making a conscious effort to keep themselves out of danger. At the same time, there was this excitement that the British – our lads – were coming. One old lady, Mrs Blanche McAskill, had been making sausage rolls ready for them for several days, cooking them and putting them in the freezer.

Nidge Buckett's ex-Royal Engineer husband Ron worked at Stanley's Public Works Department and Nidge had begged a Mini from his department, painted large red crosses on it and, with Argentinian approval, had been systematically visiting Stanley's old people and carrying out other welfare work with a team of helpers. Nidge Buckett received a decoration for this work after the war.

The Argentinians had recently ordered people in certain areas on the outskirts of Stanley to leave their homes permanently, and the West Store next to the Cathedral became the central accommodation point for many of these people. The curfew was extended and now lasted from 4.0 p.m. to 8.30 a.m. Two International Red Cross observers arrived and negotiations took place between them, the Argentinians and the local people to identify a safe 'neutral zone' into which the entire civilian population could be concentrated and which both sides would respect in the event of a final battle. The area around the Cathedral and the West Store was suggested but the community representatives told the Red Cross that the civilians did not wish to be concentrated in one place where, if the Argentinians turned awkward, the entire civilian population could be used as hostages in negotiations. Despite this, the British Government was informed that the Cathedral neutral zone had been established; but it never was and many Falklanders are deeply resentful of the International Red Cross over this matter.

This failure of communications may have led to a tragedy. On the night of 11/12 June, four British ships were carrying out bombardments in support of ground attacks being made that night. One of the targets was an Argentinian heavy gun located not far from Ross Road West in Stanley. Number 7 was the 'safe house' of John and Veronica Fowler and three other families were congregated in it. There were ten people in the house, ranging from an eighty-year-old widow to baby David Fowler, who had been born since the Argentinian invasion. John Fowler describes what happened.

We tried to get to sleep after 11.30 p.m., some of us in beds, some on the floor, some in a makeshift air-raid shelter in a room made of tea chests full of peat, a sideboard stuffed with clothes and with a plank-and-mattress roof.

There were two shells. The first burst at the front of the house; it blew away a section of fence and all the windows and destroyed the porch. It was quiet then and, foolishly, we all thought that was 'our' shell and that we had had a very, very lucky near escape. There seemed to be a lull in our immediate vicinity and, as had been our habit when we had been woken up, we decided to have a cup of tea. The kitchen is at the east end of the house, which was the direction from which the shelling had been coming and, having got our cups of tea, I became increasingly nervous about remaining in that exposed end of the house and suggested, very firmly, that we should leave the kitchen and either go back to our own rooms or go to the room which I considered safer, which was where old Mary Goodwin was still in bed – though she was awake and we were taking her a cup of tea. While the general exodus from the kitchen was going on, I decided to look into the shelter in another room, where the children were asleep – they had slept right through it all so far.

My recollections now become a bit confused. There was certainly a very loud whistling of a shell – we had got used to shells whistling over but we knew this one was going to land fairly close – and it was followed by a tremendous explosion which, in turn, was followed by a great cloud of cordite, smoke and fragments of plaster and who knows what else. Somewhere in amongst all this, I felt I had been jabbed in the leg by a red-hot poker, a burning sensation, a short sharp pain which only lasted a moment. It did not seem serious to me; my main concern was the children.

Veronica, my wife, was shouting that the house was on fire, which was reassuring because it meant she was about and able to speak. That was followed by Steve Whitley shouting that it wasn't a fire, only the water pouring down from the roof tank. That also reassured me because it meant he was all right. Then my wife came and told me she thought that Doreen Bonner and Sue Whitley were both dead.

They had died at once. Veronica and Doreen Bonner had thrown themselves on the floor at the foot of Mary Goodwin's bed and thrown their arms around each other instinctively. After the explosion, Veronica found that Doreen was already dead. The doctors later found a lump of shrapnel in the back of her head. Sue Whitley had been standing with Mary Goodwin's cup of tea and had caught the full blast.

John Fowler, who became very distressed in the telling of this story, was later annoyed over press reports that eighty-year-old Mary Goodwin died in his arms (other reports said in Veronica Fowler's arms), but Mrs Goodwin actually died in hospital four days later.

Mary Goodwin, Susan Whitley (the local vet's wife) and Doreen Bonner are the official Falklands war dead but John Fowler considers that Doreen's husband Harry Bonner should be added to the list. This middle-aged man was the deputy director of the Public Works Department of Stanley. His chief was in England when the Argentinians invaded and Mr Bonner had since broken down under the pressure of the Argentinian occupation. He was in hospital on the night of the shelling. The combination of his illness, the news of his wife's death and the sense of guilt he felt at not being with his wife on that night all led to his own death on New Year's Eve that year.

Major-General Moore's earliest plan had been that the British assault on the Stanley defences should commence on the night of 8/9 June. This was set back by two nights because of the difficulties of establishing enough stores, particularly artillery ammunition, in the forward battle area. Then a further postponement of twenty-four hours had been forced by the disruption caused to helicopter schedules by the bombing of the *Sir Galahad*, although more helicopter reinforcements arrived on 9 June in the *Engadine*. The British troops who were established in forward positions thus had a further four days and nights of waiting. It does not sound a long time but it was a hard period for the men in those forward positions. There were four units in what might now be called the front line. 3 Commando Brigade had 3 Para, 45 Commando and 42 Commando in the Mounts Vernet, Kent and Challenger areas and 5 Brigade had the Scots Guards forward of Bluff Cove. The first three of these units were enduring particularly severe conditions in the biting cold, sleet, snow and winds on Vernet, Kent and Challenger. It was the debilitating effect of these conditions which was now spurring the British commanders to start the attack on Stanley's defences as soon as possible. The drain of casualties from exposure had not yet become serious, but there must be no further delays. Many of the Argentinian soldiers were enduring the same conditions, but in long-prepared positions and with Stanley at their backs. For the British soldiers, it was a long way back even to the modest comforts of San Carlos; Plymouth, Arbroath, Aldershot and Chelsea Barracks must have seemed like places on another planet. The experiences of Marine Nigel Rees, of 42 Commando, are typical.

It boiled down to personal survival. It was very cold; sometimes we were in the clouds. The wind was horrific, always whipping across the top of that mountain. We could not dig in; we only had makeshift bivvies. The main problem was staying dry. We tried desperately to keep our feet dry. The feet are your main thing; it doesn't matter what happens to the rest of you. We had Cairngorm boots which were very good but, when wet, they retained the water and became thick and heavy and got very cold. You would take your boots off, then the socks off, put the wet socks inside your shirt next to the body and try to dry them out while you were asleep. While you were asleep, you kept your feet dry in your sleeping bag – if that wasn't wet; if it was wet – tough. Then, in the morning, you put your spare socks on and your wet boots back on, and were ready for another day.

For rations we had to go back down the mountain to the helicopter landing zone and carry the rations up in boxes. It was only a kilometre's yomp but it developed into a right pain in the you-know-what. That kilometre took up to two hours to do over rock screes and steep ground. One party lost its way in the fog and took four hours.

Patrolling and reconnaissance took place every night during the waiting period and much valuable information was gained while probing the Argentinian defences. This was an activity at which the British units excelled. Many patrols were carried out successfully and without loss; others hit trouble. Mines were the main problem, often not discernible until a man stepped on one in the dark. Marine Mark Curtis describes a patrol carried out by 4 Troop of 42 Commando.

It was at the bottom of a little slope that I stepped on the mine; 'Cuth' and the other marine had walked over it. I seemed to be thrown up in the air and fell on my right side. I took the gun off my shoulder and pointed it forward, waiting for someone to fire at us; I still thought it was the ambush. My foot started to feel numb. I tried to feel down but my trousers were all torn round the bottom. The middle of my foot had been blown off; the toes were still there, connected to my shin by a fleshy bit of skin. It looked weird. Half an inch of my heel had been ripped back. That was all there was left – the toes and the back of the heel. 'Cuth' shouted, asking what was going on – a bit of heavy language. I told him I'd had my foot blown off, but I didn't put it quite like that. Everything was quiet then. He crawled over on his hands and knees, looking for mines. He tried to bandage my leg and I gave myself some morphine. You keep it on your dog tag – like a little toothpaste tube with a needle. I couldn't get the plastic cover off and had to bite it off. I injected myself in the muscle of the thigh. It didn't seem to have any effect for half an hour; the pain had started after five minutes. 'Cuth' picked me up and carried me out.

It took seven hours of carrying to get Marine Curtis back to the British positions and a further eighteen hours for him to reach the surgeons at Ajax Bay. The remains of his foot were amputated and he was still having further operations more than two years later.

The British units suffered several casualties – all Royal Marines – during the patrol and reconnaissance period. Corporal Jeremy Smith became 42 Commando's first fatal casualty when he died after being wounded by an Argentinian shell; he had been earmarked as potential for further promotion up to R.S.M. 45 Commando suffered a tragedy when one of its patrols was operating on the night of 10/11 June. Y Company sent a recce patrol out towards Two Sisters; a section of mortars was also sent out so that it could give fire support if the patrol met trouble. But the location of the two groups became confused, the mortar section probably being in the wrong position. They were spotted by the recce patrol whose commander then checked by radio that no other patrols were operating. He also spoke with the sergeant in command of the mortar group, who said he was in the correct position 'on high ground'. The recce patrol was looking *down* on the suspect group, and assumed it must be Argentinian. They opened fire, killing the mortar sergeant, two corporals and a marine, and wounding three other men.

Special forces parties also continued to be active in their secretive way. Some items are available from this period. S.B.S. men managed to get right into Stanley Harbour and were hiding in the hulk of an old wreck, the *Lady Elizabeth*, which was in Whalebone Cove. The S.B.S. men were able to observe the whole of Stanley waterfront two miles away, and were directly under the flight path of the nightly Argentinian transport aircraft coming into the airfield only a mile distant. Civilians in Stanley say that Menéndez was concerned that British special forces were after him and that he slept in a different place every night, but that may be a rumour.

There was, however, one bold attempt to strike not only at Menéndez but at a larger part of the Argentinian command. It had been observed, possibly by the S.B.S. men in the hulk in the harbour, that Argentinian officers congregated at the large building on Stanley waterfront which housed the Post Office and the Town Hall. A Wessex helicopter of 845 Squadron was ordered to attack that end of the building where the Argentinians met soon after dawn on 11 June, just a few hours before the British offensive on the Stanley defences started. A message was sent by roundabout means, advising the civilian Post Office staff not to go to work too early that morning, but the message did not get through. The Wessex was piloted by Lieutenant Peter Manley, with Petty Officer Arthur Balls firing the two AS12 wire-guided missiles. Here is Balls's account of this operation.

I found the missile sight had misted up; I don't think it had ever been used so early in the day and the cabin heating had misted it up. We tried various methods to clear it and actually landed near Mount Round, behind the enemy lines, and made contact with the S.A.S. We eventually cleared it sufficiently and took off again. We were intending to hover over some lakes north of Stanley, using the dark mountains behind us as a backdrop, and fire from there – about 5,500 metres range, 6,000 maximum. But we couldn't find the ponds; they were dried up or it was still too dark. We could see Stanley, but only some of the more prominent buildings with difficulty. I saw Government House and the Fuel Tanks west of town and the jetty. I aimed the first missile at the estimated position of the Town Hall. After the thirty seconds' flight time, I saw a puff of smoke and thought I had hit the target. There was no Argentinian response.

I fired the second missile but it only flew for five to ten seconds; there was a defect in the missile or the wire snapped and it fell well short. I made all the switches safe, got rid of 6,000 yards of wire, looked up and saw shells exploding around us. I realized we had been under fire for several seconds, with the pilot holding the helicopter in perfect steadiness for me. We got out fast, back to Teal, singing and shouting all the way.

The missile had missed the Town Hall by only a few feet, flew across the road and struck the corner of the roof of the Police Station which was occupied by Argentinian military police. The new chief of military police, Major Roberto Berazay, emerged with his coat ripped but there were no casualties.

The British special forces' task was nearly finished on East Falkland but parties were still active on West Falkland. Argentinians of the 5th Regiment discovered an S.A.S. observation post near Port Howard and surrounded it, trapping Captain Gavin Hamilton and his signaller. The two men attempted to fight their way out but Hamilton was injured. He sent his signaller on, remaining behind himself to give covering fire until he was killed. The Argentinians claimed to have taken a British sergeant prisoner in this action, but this was not admitted by the British. The gallant Captain Hamilton, a former Green Howards officer, was awarded a posthumous Military Cross.

One subject which cannot be spared much space but which was so important to the success of the campaign is the herculean efforts being made by the British supply services. The hills in front of Stanley were at the end of an extended supply line stretching right back past Ascension Island to Britain. Every element along that line can claim a share in the victory which the British forces were about to achieve – the logistic units at Teal Inlet, Fitzroy and San Carlos, the helicopter crews and the men operating all the other means of getting supplies forward, the supply ships, the long-range Hercules crews now regularly flying sixteen-hour sorties from Ascension to drop vital stores to the task force, the exhausted Victor tanker crews at Ascension, the R.A.F. transport squadrons operating the daily shuttle down to Ascension from Lyneham and

Brize Norton, the supply depots in Britain. I asked Commander Brian Goodson, Fleet Supply Officer at Northwood who had overall responsibility for Operation *Corporate*, what was the most serious shortage encountered by the task force. After a long pause for thought, he could not recall any serious shortage, though there was 'some anxiety over the range and depth of some of the avionic spares, particularly for the Harriers'. It is a measure of the efficiency of the British supply services and of the establishment of superiority at sea that, while the Argentinians around Stanley, only 400 miles from home, were often hungry and short of certain types of ammunition and had weapons which would not work for want of spares, the British forces in front of Stanley, 8,000 miles from home, never went short of the essentials – although, in another Falklands-type campaign, they would like to have better boots.

Finally, as we arrive on the eve of the Stanley battles, remember again the ships of Sandy Woodward's task group which were still supporting the land forces in a variety of ways. The crews of many of these ships had now been at sea for nearly three months and had been in the operational area under threat of various forms of attack for six weeks.

Major-General Moore – now established with a small forward headquarters at Fitzroy – had finalized his plans for the largest battle to be carried out by British land forces since 1945 and the largest operation commanded by a Royal Marines general since the invasion of Madagascar in 1942. Moore had at his disposal the equivalent of seven battalions of infantry: 42 and 45 Commandos, 2 and 3 Para, 1st Welsh Guards (their losses on the *Galahad* made up with two companies of 40 Commando), 2nd Scots Guards and 1/7th Gurkhas (less one company back at Goose Green). The infantry would be supported by thirty 105-mm guns from 7, 8 and 79 Batteries of 29 Commando Regiment and 29 and 97 Batteries of 4 Field Regiment. Eight light tanks of the Blues and Royals were also present, though the steep and rocky terrain of some of Moore's objectives would inhibit their use.

Military commanders have often been forced to choose between the broad and the narrow front advance. Jeremy Moore and his three brigadiers – Julian Thompson, Tony Wilson and John Waters (Moore's deputy) – had discussed this problem at length.

A narrow-front attack would thrust through the southern peaks to Stanley, bypassing the Argentinian positions to the north. This route would require the taking of Mount Harriet and Two Sisters by one of the brigades in a first bound and Tumbledown Mountain and Mount William by the other brigade in a second bound, leaving only the smaller and much weaker feature of Sapper Hill in front of Stanley. There was much merit in this policy. The British units would have been spared from two set-piece attacks, on Mount Longdon and on Wireless Ridge; the limited amount of artillery ammunition available could be concentrated on fewer objectives with a consequent greater effect and a further possible reduction in British casualties. But an insurmountable snag was encountered when the plan was thought through. The British artillery would need to be moved forward by helicopter from the gunline behind Mount Kent to a large bowl of open ground in front of Mount Kent before the second phase of the attack. The only reliable helicopter route for this lift – and for all the other supplies needed for the second phase of the battle – was round the north end of Mount Kent *and that route was overlooked by Mount Longdon*. The narrow-front option thus disappeared.

Moore's eventual plan was for two consecutive nights of attack starting on 11/12 June, with 3 Commando Brigade – whose patrolling was more advanced – attacking Mount Longdon, Two Sisters and Mount Harriet on the first night and with both brigades attacking Tumbledown Mountain, Mount William and Wireless Ridge on the second night. By then, it was hoped, the Argentinians would surrender; radio appeals for them to do so were already being broadcast regularly, though not yet bringing any response. If the Argentinians still fought on, Sapper Hill would be attacked on the third night and then Stanley would have lost all of its western defences. All of the attacks would be by night, when the British troops could make use of their superior tactical skills and avoid daylight movement over what one commander called 'that bare-arsed land'.

As the world now knows, Jeremy Moore's plan proved to be successful, but it is worth considering what might have happened if the Argentinian defence had proved stronger and if the British troops had been held up in the mountains. Moore was committing most of his strength to these attacks; his only reserves were half of 40 Commando at San Carlos and a company of Gurkhas at Goose

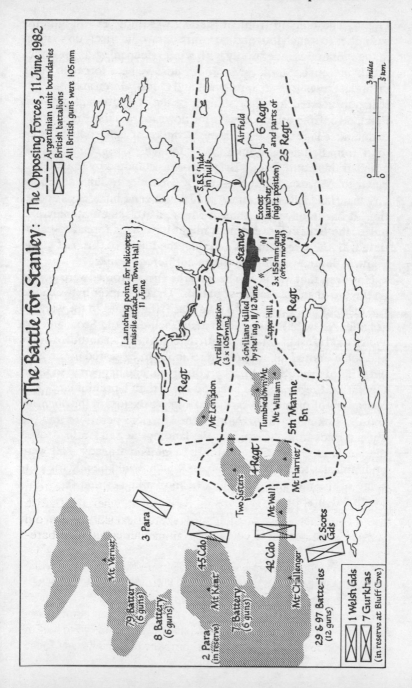

The Battle for Stanley: The Opposing Forces, 11 June 1982

- - - - Argentinian unit boundaries
British battalions
All British guns were 105mm

S.B.S. 'hide' in hulk

Airfield

6 Regt

and parts of 25 Regt

Exocet launcher (night position)

Stanley

3 x 155mm guns (often moved)

3 civilians killed by shelling, 11/12 June

Sapper Hill

Launching point for helicopter missile attack on Town Hall, 11 June

Artillery position (3 x 105mm)

7 Regt

Mt Longdon

3 Regt

Tumbledown Mt
Mt William

5th Marine Bn

4 Regt

Two Sisters

Mt Harriet

Mt Vernet

3 Para

79 Battery (6 guns)

8 Battery (6 guns)

45 Cdo

Mt Kent

2 Para (in reserve)

7 Battery (6 guns)

Mt Wall

42 Cdo

2 Scots Gds

Mt Challenger

29 & 97 Batteries (12 guns)

1 Welsh Gds

7 Gurkhas

(in reserve at Bluff Cove)

3 miles
5 km.

Green. The units in front of Stanley were not yet losing many casualties to trench foot and exposure but would surely do so if left in the mountains indefinitely with winter deepening. If his forces 'stuck' in front of Stanley, Moore would have been forced to bring up his last reserve from San Carlos and Goose Green and leave his rear unprotected from the danger of the Argentinians on West Falkland or from other forces being flown in from the Argentinian mainland to land at Goose Green. No more British troops were *en route* from Britain, although the 1st Queen's Own Highlanders – a well-trained unit from 1 Brigade – was standing by to be flown down to Ascension and then on by Hercules to land at Goose Green airfield, although that would have been a hazardous operation. It has to be stated that, if Moore's attacks on Stanley had failed, the British Government might have been forced to negotiate with the Argentinians and then the future of this part of the South Atlantic would have taken a different course.

However, that was not to be, and the British forces were poised on the eve of an epic series of battles. The encounter between the two main forces was about to take place; the climax of the war was at hand. A few battalions of British (including the Scots and the Welsh) and Gurkha infantry, with their artillery, a handful of tanks and some limited air support were about to be tested in the age-old skills of the battlefield. Everything that had happened so far in Operation *Corporate* had been no more than a preliminary – the operations of Admiral Woodward's ships, the loss of the *Sheffield* and the sinking of the *Belgrano*, the Harrier operations and the Vulcan raids from Ascension, the landings at San Carlos – even the battle at Goose Green and the *Galahad* tragedy – all were preliminaries to the coming battles which would decide the outcome of the war and the future of the Falklands and the whole South Atlantic area.

CHAPTER TWENTY-ONE

The Battle for Stanley

The detailed planning of the first night's attacks was carried out by Brigadier Julian Thompson and his 3 Commando Brigade Headquarters. Each of Thompson's units was allocated the Argentinian position directly facing it. Running from north to south, 3 Para's objective was Mount Longdon, 45 Commando's Two Sisters and 42 Commando's Mount Harriet. There was still some degree of optimism about the poor fighting qualities of the Argentinians and all units were ordered to be prepared to press on to the next high ground if their main objective was achieved quickly. These exploitation objectives were, again running from north to south, Wireless Ridge, Tumbledown Mountain and Goat Ridge. In retrospect, the exploitation aspect of the plan appears to be unrealistic and it was really only a contingency in case there was a wholesale rout of the Argentinians on the first assault. To balance this, some lessons in prudence had been learnt from the earlier action at Goose Green. No battalion was given an objective anything like as large as that given to 2 Para at Goose Green; much more patrolling and reconnaissance was carried out; plans were made for any attack which met trouble to 'go firm' before dawn and avoid the daylight extension of the battle in which 2 Para had suffered so many casualties. The five batteries of British guns had 11,095 shells available for this first night's battle. Admiral Woodward was providing four warships to give additional support: *Avenger* was allocated to 3 Para, *Glamorgan* to 45 Commando, *Yarmouth* to 42 Commando and *Arrow* for special forces' work; the combined ammunition stock of these ships was approximately 1,400 rounds. All of the attacks would be 'silent'; that is, the artillery would not open fire until the infantry advanced to contact and called for support.

The coming night's battles would be fought over two sheep-farming properties. The two Royal Marine objectives – Two

Sisters and Mount Harriet – formed part of one of the Falkland Island Company's smaller sections, the Port Harriet–Two Sisters property. Mount Longdon was part of a large Falkland Islands government-owned property which was leased to a local sheep-farmer, Mr Richard Hills; the later objectives – Tumbledown Mountain, Mount William, Wireless Ridge and Sapper Hill all belonged to the same property.

3 Para's attack started first. Lieutenant-Colonel Hew Pike had planned a standard assault. He decided that Mount Longdon could not easily be approached from a flank and the feature was to be attacked frontally at its western end. A Company on the left would attempt to seize a feature to be known as WING

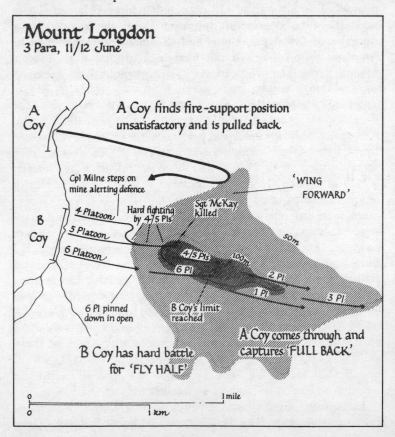

Mount Longdon
3 Para, 11/12 June

A Coy

A Coy finds fire-support position unsatisfactory and is pulled back.

Cpl Milne steps on mine alerting defence

'WING FORWARD'

Sgt McKay killed

B Coy

4 Platoon

Hard fighting by 4/5 Pls

5 Platoon

6 Platoon

4/5 Pls

50m

100m

6 Pl

2 Pl

1 Pl

3 Pl

6 Pl pinned down in open

B Coy's limit reached

B Coy has hard battle for 'FLY HALF'

A Coy comes through and captures 'FULL BACK'

0 ————————— 1 mile
0 ————————— 1 km

FORWARD which would then act as a fire-support base for an attack by B Company on the main Longdon feature, whose two parts were designated FLY HALF and FULL BACK. C Company would move up as a reserve and as the exploitation company if there was a swift Argentinian collapse. The Argentinian defenders of Longdon had been correctly identified as belonging to 7 Regiment, a conscript unit recruited in the province of La Plata. But 7 Regiment was holding Wireless Ridge as well as Longdon and there were probably only about one and a half companies reinforced with specialist snipers and commandos on Longdon itself.

The paras moved off after dusk for the approach march to the Start Line. Lieutenant-Colonel Pike and his Tactical Headquarters were given a lift by civilian Land-Rovers, one of which was driven by a woman, Trudi Morrison, perhaps the first time a battalion commander had been driven into action by a farmer's wife. The British artillery were firing no more than the normal evening harassment. A and B Companies reached their Start Lines on a little stream about fifteen minutes late. Corporal Ian Bailey was a section commander in B Company.

We were only on the Start Line a few minutes. I went round the lads and checked everybody and had a joke with my mates. The lads were quiet, each man whispering to their own very good friends, having a last drag. It was a time for being with your own mates. They knew some of them were going to get killed. For some reason, most of them fixed bayonets; I put mine on and looked round to find all the others were putting theirs on too.

We stepped over the stream and set off. It was a very clear night, cool, but it didn't feel cold; there was too much adrenalin flowing. We knew we had got a punch-up on our hands. It was uphill, a fairly steep gradient, lots of rocks, tufts of grass, holes where you could break your ankle easily – just like a good training area. You could see 200 or 300 metres ahead of you. As we went up, we were funnelled together into a space between the main Mount Longdon and a large separate rock. At one time, we were shoulder to shoulder, so we tried to spread out, some men waiting while others moved on faster. That was only about twenty feet from their main position but that first position turned out to be empty.

The silent approach ended when Corporal Milne, a section commander in B Company's left forward platoon, stepped on a mine and severely injured his leg. The explosion, and Corporal Milne's scream, alerted the Argentinians and the battle began.

It is not going to be easy to do credit to every sub-unit in every battalion action. It will help on this occasion if the action of A Company, attempting to take the subsidiary position at WING FORWARD and establish a fire base there, can be disposed of quickly. WING FORWARD turned out to be a poor position. It was not occupied by the Argentinians but it was overlooked by the main Longdon heights and Major David Collett's company was fired upon by small arms and mortars as soon as they advanced on to it. The paras then found that the position was almost useless as a fire base, because of its inferior height. One man was killed and two were injured, one of whom died later. Colonel Pike ordered Collett to withdraw from the feature an hour and a half later and A Company was brought round into the main area of action where, in the interval, Major Mike Argue's B Company had been involved in fierce fighting.

That fight in the rocks on the western end of the mountain proved to be the core of the Longdon battle. The early mine explosion drew fire before B Company reached the main Argentinian positions and the paras found themselves squeezed into a series of steep and narrow rock channels on the face of the hill, called by one officer 'bowling alleys', down which the Argentinians could fire their weapons and toss grenades. Some British artillery support was possible but could not be brought close enough to engage the Argentinians firing on the paras.

The three platoons were faced by well-constructed and stoutly manned defences, including at least two heavy 0·5-inch machine-guns, many other automatic weapons and snipers, the many 'signatures' of whose infra-red night sights could be detected. Corporal Ian Bailey, leading a section in 5 Platoon, resumes the story.

Then we got the first grenades; they were just bouncing down the side of the rock face. We thought they were rocks falling, until the first one exploded. One bloke caught some shrapnel in the backside. He was the first one in the section to get injured, not badly but enough to put him out of the fighting. The small-arms fire followed soon after. People were getting down into cover again then. Because we had got funnelled, we weren't really working by sections now; the nearest private soldier to you just stuck with you. Corporal McLaughlin, the other leading section commander, was ahead of me now. He and his men were getting the small-arms fire. It was keeping his group pinned down but no one was being hit in either of our sections. There was a lot of fire on the right, where 6 Platoon was going up but it was quiet on the left where 4 Platoon was coming up after hitting the minefield.

My men started firing their '66s' [hand-held 66-mm anti-tank rockets]. Whoever was in the best position to spot targets fired; the others passed spare rockets to them. It was a very good bunker weapon; there wasn't going to be a lot left of you if your bunker or sangar was hit by one of those. We could see their positions by now, up above us, possibly thirty feet or so away, we could even see them moving, dark shapes. Their fire was sparse to start with but then it intensified. Some of them were very disciplined, firing, moving back into cover, then coming out again and firing again or throwing grenades.

The next cover to get forward to was in some rocks with one of their positions in the middle of it. Corporal McLaughlin's GPMG gave us cover and we put about four grenades and an '84' round into the position, which was a trench with a stone wall around it and a tent, which was blown over. We went round and on, myself and whichever 'toms' [private soldiers in the Parachute Regiment] were available, and the two men with the '84' launcher. It was all over very quickly. We ran across, firing at the same time. Just as we went round the corner, we found one Argentinian just a few feet away. Private Meredith and I both fired with our rifles and killed him. The rest of the post – two men – were already dead, killed by the grenades or the '84' shrapnel. We put more rounds into them, to make sure they were dead and weren't going anywhere; that was normal practice.

Meanwhile, on the left, Lieutenant Bickerdyke's 4 Platoon had become pinned down by one of the heavy machine-guns and was largely out of contact with the company commander. They were released from this position by Lance-Corporal Lennie Carver and Privates Gough and Grey of 5 Platoon who put the machine-gun out of action. The advance route on the left was still dominated by fire so Bickerdyke edged 4 Platoon to the right and two of his sections became intermingled with 5 Platoon. The progress of these men was now blocked by a second heavy machine-gun which caused some casualties. Lieutenant Bickerdyke was hit in the thigh, the impact of the bullet being so strong that he was thrown over backwards. Lieutenant Cox, of 5 Platoon, was pinned down in the same area, so that it was left to the N.C.O.s of 4 and 5 Platoons to carry on the fight. When he was hit, Lieutenant Bickerdyke called out to his platoon sergeant, 'Sergeant McKay; it's your platoon now.'

The next phase of the action was thus led by Sergeant Ian McKay, with some of his own men of 4 Platoon and with Corporal Bailey's section from 5 Platoon. Corporal Bailey again.

Ian and I had a talk and decided the aim was to get across to the next cover, which was thirty to thirty-five metres away. There were some Argentinian positions there but we didn't know the exact location. He shouted out to the other corporals to give covering fire, three machine-guns altogether, then we – Sergeant McKay, myself and three private soldiers to the left of us – set off. As we were moving across the open ground, two of the privates were killed by rifle or machine-gun fire almost at once; the other private got across and into cover. We grenaded the first position and went past it without stopping, just firing into it, and that's when I got shot from one of the other positions which was about ten feet away. I think it was a rifle. I got hit in the hip and went down. Sergeant McKay was still going on to the next position but there was no one else with him. The last I saw of him, he was just going on, running towards the remaining positions in that group.

I was lying on my back and I listened to men calling to each other. They were trying to find out what was happening but, when they called out to Sergeant McKay, there was no reply. I got shot again soon after that, by bullets in the neck and the hand.

Sergeant Ian McKay had carried on alone to attack the machine-gun position with grenades; he put it out of action but was killed in the process. He was awarded a posthumous Victoria Cross. Corporal Bailey, who was typical of some excellent junior N.C.O.s I interviewed in several units, received the Military Medal.

Sergeant McKay's action had enabled the remnants of 4 and 5 Platoons to get into a good position astride the main ridge from which they could bring fire to bear on further Argentinian positions. The company commander then ordered the leading troops to withdraw slightly, while artillery fire was called down on these. The action continued and steady progress was made. Lieutenant Cox came up with the forward artillery and naval-gunfire observers and some reorganization took place. Then a voice came out of the dark, '*Hey, amigo!*', and a fresh Argentinian position opened fire. One man was killed and Corporal

Graham Heaton was injured before the position was silenced. This is Heaton's description of being hit.

We were pulled back by groups; I had come back with five men and Corporal McLaughlin was covering me with his group. I got back up on to the hill, got my section into position and leaned forward to shout to McLaughlin to follow me up. But I never had a chance; the old slow motion began. Everything goes into slow motion when you are hit. I saw the tracer coming towards me from my right. It seemed like five minutes before it hit me, but that was the slow motion. I was thinking about moving to avoid it but I didn't have time. Then the rounds hit me and knocked me on to my face. It felt just like someone sticking a red-hot poker into my leg. I lay on my face and tried to call to the other blokes but it had all gone cold then and I could only move my left leg.

There was no pain; the right leg was really cold, just as though it had been stuck in ice for weeks. I thought I was going to die because of what I had heard from the others, that the Argentinians liked to injure one man and then kill the one that went to help him. I remember two blokes pulling me, dragging me up the hill, and one of them saying, 'Get his webbing off; he's too heavy.' I managed to undo the webbing and one of the lads pulled it off and got me away all right without being fired upon.

(I was most impressed with Graham Heaton's attitude. His foot was amputated after the battle but then gangrene set in and the knee had to go as well. I met him at the Parachute Regiment Depot prior to release, facing, as he put it, 'competition with three and a half million unemployed, who all had two legs'.)

While 4 and 5 Platoons had been fighting so hard, an unlucky 6 Platoon had been pinned down in the open on the right, mostly by the fire of Argentinian snipers with night sights who had been bypassed by the main attack. The 6 Platoon men had been able to remain in radio touch with the other platoons and had been able to give a little covering fire to the main attack but had not been able to get forward themselves. Five men in the platoon were killed, mostly picked off while trying to help the wounded, and many others were badly injured and forced to remain in the open, untended, for several hours.

But the B Company attack was now over and the first half of the ridge – FLY HALF – was secure. The success was mainly due to Sergeant Ian McKay and some very good section commanders. Night fighting in rocks must rely heavily on such junior leaders; they suffered heavy casualties. Ten paras died in B Company's action and, with the wounded, the company lost more than half of its strength.

The first half of the ridge now being taken, A Company came through to take the second part, FULL BACK. This next objective was first well shelled and Captain Adrian Freer, A Company's second-in-command, set up a fire-support base of machine-guns on an outcrop of rocks covering the next bound. 1 and 2 Platoons, commanded by Second Lieutenants John Kearton and Ian Moore, were then infiltrated through a narrow gap in the rocks to deploy silently before continuing the attack along a much flatter and more open terrain than B Company's old battleground. Ian Moore, an Australian, describes the action of his platoon.

We took all our webbing off; we needed to crawl, and British equipment is too bulky to crawl among rocks with. We only kept ammunition, grenades and as much ammunition for the machine-guns as possible. B Company gave us all their belt ammunition, grenades and dressings. Only the platoon signaller had any bulky equipment. The tracer from the two Argentinian machine-gun arcs did not meet in the gap but twenty metres back, so we could just get through. Eventually, the whole platoon was through and started moving forward on a thirty-metre frontage. Communications were by shouting.

The system worked well. A man called out, then lit up a cigarette behind a rock to signify his location and called for fire on a bearing from him. Then he gave a correction, left or right, up or down, almost like mortar-fire corrections. As soon as an enemy position was spotted, all available fire was brought down in this way. The machine-guns did a very good job, often tearing the walls of the sangars apart with 200 or 300 rounds of fire. The Argentinians had made the mistake of filling in the walls with too much turf. When a section was ready to go on with grenades, we called back to the fire-support group to stop; then the men went in in pairs, one covering the other.

After about an hour of this, we had removed most of the first group of positions – about a hundred metres in depth, about eight positions destroyed in all. My platoon only had one man wounded during this time – hurt by his own grenade. He threw it, then looked up to see what effect it had and got himself wounded in the shoulder.

After that first hundred metres, taken by first-class infantry tactics at little cost, the Argentinian defence steadily weakened and it required only a further three quarters of an hour to capture the last 800 metres or so of the ridge. Not a single para was killed in the action on FULL BACK. Many Argentinians surrendered; others ran off in the gloom, either back to Wireless Ridge or down the side of Longdon from where they came in next day to surrender or were shot or mortared as they tried to escape.

3 Para had won a clear victory in a difficult battle but at heavy cost. The Argentinians had fought well, most of the opposition probably coming from the regular N.C.O.s of 7 Regiment and the commandos and snipers sent in as 'stiffeners'. 3 Para lost eighteen men taking Mount Longdon. It was the costliest battle of the war for the British. The senior-ranking casualty was Sergeant Ian McKay, V.C.; two of the dead – Private Ian Scrivens and Jason Burt – were only seventeen years old. At least thirty-five paras were wounded. 3 Para claimed more than fifty Argentinians dead and about fifty prisoners.

The commanders of 42 and 45 Commandos, attacking further south, had asked for a later Zero Hour to that of 3 Para so that their final fighting and consolidation could take place in the half light before dawn. We will deal next with the attack of Lieutenant-Colonel Andrew Whitehead's 45 Commando on Two Sisters, two miles south-west of Mount Longdon and in the centre of the 3 Brigade's front. Zero Hour was intended to be 11.0 p.m. local time (02.00 Zulu). Colonel Whitehead's plan contained some similarities to that of the 3 Para attack on Mount Longdon, the seizure of one peak – the south-western one – to provide a fire base and then an assault on the main peak. The topography of Two Sisters gave 45 Commando an advantage over the 3 Para attack; the 'subsidiary' peak was actually higher than the main bulk of the objective and it should have made a very good fire-support base. It was believed that only one strong company of Argentinians was defending the whole objective, with a standing night patrol forward on the south-western peak. The Royal Marines were also fortunate in that Two Sisters had not been included in the original ring of defences around Stanley when the Argentinians had believed that a British attack would come from a landing on the south coast. When the Argentinians had been forced to swing their attention to the overland attack from the west, the Argentinian unit in the Mount Harriet and Mount Wall sector had been forced to take over Two Sisters at a late date. The defences here had thus been more hastily prepared and there were hardly any minefields. As on Mount Longdon, Two Sisters was defended by an Argentinian company supported by mortars, heavy machine-guns and snipers. The company was from the 4th Regiment, a mainly conscript unit from Corrientes Province. The local commander was Major Ricardo Cordon.

45 Commando's attack was delayed by one and a half hours because Captain Ian Gardiner's X Company, making the subsidiary attack on the south-western peak, experienced serious difficulties on the approach march. The company was carrying Milan anti-tank launchers, superb support weapons, but they and their rounds of 30-lb ammunition were very heavy loads to carry forward over the tussocky and rocky terrain in the dark. It

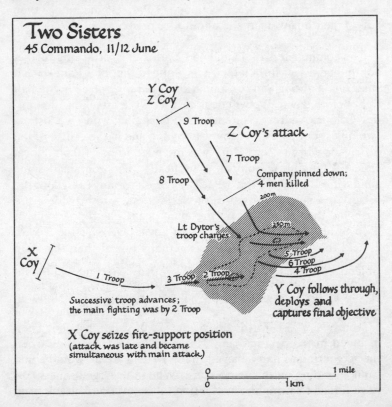

Two Sisters
45 Commando, 11/12 June

Y Coy
Z Coy

9 Troop

Z Coy's attack

7 Troop

8 Troop

Company pinned down;
4 men killed

200m

150m

Lt Dytor's
troop charges

5 Troop
6 Troop
4 Troop

X
Coy

1 Troop   3 Troop 2 Troop

Y Coy follows through,
deploys and
captures final objective

Successive troop advances;
the main fighting was by 2 Troop

X Coy seizes fire-support position
(attack was late and became
simultaneous with main attack)

0   1 mile
0   1 km

took X Company six hours, instead of the planned three, to reach their Start Line. Captain Gardiner writes:

We were at least two hours late. We were 150 very fed-up and tired men – and that was before the real work began. It is miraculous where the reserves of energy can come from. I explained briefly to the Colonel on the radio what the position was. To his everlasting credit, he put me under no unreasonable pressure and said simply, 'Carry on as planned; I will do nothing until I hear from you.' As a result of his patience and understanding, I was able to turn round to my troop commanders and say, 'Put the last six hours right behind you, make your final preparations in your own time, and when you are completely ready, let me know and we will go.' Ten minutes later, 150 men were as good as new and the assault began.

Half of this subsidiary feature was occupied before any opposition was met. Mortars, artillery and the Milans then all came into action, although the artillery, needed elsewhere, soon stopped and the mortars too, their base plates sunk into soft ground. Lieutenant Chris Caroe's 2 Troop, supported by the rest of the company, then assaulted and cleared the peak. Only one man was injured, by an Argentinian artillery shell.

X Company's delay had forced Y and Z Companies to wait for a long time on their adjacent Start Lines well to the north. Because of the delay to X Company, Colonel Whitehead ordered these two companies to start their attack at the same time as X Company. The result of this was that the main attack did not have the benefit of the fire support base which should have been established on the south-western peak, in exactly the same way as 3 Para's attack on Longdon had been denied its close fire support. Z Company moved off on the left, intending to assault the first of two peaks in the main position; Y Company would then hook left, round the back of this peak and attack the final objective. The leading troops of Z Company – Second Lieutenant Paul Mansell's 7 Troop and Lieutenant Clive Dytor's 8 Troop – then advanced to within 400 yards of the top of their objective without encountering any mines or any Argentinians, but then one of Dytor's section commanders spotted the Argentinians. This is Clive Dytor's account.

I moved over to him in the dark. He could see people moving on the ridge. I didn't believe him but I borrowed one of his riflemen's night sight and, yes, I could see them. I saw one man talking to another head – it turned out later that it was the head of one of their Browning 0·5-inch machine-gunners. From the section on my left, I could hear men saying they could see them as well. The first thing was to stop the marines firing. I could hear them saying, 'I can see him, I can see him. Let's kill him.'

I talked to the company commander on the radio. There was a bit of an argument then. The company commander must have reported to the C.O. and then I heard an order for me, 'From Nine. Move forward.' 'Nine' was the C.O. I said, 'No,' because I knew they would see us as soon as we stood up and we would lose a lot of blokes. The conversation went on for some time. I wasn't speaking directly to the C.O.; we were not on the same net. Then, something happened which started the whole fire-fight off. Something like a little flare came whizzing over the top of their positions, not a proper flare, just a little light. I immediately told my two forward sections to engage, which they did very willingly. Everyone opened up; it was the signal for the whole two companies to open fire.

Marine Steve Oyitch, in the rear troop, 9 Troop, also saw the flare and took part in the ensuing fire-fight.

It had been quiet and I was thinking, 'Thank Christ; there's nobody there; they've all buggered off again', when this Argie up on the top stepped out, lit a flare, held it above him, waving it around, and then threw it down and he disappeared. I could see him quite plainly. The next moment, a whole wall of tracer came down on us.

The fire-fight went on for an hour. We were laughing and joking about it; we were in dead ground and their fire was passing just above our heads the whole time. I was behind a rock and I kept pushing my rifle round it and getting a few rounds off. I could see their trenches and positions through my night sight; they had them as well, much better than ours, but their cover was good and I don't think our fire was doing much good either.

Then, after about an hour, their artillery started coming in on to us. Our artillery started about the same time; they were soon smack on the Argentinian positions. Their artillery was getting close and 7 Troop, the left-hand point troop, started getting casualties. Then fire moved down on to my troop; mortar fire. You couldn't hear them coming but you could see the red glow of the fuze coming through the darkness – and that's when I started to get frightened. Our section was in a circle – about ten yards wide – and one bomb landed right in the middle. It lifted me right up and threw me straight back on the deck. The man in front of me was wounded in the arm and chest and Corporal Burdett was badly wounded in the foot. That's when Gordon Macpherson was killed; he was just lifted and thrown into the rocks and must have died instantly. That just about put an end to my section's usefulness in the battle. Wounded men from other troops were brought back to us and we, in effect, became the casualty post.

Oyitch's section commander, Corporal Julian Burdett, was awarded the Distinguished Conduct Medal for carrying on though badly hurt, and for helping the other wounded and transmitting artillery fire corrections on his radio.

The attacking troops were in that classic infantry dilemma again, pinned down by defensive fire but needing to move. This time it was the young Welshman, Lieutenant Clive Dytor, who broke the deadlock. This is his account. ('Pepperpotting' is a tactical move in which individual men, or small groups of men, take it in turns to dash forward or stop and fire while others move, all in short bounds.)

It came to a point where I realized it was a stalemate and I actually remembered, at that point, a piece from a book I had read once – in a book called *The Sharp End* – a bit about the Black Watch in the Second World War. They were pinned down; they had gone to ground and wouldn't get up. I think the chapter was about leadership. The adjutant had got up and waved his stick and said, 'Is this the Black Watch?' and been killed immediately, but the whole unit had got on then, surged forward. I remember thinking about that and then, before I knew it I suppose, I was up and running forward in the gap between my two forward sections. I shouted, 'Forward everybody!' I was shouting 'Zulu! Zulu! Zulu!' – for Z Company. I talked to my blokes afterwards; they were amazed. One of them told me that he had shouted out to me, 'Get your fucking head down, you stupid bastard!' I ran on, firing my rifle one-handed from my hip and I heard, behind me, my troop getting up and coming forward, also firing. The voice I remember most clearly was that of Corporal Hunt, who later got a Military Medal. I think what happened was that Corporal Hunt was the first man to follow me, his section followed him, the other sections followed, and the troop sergeant came up at the rear, kicking everybody's arse.

So 4, 5 and 6 Sections came up abreast, pepperpotting properly. I could hear the section commanders calling, 'Section up, section down.' It worked fantastically; it was all done by the three section commanders and the troop sergeant at the rear shouting to keep everybody on the move and the hare-brained troop commander out at the front.

That assault up that hill was the greatest thrill of my life. Even today,

I think of it as a divine miracle that we went up, 400 metres I think it was, and never had a bad casualty. Only one man was hurt in the troop, with grenade splinters from a grenade thrown by a man in his own section. When we had been waiting on the Start Line, I had prayed that the Lord would give me the strength and courage to lead my men and do with me what you will and He did just that.

Marine Oyitch, left behind with the casualty group, heard his comrades charging.

We could hear them calling out, 'Commandos, Royal Marine Commandos!'; that was to let the Argies know who was going to go in and kill them. If they chose to mix with the best in the world, they were going to get burned.

The rest of Z Company joined in what might be called 'Dytor's Charge' and the company's objective was taken. Clive Dytor was awarded the Military Cross. Despite the mass of rifle and machine-gun fire poured down by the Argentinians, not one marine had been wounded by Argentinian bullets in either the big fire-fight or the recent assault. Colonel Whitehead came up and there was much patting on backs before Y Company came round and started the final assault. The company's three troops deployed in a line and then swung in an arc to assault the southern side of the final feature. The main Argentinian resistance had already been broken. Y Company suffered no casualties in their attack and captured their entire objective, right on to the Argentinian mortar positions at the eastern end of the ridge. The fight for Two Sisters was over.

45 Commando had taken one of the Argentinians' main defensive positions with the loss of only four men killed and ten injured (one of whom was an artillery forward observation officer). The four fatal casualties all died because of shelling or mortaring when Y and Z Companies were pinned down. Y Company's only dead marine was Michael Nowak, known to his friends as 'Blue'. Y Company's flag now contains a small blue square in memory of the company's only death in the Falklands campaign. Forty-four Argentinians were taken prisoner and 'about ten' bodies were buried (nine were recovered after the war). The rest of the Argentinian garrison fled. There had been particularly good cooperation with the artillery. 8 Battery of 29 Commando Regiment had provided most of the support and the

battery command post was delighted at the full use made of their services; approximately 1,500 rounds had been fired.

The final land action of the night was the attack by 42 Commando on Mount Harriet. It was thought by the British commanders that this would prove to be the most difficult objective of the night because of the combination of the length of the approach march required, the nature of the ground both during the final approach and on the objective, and the presence of minefields; a post-war British mine-clearance map shows nine minefields to the west and south of the mountain. Mount Harriet had always been part of the main Argentinian defences covering Stanley from a local landing on the coast and there had been plenty of time to lay these minefields and to construct defensive positions. The main Argentinian unit on the mountain was C Company of 4 Regiment but there were probably elements of several other units present.

Thanks to some particularly skilful patrolling, Lieutenant-Colonel Nick Vaux knew the location of most of the minefields and also the fact that the strongest Argentinian defences were on the western end of his unit's objective. He was determined not to make a direct assault on these defences but to attack from a flank. A patrol had discovered a mine-free route which could be used right round the southern flank of the mountain. Vaux decided not only to use this but took the bold step of sending the two assaulting companies as far as the south-eastern side of the objective in order to attack almost from the Argentinian rear.

This audacious plan required K and L Companies to make a four-mile approach march to their Start Line over open ground, the two companies threading their way through minefields and hoping not to be detected and lose the element of surprise. This part of the operation went smoothly. Meanwhile, artillery fire was falling on the western defences of Mount Harriet and captured Argentinian weapons were being fired as though Argentinian outposts were in action against a British force attacking from the west. To reach their Start Line, the two Royal Marine companies had to cross the boundary into the area held by 5 Brigade. It had been decided that the Recce Platoon of the Welsh Guards would secure the Start Line area and guide the marines in the final approach to it. But the Welsh Guards could

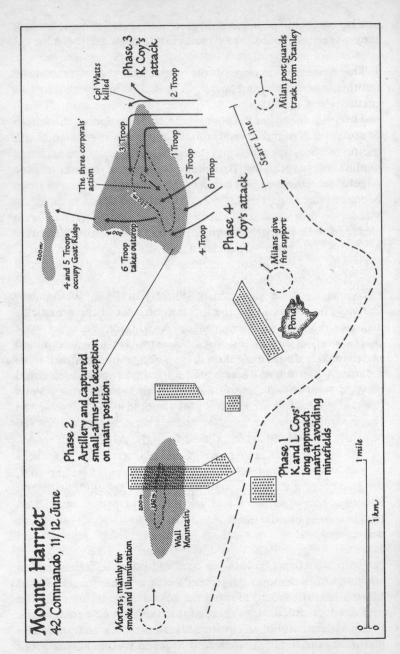

Mount Harriet
42 Commando, 11/12 June

Mortars; mainly for smoke and illumination

Wall Mountain
200 m
150 m

Phase 1
K and L Coys'
long approach
march avoiding
minefields

Phase 2
Artillery and captured
small-arms-fire deception
on main position

Pond

Milans give
fire support

Phase 4
L Coy's attack

Start Line

Milan post guards
track from Stanley

4 Troop

6 Troop takes outcrop

6 Troop

5 Troop

1 Troop

The three corporals' action

3 Troop

150 m

200 m

4 and 5 Troops
occupy Goat Ridge
200 m

Cpl Watts killed

Phase 3
K Coy's attack

2 Troop

1 mile

1 km

0

not be found at the appointed time and an hour was lost until they were discovered, not deployed tactically and keeping the area secure as agreed, but sitting nonchalantly against a fence, talking and smoking, 'totally unprofessionally' says one marine; another marine wanted to shoot them for compromising the security of the Start Line.

The plan called for Captain Peter Babbington's K Company to spread out, with 2 and 3 Troops leading on a frontage of 150 metres and making a silent advance up the mountain until contact was made. This move started forty minutes late. Marine Nigel Rees describes that advance up the south-eastern slope of Mount Harriet.

I was surprised that it all went so smoothly; everyone seemed to know his job. On exercises, there was always a hiccup of some sort or another; someone always falls over or coughs on an exercise but, on this occasion, there was complete and total silence. I think it was the adrenalin; everyone's hearts were going ten to the dozen.

This is where the eerie bit comes in, the bit I remember most about the attack when I think about it now. We knew there were enemy there who were going to try and kill us, but we couldn't quite believe that they were actually there, because they hadn't opened fire. It was perfectly clear – everything silent. We could see into the crags in the mountain. Why couldn't they see a bloody great long line of blokes yomping towards them – albeit quietly? It lasted about fifteen minutes from the Start Line before we met them. We walked past two or three of their bivvies – a big pile of kit next to them. They were empty, but the chaps in 2 Troop on our right said they got within five metres of bivouacs with Argentinians in them – and they still didn't know we were there. We were really behind their lines, taking them from the rear.

I heard one single shot ring out and heard one of our corporals shout out, 'Who fired that shot?', at the top of his voice and then the shit hit the fan and the bloody 'spics' realized what was happening and rounds went everywhere.

The British firing had started on the radioed orders of Captain Babbington. As soon as this happened, Babbington turned to the artillery observer alongside him and called for shell-fire on the left-hand (western) side of the company objective, while his right-hand troop – 2 Troop – cleared the eastern end. The guns had been steadily tracking a hundred metres ahead of the advancing marines and accurate fire fell with a crash on the Argentinians. 2 Troop found itself in an area of broken rock and gullies and cleared four mortar posts and other Argentinian positions there. Some of the defenders were huddled in their sleeping bags but others fought hard. There was a lot of firing and shouting, 'just like a football ground during a particularly wild match'. The task took nearly an hour – about twenty minutes of 'killing time' followed by more than half an hour of sorting out the prisoners and 'packing them off in big gaggles to the sergeant-major'. Six Argentinians were killed and more than twenty captured. One marine died. This was Corporal Larry Watts, a section commander who was clearing an Argentinian bivouac whose three occupants were armed with rifles. Watts was so close that he was actually pushing a rifle away with his hand when he was shot in the neck. He fell back, bleeding badly, but returned to the attack, firing his rifle, only to be shot again, the bullet passing through his radio set and then through his heart. The three Argentinians surrendered soon afterwards.

Nos. 1 and 3 Troops had meanwhile started to work their way westwards through the rocks along the other half of K Company's objective. There was a lot of noise and much firing. The Argentinians had good flash eliminators on their Belgian FN rifles but their machine-gun positions were revealed because they used too much tracer. As was noted again and again in the Falklands, the Argentinians consistently fired too high in these night battles. The use of well-practised tactics and good radio discipline took the marines steadily forward. Sections 'pepperpotted' their way along the ridge until held up, and then called for supporting fire. All troop and section commanders were on the company commander's radio net – thirteen call signs in all. Star shells were fired to light up the target and, when high-explosive fell, the nearest radio set called for final corrections. Milan heavy anti-tank rockets from the fire-support base 800 metres away came into action and the smaller anti-tank rockets now being used in standard fashion as 'bunker busters' were particularly useful. A complication was introduced when a phosphorus grenade set fire to a large store of bandages and wound dressings. Far away in Stanley, a report of the battle reached the Argentinian headquarters. Brigadier-General Joffre, who was handling the overall Argentinian defence, was told that the eastern end of Mount Harriet had fallen and he ordered his artillery to fire on the feature, using the blazing store as an aiming point. Menéndez, who was watching, noted in his memoirs that Joffre looked at him with an expression which seemed to say, 'May God forgive me if any Argentinians are still fighting there.' Argentinian shells thus fell at the same time as British shells, causing casualties to both sides and creating problems when fire corrections were called by the Royal Marines.

K Company captured most of its objective without suffering any further fatal casualties. A saddle was reached, leading to a second ridge of high ground. Most of this would be L Company's objective but K Company was required to cross the saddle and secure the first part of the next ridge. There now occurred another good example of the initiative showed by junior British N.C.O.s in the Falklands, although an officer and some marines also took part. The leading elements of K Company were pinned down by a machine-gun which was dominating the saddle that was the company's final bound. Three corporals worked together to overcome this obstacle. They were Steve Newland from Ipswich, in 1 Troop, and Mick Eccles from Sheffield and 'Sharkey' Ward of London in 3 Troop. This account is provided by Mick Eccles.

The machine-gun was on the edge of the next rise, about eighty metres away. The O.C. called down artillery fire on to it and this had some effect; we heard screaming, foreign, in pain, but it didn't stop all the fire. I tried to get the men of my section round to the right. Corporal Ward was trying to do the same to the left. But we were all pinned down as soon as we moved. The troop commander was back organizing the prisoners. There were more prisoners than marines in 3 Troop; they were a nuisance, milling around like sheep.

Sharkey Ward and I stopped, had a fag, and a rethink. We were in touch, shouting to each other. Then Corporal Newland came into the act. He had heard our messages to the O.C. He could see the machine-gun and said he would deal with it, although he had already reached his limit. We all thought he was going to do it with his whole section; that's what we would have done, but he moved up on his own and got to within ten yards or so of the machine-gun that was troubling us. He opened up on it with his rifle and took it out, that is hit the guy behind the machine-gun and killed him. But, while he was moving up to the gun, he was hit, shot through both legs by a rifleman behind the machine-gun post. He came up very calmly on the radio and reported that he was hit and that there were more than he could deal with. 'I'm going down for a fag,' he said.

Sharkey and myself decided we had to do something. We whacked two 66s into it; we guessed the range at seventy to eighty metres. Sharkey's went right into the hole where the enemy was. I was ready to fire mine but he had the range so well that I gave it to him and he whacked exactly into the right place again. We heard a lot of shouting and some screaming; we were told later that it was a captain who was badly wounded. We had heard him shouting earlier, leading the others, and we had aimed at the area we thought he was in.

We decided we had to get across there while the confusion was raging. Sharkey went first, to a rock half-way across the saddle. As soon as he hit this, he stopped and I pepperpotted past him and reached the far side of the open ground and was in the rocks below where the Argentinians were. Sharkey came up on my right, a few yards away. We stood with our backs to the rocks and could hear them just above us, about five feet above us. I took an L2 grenade out and showed it to Sharkey. He'd got his out; he nodded. We intended to throw them over the top but then Lieutenant Heathcote, the troop commander, came across the open space with a machine-gunner, sussing the situation out. He stopped us throwing the grenades; I thought he was the unit padre. We had a small conference; what we said to him isn't really printable but the troop commander was right. He said he thought they were surrendering.

He called out, '*Arriba las manos*' – that's how we'd been told to call for them to put their hands up, but I don't think they understood him and we had given away our positions. So Sharkey and I went straight over the rock wall, kind of aggressively. Lieutenant Heathcote was over there with us and Marine Barnett, the machine-gunner. The Argentinians panicked. One looked as though he was going to throw a grenade but Barnett shot him and they came in from all directions to surrender then. We had thought there would only be four or five at the most but there were seventeen of them alive and a few dead.

The three corporals were all awarded the Military Medal.

K Company had now taken its entire objective, at a cost of only one man killed and less than ten injured. Their battle had taken five hours. L Company started its attack on the eastern half of Mount Harriet. As so often in the Falklands battles, the initial fighting was followed by a weakening of resistance by the remainder of the defenders. Confusion, steady battering by artillery and mortars, and the realization that the British had attacked from the rear contributed to this on Mount Harriet. But L Company still had much fighting, with Argentinian snipers and machine-gunners fighting determinedly as usual. There is no major story to tell here. The British fought their way steadily through the objectives. One determined Argentinian sniper just below the highest part of the mountain held out long after other resistance in that area had ended. He hit 6 Troop's commander, Lieutenant Pusey, and the troop sergeant took over. There was a humorous moment when Sergeant McIntyre called out to a marine to get into position and throw a grenade. The marine moved but then nothing happened. The sergeant's shout of 'Well, throw it!' brought the plaintive reply, 'I haven't got a grenade.' The sniper was eventually silenced by an 84-mm Carl Gustav rocket round fired at fifteen yards range. L Company suffered no fatal casualties but two officers and at least six marines were hit. (One of the casualties was Lieutenant Ian Stafford, an Argyll and Sutherland Highlander whose regiment has had an officer attached to a marine commando unit since the Second World War, when marines from the sunken ships *Prince of Wales* and *Repulse* – both manned from Plymouth Naval Barracks – joined with the remnants of the 2nd Argyll and Sutherland Highlanders to form a composite unit called the 'Plymouth Argylls' in the defence of Singapore.)

The Argentinian defence finally collapsed and a mass of prisoners was marshalled into two lines. Some marines, in the immediate aftermath of battle, started to abuse and hit the prisoners but Company Sergeant-Major Cameron March restrained them, saying that the Argentinians were good soldiers who had fought properly. Some Argentinians fled to a nearby outcrop of rocks and opened fire from there but, after British artillery had shelled the ridge, Lieutenant Jerry Burnell led 5 Troop in an attack and quickly cleared it. A further feature – Goat Ridge – which was not defended by the Argentinians, was then occupied and 42 Commando had thereby taken all of its objectives.

The battle had lasted for eight hours. Preparations were made to resist an Argentinian counter-attack and souvenir hunting was carried out 'in the most determined manner'. Mount Harriet was a brilliant success, described by one senior officer as 'a most consummately successful operation'. 42 Commando's bold attack from the Argentinian rear had enabled them to capture this strong position with the loss of only one man killed and about twenty injured. It is believed that twenty-five Argentinians died and the massive number of 300 were herded off as prisoners.

So ended the first night of attacks on the main Argentinian positions. All three objectives had been captured, but the fighting had taken so long that no further exploitation forward was possible. The British had lost twenty-four men killed and about sixty-five men were wounded. Comparisons must be made between 3 Para's hard fight in their frontal attack on Mount Longdon, which cost nineteen dead, and the two Royal Marine attacks which used flanking approaches and which cost the lives of only five men. After the Falklands war, when two Victoria Crosses were awarded to members of the Parachute Regiment, the media descended in great numbers on Aldershot, only thirty miles from London, and the paras received the lion's share of the post-war publicity. The Royal Marines, 200 miles away at Plymouth or nearly 400 miles away at Arbroath, received less attention. The marines were quietly resentful that their skill and professionalism in the Falklands went largely unnoticed. A philosophy of direct action, speed and aggression is doubtlessly required in parachute operations but the more careful preparations of the marines and their search for the indirect approach were well suited for the Falklands fighting. I hold no particular brief for either colour of beret and have been shown much help and friendship by units of both, but I feel that these comments should be made.

Some of the Argentinians had fought well. Approximately eighty-five were killed on this night and 400 taken prisoner. But, once forced out of their fixed defences, the Argentinians had been unable to mount any sort of counter-attack. Brigadier-General Joffre, in the central command post at Stanley, had ordered several counter-attacks to be made but none had even got started. 3 Regiment, stationed near Stanley, was ordered to send a company forward to join in a counter-attack being organized, but the lorries allocated for the move up the track out of Stanley never started, a hard frost being blamed. The best the Argentinians could do was to establish a series of blocking positions so that the British did not come rampaging on further towards Stanley. The Argentinians had been clearly outfought in this, the most important few hours of the war.

Dawn found the British troops consolidating their objectives and dealing with the casualties. Royal Marines on Two Sisters lit a peat bonfire and one marine lay alongside a badly injured Argentinian for several hours to keep the casualty from losing too much body heat. One man who was helping with Argentinian prisoners on Mount Harriet was Kim Sabido, a reporter for Independent Radio News who had actually accompanied L Company of 42 Commando in its attack. Sabido is believed to be the only reporter allowed to move beyond Battalion Head-quarters and join an assault in a Falklands battle, this permission probably being given because he spoke Spanish. His presence now proved very useful in the tending of the Argentinian wounded. Army Air Corps helicopters evacuated the wounded, one Gazelle pilot noting how 'the crewman and I felt strange when we picked up two wounded private soldiers – one British marine, one Argentinian – and saw them sitting in the back like bosom buddies. The "Ginty" had his Catholic worry-beads. They had no common language but they ignored us and nattered away to each other and helped each other out at the end of the flight.'

Two good accounts from men of 3 Para describe the appearance of Mount Longdon immediately after the battle. The first is by Second Lieutenant Ian Moore.

The mist and frost caused everything to have a grey appearance. I found that the platoon had passed eight to ten sangars without realizing it and come right across the arc of a 50-calibre. The machine-gun had been knocked out by artillery; the gunner's hands were still on the gun. He was dead, probably by concussion. The other positions were all empty, just the occasional dead body.

We found their positions very difficult to spot because of the natural qualities of the ground, a mass of stone outcrops all over, only a few of which were Argentinian positions. But we quickly learned to identify their positions by the mass of empty cans thrown out – and their shit. A British soldier would have buried or carried away every trace of his occupation.

Then the party started. At the end of the ridge we found an Argentinian quartermaster bunker, full of tinned stuff – peaches, some good beef in sauce, tons of corned beef – new clothing, big cheeses, and lots of medical equipment. It was all dug into a side of the hill, seventeen metres deep – it was just like an Aladdin's Cave. You can tell a para anywhere. As soon as he gets to a position, he starts sniffing around to see what he can get his hands on. They were stuffing the booty down the front of their smocks. Our priority was cigarettes, but there were none. There was plenty of chocolate, however, and two bottles of cognac, which were a special surprise. We gave the C.O. one when he came up.

Colour-Sergeant Brian Faulkner came on to Longdon to help sort out the wounded.

It was a typical battlefield mess – entrances to sangars blown away – bodies strewn everywhere; you could see many of their bodies shot as they ran away. What I remember most vividly is the way our bodies were marked by an upturned rifle, stuck into the ground by the bayonet, with a helmet or other identifying mark hung on it. We couldn't reach all of our bodies for several hours because of the shell fire. We didn't bother with theirs; they were just like dummies to us.

Some of the Argie wounded had been injured by phosphorus grenades – severe, deep, burn wounds, very painful. They screamed, were very upset. One or two had bayonet wounds – very unusual in a modern battle – and some were even physically mauled, literally from hand-to-hand fighting with rifle butts or anything that had come to hand.

The Argies had fought very well.

The successful British units would now have to endure forty-eight hours of shelling and further casualties. But Stanley could easily be seen, only four miles away, and possession of the high ground allowed the British artillery to exercise its power. The artillery and mortar controllers had a feast of opportunity in the next two days; 3 Para called it 'a mortar fire-controller's dream' and could not engage all the targets offered.

The R.A.F. had carried out another Vulcan – Victor Black Buck operation during the night. Vulcan XM 607, which had carried out the first raids, returned to carry out another bombing raid. It was flown by Flight Lieutenant Martin Withers and his 101 Squadron crew who had flown Black Buck One six weeks earlier.

The raid did not go according to plan; indeed an embarrassing story started circulating that the bombs were dropped unarmed and did not explode. Two people told me this version, blaming the mistake on 'finger trouble', the incorrect setting of switches in the Vulcan. The bombs were fitted with a radar-altimeter-type nose fuze, set to detonate the bombs fifty feet above the ground. The object was not to crater the runway but to hit 'soft targets' in the airfield area – troops, equipment, or any of the remaining Pucarás. The unfamiliarity of the nose fuzes, and some ambiguously worded instructions, led to the nose fuzes being unarmed when the twenty-one bombs left the Vulcan. But the bombs were fitted with a reserve fuze in the tail, set to detonate on impact with the ground if the primary fuze failed. Because the war was nearly over, no investigation seems to have taken place over what became of the bombs, but the Vulcan crew were convinced that the tail fuzes were armed and, although they did not see the flashes of explosions, a prolonged 'crummpp was heard/felt' at exactly the right time interval, indicating that the bombs had exploded on impact, but it is not known where.

The last piece of action on this night was the result of a long-prepared retaliation by the Argentinians against the British warships which came in each night to bombard shore positions. The Argentinians had been unable to drive away the ships with gunfire and the British ships always retired before dawn could allow the few remaining locally based aircraft to attack. The Argentinian Navy had removed some naval Exocets – the MM38 version – from a frigate at a mainland port and a marine officer, Captain Pérez (probably Julio Marcelo Pérez in the prisoner-of-war list), and two civilian technicians had produced a makeshift land-based Exocet launcher with two firing tubes mounted on a trailer, together with an improvised computerized firing system. This equipment had been landed at Stanley in a Hercules transport aircraft late in May. The trailers were concealed in Stanley by day, but each night they were taken out on the airport road – the only road on the Falklands level enough – in an attempt to get a shot off at the British bombardment ships.

The first firing, on the night of 27/28 May, had just failed to score a hit. *Glamorgan*, *Alacrity* and *Avenger* were on the gunline south of Stanley when two Exocet firings were attempted. One missile failed to leave the launcher but the second locked on to the frigate *Avenger*. But the missile flew too high and passed just over the ship. That first firing of the land-based Exocet had alerted the British to this new danger and the bombardment ships always approached and left their firing area by a route which would leave a tongue of land, Sea Point, between the ships and the Exocet firing location.

Now *Glamorgan* and *Avenger* were on duty again, supporting the land battles, intending to retire at 2.45 a.m. local time in order to be well out of the area before dawn. Captain Mike Barrow, in *Glamorgan*, decided to delay a further half hour because 45 Commando, the unit he was supporting, had been badly delayed in its attack on Two Sisters and still needed help. When *Glamorgan* finally left, she miscalculated slightly the range of the Exocet and just came within its range. At 3.36 a.m., an Exocet was seen both visually and on radar. The missile performed well. *Glamorgan* had time only to start turning her stern towards the Exocet and hurriedly fire a Sea Cat missile, but this failed to stop the Exocet which struck the ship's side near the stern, *Glamorgan* escaped the worst effects of the hit; the missile did not strike the hull directly but hit just at the junction of the hull and the upper deck, carving a groove along the deck before just penetrating the deck plating and exploding opposite the helicopter hangar doors. If the Exocet had struck just a foot or two lower, it would have exploded well inside the ship and blown the stern off. This narrowest of escapes was attributed to the fact that *Glamorgan* was still heeling over to port in her turn and this lowered the port side of the ship sufficiently to cause the Exocet to strike just above the hull. Ironically, if *Glamorgan* had delayed a further half hour before leaving the firing area, she might not have been hit. The Argentinians normally started packing up their equipment at 4.0 a.m.

Much of the blast effect of the Exocet explosion fortunately expended itself upwards into the open but there was serious damage in the hangar area, where eight men were killed and the Wessex helicopter was destroyed. Underneath the explosion was the ship's galley which, twenty minutes later, would have been crowded with half of the ratings starting their breakfast. Four cooks and a steward were killed here and other men were injured, the flying pieces of razor-edged Formica wall covering causing many of these injuries. (This shattering Formica was an unexpected hazard and much of it was replaced in British ships after the Falklands war). There was much minor damage caused by blast and the shock effect of the explosion, and a fierce fire raged in the hangar area for some hours. But there was no damage in any vital area and *Glamorgan* was able to steam back to the task force. The thirteen dead were buried at sea later that day; they were the last naval deaths of the Falklands war. The Argentinians never got the chance to use their Exocet launcher again; it was captured with three unused missiles a few days later.

Thus ended another period of intense action. That night had seen the breaching of the Argentinian defences around Stanley, a Vulcan raid from Ascension, the deaths of three civilians in Stanley through shell fire, and the Exocet hit on the *Glamorgan*.

Major-General Moore's intention had been to renew the battle on the following night, that of 12/13 June, with attacks on Tumbledown Mountain, Mount William and Wireless Ridge. But this immediate resumption of the battle proved to be impossible because the Scots Guards and the Gurkhas, due to attack Tumbledown and Mount William, could not complete their preparations, particularly their reconnaissance and observation, and the commanding officers asked for a postponement. Moore granted a twenty-four-hour delay. The intervening period was used to move forward some of the British artillery batteries and to bring up more ammunition. The artillery fired on many Argentinian targets but the Argentinian artillery was also in action and four men were killed on Mount Longdon by shell fire, three paras and a R.E.M.E. man.

The Argentinian Air Force made their last efforts to help their comrades on the ground. There were two air raids on 13 June. Skyhawks made a daylight raid on 3 Commando Brigade Headquarters near Mount Kent and on 2 Para's positions near Mount Longdon. The British suffered no casualties in these raids but three helicopters were damaged and some delay imposed upon 2 Para's preparations for their next battle. A Rapier post of T Battery fired two missiles and hoped that one might have hit a Skyhawk, but the Argentinians returned safely to their base. The very last Argentinian air effort was made that evening when an unknown number of Canberras attempted another raid, but Admiral Woodward's ships were still on guard and H.M.S. *Cardiff*, patrolling north of Stanley, broke up this raid and shot down one Canberra with a good, long-range Sea Dart shot. The pilot ejected, parachuted into the sea, was blown ashore in his dinghy and became a prisoner of war two days later, but his navigator died.

The British aircraft were very active during this period, particularly the R.A.F. GR3 Harriers. The R.A.F. pilots had been forced to face Argentinian ground fire on many of their missions but, on the evening of the 13th, technology enabled them to carry out some highly accurate bombing without ever coming within range of ground fire. This was the first British use of the laser-guided bomb. Twenty expensive laser kits were available and Wing Commander Peter Squire and Squadron Leader Jerry Pook were attacking targets in the Tumbledown–Mount William area. The Harriers came in over Bluff Cove, lifted their noses, and 'tossed' their 1,000-lb bombs into the air. The laser guidance system then took over and the bomb flew down the laser 'cone' to its target. These attacks were carried out under the direction of two forward air controllers – Majors Anwyl Hughes and Mike Howes, who were on Two Sisters and were under Argentinian shell fire at the time. Four bombs were dropped and two are believed to have scored direct hits, on a company headquarters and on a machine-gun post (which the press later upgraded to 'a brigade headquarters' and 'right down the barrel of a 155-mm gun'). The effects of these attacks were considered to be devastating. The Argentinians, instead of being fired upon by the standard artillery shell, suddenly experienced the effects of a well-aimed bomb more than forty times heavier than a shell and delivered by a plane they never saw.

The evening of Sunday, 13 June, arrived and three battalions of the British forces made ready to resume the assault on the Stanley defences. The main effort would be made in the south by 5 Brigade, making its first attacks of the war. The 2nd Scots Guards would start by assaulting the formidable Tumbledown Mountain, with the 7th Gurkhas following through on the same night if possible and taking the adjacent Mount William. In the north, 2 Para had been transferred back to 3 Commando Brigade and was to attack Wireless Ridge. The attacks of this night were to be supported by five batteries of artillery, four warships and the eight tanks of the Blues and Royals which had arrived after motoring overland from San Carlos.

The Scots Guards attack on Tumbledown started first. It was a difficult objective. The long, narrow, rocky ridge had an almost sheer drop to the north, which ruled out any flanking attack from that direction. The ground to the south swept away more gently but was completely open and the Argentinians had sited their defences to dominate this ground. Mount William, to the south-east, also overlooked this southerly approach and any attack there would be enfiladed. (See the map on page 360.) For these reasons, Lieutenant-Colonel Mike Scott decided that his battalion would attack directly from the west, with three companies each taking successive bounds along the ridge. This attack would, however, be preceded by a major diversion mounted by a force of Scots Guards and four tanks which would move along the track south of Tumbledown and simulate a major advance towards Stanley along that track.

Many people would be watching with interest how this battalion, trained as lorried infantry and sent to the Falklands almost directly from ceremonial duties in London, would perform in a mountain battle at night. Colonel Scott had ordered that steel helmets were not to be worn, though they were carried strapped to the guardsmen's equipment for the expected artillery retaliation next morning. He says:

> I believed that the guardsmen would feel better morale-wise with their berets and cap badge. It would also help with identification; anyone seen with a helmet on would be an Argentinian. I also decided not to use the complicated Nato password and response system, which I believed would be too ponderous for close combat in the dark with the guardsmen's blood up. We would simply use 'Hey Jimmie' and the Gurkhas who were following us would use 'Hey Johnnie'. We knew that the Argentinians were incapable of pronouncing a 'J'. The system worked well.

The Argentinian garrison was provided by the 5th Marine Battalion, probably the best major Argentinian unit in the Falklands. There were two companies on Tumbledown itself and other troops on Mount William and to the south near the Stanley track. The Argentinians regarded Tumbledown as the key to the Stanley defences. The commander on Tumbledown itself was an army officer, Lieutenant-Colonel Martin Balza, whose forty-eighth birthday it was. A tape-recorder was set running at the Argentinian control post at Stanley so that messages concerned with the night's action could, according to Menéndez's memoirs, 'be recorded for posterity'. When early shelling fell on Tumbledown, Lieutenant-Colonel Balza came up on the air to Brigadier-General Joffre, 'Look, General, the dear little English are celebrating my birthday,' and other officers immediately came on and wished him a happy birthday. Unfortunately, the recorder broke down later and most of the night's action was missed.

The Scots Guards diversionary move went in thirty minutes before the main attack. The force was composed of about thirty guardsmen – members of the Recce Platoon, drivers, clerks and such like – and some Royal Engineers to clear mines; it was commanded by Major Richard Bethell, an ex-S.A.S. officer who was now the commander of Headquarters Company. Lieutenant Mark Coreth's troop of tanks – two Scorpions and two Scimitars – were also involved. Many people mention the clear and calm night but it was bitterly cold and there would be snow showers later. This preliminary action was a hectic affair and deserves to be described in some detail. Corporal of Horse Paul Stretton and Trooper Pete Fugatt provide a joint account of the tank crews' experiences.

We were behind the Scots Guards diversion group, ready to give fire support to them, but communications to them went; their radios were too weak. So Lieutenant Coreth decided to move further along the track. We drove up through some artillery fire.

Then Lieutenant Coreth reached a crater in the road and assessed it as a hole made by a stray artillery round and he decided to go round to the left. There was a big flash and a 'wallop' and his tank was lifted up three or four feet. The engine came out of the mounting and the right-hand track was ripped in half. We believed it was an American 15-lb anti-tank mine and that the crater was a demo job, with a minefield to both left and right of the track.

His crew jumped out and ran back to the second tank, which was twenty metres back, and climbed aboard. The mine explosion resulted in heavy shelling and we made a 'tactical withdrawal', that is we reversed as fast as possible back up the track. Communications then came back with the Scots Guards diversion party who were now in trouble. We were able to help by firing on Mount William, which was the Argentinian mortar position area, and putting up flares to help the Scots Guards find their way back to us.

Lance-Sergeant Ian Miller was with Major Bethell's main party.

We were formed into two assault groups and a fire-support group. We moved to a position one and a half or two kilometres south of Tumbledown. We were in single file and came upon the position we had been warned of – a set of bunkers on a small rise of sandy gravel.

We spotted the bunkers but the Argentinians did not see us. Most of them were probably asleep but we were certainly within sight of an armed sentry who we could see. We stopped and deployed; it was all done very quickly. Drill Sergeant Wight's group got to within three yards of a covered bunker, a good strong position, so close that we believed it must be empty. Then that Argentinian sentry must have woke up, saw a six-foot-three-inch Guards Drill Sergeant in front of him and fired three rounds, three single shots in rapid succession. It was so close that the other men had ringing in their ears. Drill Sergeant Wight went straight down on the first shot, killed at once. Lance-Corporal Pashley, one of the R.E.s, was hit in the throat and died soon afterwards and a guardsman was wounded in the arm.

A fierce fight around those bunkers followed; it lasted ten minutes at most – mostly grenades. We stopped when there was no more return fire. Our casualties were two dead and seven wounded and we were told later that there were ten dead Argentinians.

Major Bethell now had 30 per cent casualties, including himself wounded, and had fulfilled his task. We moved back round their mortar shelling. It was a difficult move in the dark. Every man was either carrying or supporting a wounded man or carrying spare weapons. But our spirits were high. The uninjured were relieved that they were not hit and the wounded were relieved that they were not dead.

But then we hit a minefield. Myself and a young guardsman stood on mines at the same time – small cigarette-packet-sized mines, possibly scattered by helicopter and very difficult to spot. My foot was literally blown off.

The survivors of the diversion force were eventually extricated from the minefield, Corporal Paddy Foran of 9 Para Squadron R.E. being awarded the Military Medal for helping to deal with the mines.

A mile to the north, the main attack made an excellent start.

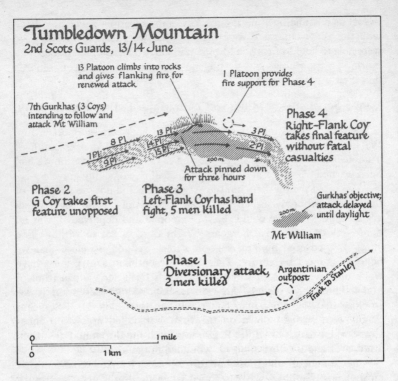

Major Dalzell-Job's G Company made a silent approach on the western end of the main ridge. (Dalzell-Job was known in the battalion as the 'Plockton Flyer' because of his home in the Highlands and his individual running style.) The company had been warned that they were likely to encounter two machine-gun posts at the far end of their objective but these were empty; the diversionary attack may have prevented this forward position from being manned. The three platoons moved carefully but steadily through the rocks, reached the limit of their objective and had taken one third of Tumbledown Mountain without being detected by the Argentinians.

Major John Kiszley's Left Flank Company* moved off on the next bound, across a saddle of ground and into the next outcrop

---

* Guards battalions traditionally have a company system based on the old 'forming of line' principle of 'tallest on the right, medium on the left, smallest in the centre'. Left Flank Company thus contained the guardsmen of medium height.

of rocks. Again, all went well for thirty minutes or so but, then, there was 'one long continuous hail of bullets'. As usual in the Falklands night battles, it was the section commanders who were in the forefront. This is Lance-Sergeant Tam McGuinness.

They were right across the front of us, in the crags. We found out afterwards that they had very good night sights, better than ours – American-made I think they were. Most of them were using single shots. It was very confusing for us when they opened fire because we did not expect a fight – right up to the last minute. Right from the word go, when we joined the *QE2*, we thought in our hearts that we'd get down there and there would be no fighting, they would throw their rifles down. We were too confident; I think that 50 per cent of us felt like that. There was no real sense of fear but it was a shock when they fired at us.

Lance-Sergeant Ian Davidson was another section commander in the same platoon.

Everyone quickly had their heads down behind the nearest rock. I only had a small rock, big enough for my head only, and I could hear bullets striking the rock in front of me and feel chips of stone hitting my beret. I thought, 'Oh God, I'm going to get hit in the shoulders.' It only took me a matter of seconds to decide to move. I had to find out where everybody was. People were shouting that there were casualties. There was no radio communication, only utter chaos at that stage; you had to do everything by shouting. Sergeant Simeon was hit and dying and Guardsman Tanbini was dead. There were two young lads who thought they had been injured; they were screaming blue murder but it was only bullets hitting the magazines in their pouches; they were only bruised.

I moved further up behind one of the lads in front of me; he had got a great rock and it was better shelter. We were there for hours.

13 Platoon had been badly hit. Company Sergeant-Major Bill Nicol had gone to help the wounded.

When we came under fire, everyone went to ground and was returning fire. There was a staggering amount of noise. I had gone off 'floating' around the left leading platoon, doing what I saw as a company sergeant-major's job, giving the boys encouragement – not that they needed it. They shouted for me and I went across to Tanbini and tried to pull him back into cover; if I had tried to lift him, we would have both been exposed and hit. I suggested that he tried to push back with his feet, while I pulled him, but he said, 'Sir, I've been shot' – typical guardsman, the way he addressed me as 'Sir' – and then he died.

Someone else was screaming for me then. It was the platoon sergeant, he had been badly shot in the thigh. I jumped up and ran across to him and, as I got to him, I was hit. I was just about to kneel down beside him when the bullet hit the centre of my rifle which was across the front of the centre of my stomach in the approved manner, ready for action. If I hadn't been holding that rifle in the manner in which I had been teaching people for years, I would be dead by now. The bullet ricocheted off the barrel and went through my right hand. Tanbini, John Simeon and I had all been shot in one line by the same sniper, I think. I had just received a letter from my wife to say she was pregnant and this went through my mind. I thought I was going to be next. There was nothing I could do about it. That sniper was good; I would like to have met him.

(C.S.M. Nicol's wife gave birth to twin girls after the war.)

Most of the two leading platoons were pinned down but the resourceful Lance-Sergeant McGuinness found that his section was not being fired upon and he started to move forward cautiously.

I saw three sangars fifty metres away to the left and asked the platoon commander by radio for the '84' team to come up. It was snowing by now. They fired at the first sangar but missed; the snow was blurring his sights. I took it off him; I had done a 'Skill at Arms Course' at Warminster so could judge the distance without a sight. I fired three rounds at those sangars and scored three hits. I think we blew the sangars and possibly the men over the back of the steep side of the ridge. On the third shot, I actually hit an Argentinian who had been putting his head up and down. He just distintegrated.

But this action did not open the way forward for the rest of the company; there were too many other Argentinian positions blocking the way and the Scots Guards attack had come to a complete halt. There now commenced a long delay, Left Flank Company being pinned down in the open and with no chance of a flanking movement developing on this long narrow ridge. The Argentinians brought down mortar fire and more casualties were suffered. There were serious problems in obtaining supporting fire. The machine-gun fire-support base was operating at extreme range and at a very narrow angle from the rear. The forward artillery observer with the company became temporarily lost and out of touch in the dark. The battery commander with Colonel Scott's headquarters tried to 'spot' for supporting fire but the targets were too far forward for him to do this satisfactorily.

Then, when the artillery was ready to fire, it was found that one gun was firing 'rogue' and it required a lengthy period of trial and error to identify which gun it was. All these problems resulted in Left Flank Company being held up for more than three hours. Lieutenant Anthony Fraser was the commander of 14 Platoon, the depth platoon in the company, and he describes this period.

After nearly four hours of this they were taunting us – the odd word in English – calling us to come on and I think they mentioned 'surrender'. They obviously felt completely on top.

At one stage, Sergeant Dalgliesh of 15 Platoon shot a man in the head, using a night sight. The Argentinian screamed for at least half a minute – a real caricature, pantomime of a noise. I thought it was a joke, but I found out afterwards that he was dying. Major Kiszley was heard to shout out, 'Who did that?' No one answered so he shouted out again. Eventually, Dalgliesh answered sheepishly and Major Kiszley said, 'Well done!' That, and the screaming, did our morale a lot of good.

The morale of my platoon had been very low. The cold was unbelievable – we had been there nearly four hours and it was taking all our time to keep in touch with our bodies. Most of the men had nothing to do. I felt that the combination of cold, uncertainty and the general awareness that we were stuck led to the group ego shrinking and shrinking and shrinking. At that stage, I thought we had blown it – holding up the whole brigade attack – and that those people who had said we would be no good, coming off public duties to this job, were right.

We eventually took shelter in the rocks to the left. The whole platoon got up and ran but then all fell over after a few paces; our legs were numb. There was much laughter and we got up again.

The crisis eventually passed and this difficult situation was sorted out, the long winter night allowing time for all the elements involved to solve their problems. Colonel Scott says, 'It had been a rough baptism of fire for us but, once over that, we pulled ourselves together and young leaders – corporals, sergeants and platoon commanders – became very professional'. As an important preliminary to the next move, part of 13 Platoon had managed to climb into the rocks on the left-hand side of the ridge (the northern and highest side) and had then moved forward to establish themselves in a superb fire position over-looking the main Argentinian positions. The platoon lined up, machine-gunners, riflemen and rocket launchers in separate groups. All this was achieved without the platoon's presence being detected by the Argentinians below.

The artillery plan was now ready. It had been agreed that immediately after the fourth salvo the rest of the company would rush forward. The plan worked perfectly. Lance-Sergeant McGuinness describes the effect of 13 Platoon's fire support from the rocks.

It was like a shooting gallery. I took the left-hand side with all the anti-tank weapons – ten '66s'. Mr Stuart, the platoon commander, let the company commander know where we were and asked for permission to open fire. We all opened up at once – with very good effect; we heard some screaming. We only had six night sights. I was picking my targets with the sight, putting it down, then firing the '66' and telling the men without sights to fire when the '66' hit. We definitely took them by surprise and could see them dodging around, trying to get into cover. This allowed the rest of the company to go forward; we could see them and kept moving our fire just ahead of them. It felt really good. We knew in our hearts that, if it hadn't been for us, the company might have been wiped out because they had been under heavy fire continuously.

With Major Kiszley well to the fore, 14 and 15 Platoons advanced through the Argentinian positions. The Argentinian marines fought hard and there were casualties on both sides. But the effect of the supporting fire from up in the rocks, the now well-directed artillery fire and, above all, the dash and infantry skills of the guardsmen overcame the defence and 800 yards of ground were covered without any further hold-up. Only Major Kiszley and six other men of the leading platoon and company headquarters reached the ledge of rocks which marked the end of the company's objective; a burst of machine-gun fire from the next part of Tumbledown immediately wounded three of these men, including Lieutenant Mitchell, commander of 15 Platoon. More men came up quickly, however, and consolidated.

The Tumbledown battle was far from over. Major S. Price's Right Flank Company followed up the recent advance and prepared for the next bound, somewhat nervous after their long wait. It took a further hour for this company to prepare its own attack. 1 Platoon was placed as high up in the rocks to the left as possible, to provide fire support, and 3 Platoon then started the main attack which, because of the slightly more open nature of the ground here, could take the form of a shallow right hook. There was no artillery support, because of the still wayward 'rogue' gun, because of shortage of ammunition and because of other requirements. 3 Platoon moved silently, with frequent stops to examine the ground ahead with night sights. The first Argentinian was seen, leaning on a rock, and was fired upon with a 66-mm rocket which exploded against the rock and killed the man instantly. Sergeant Bob Jackson describes subsequent action.

We moved up by half sections then, fast, pepperpotting and buddy-buddying, taking out trenches as best we could. It went very well. There was a gully with three positions in front of it – poor sangars, hastily prepared. We took those out instantly. I think there were only six or seven Argentinians. Three or four were killed; the remainder threw their hands up and surrendered – but they weren't scared or panicky; they were good troops.

The platoon went firm there and the company commander moved up. He decided we had got to take out two machine-gun positions in the high rocks before we could move. That was when I left my rifle and went over the rocks.

The citation for Sergeant Jackson's Military Medal reads:

Discarding his rifle and armed only with grenades, Sergeant Jackson clambered forward under fire over wet and slippery rocks towards the foot of the enemy's position forty metres away. Having climbed fully fifteen metres up into the rocky crags, single-handed he attacked and destroyed the enemy's positions with his grenades.

Guardsman Andrew Pengelly earned a Military Medal for a similar action. The Argentinian defence was now crumbling and scattered groups of guardsmen were able to move steadily forward. Lieutenant Robert Lawrence, 3 Platoon's commander, was badly wounded in the head – wearing only a beret – and five other men were injured, but no member of Right Flank Company was killed. Twelve Argentinian prisoners were taken. The final limit of the ridge was reached and the 2nd Battalion Scots Guards had won the Battle of Tumbledown Mountain, eleven and a quarter hours after crossing their Start Line.

It was a significant victory, albeit one which had taken much longer than planned. The key feature of the Argentinian defences around Stanley had been taken and the best of the Argentinian units defeated. Most of the remnants of this unit were seen streaming away towards Stanley, pursued by fire from all of the Scots Guards' weapons. A few prisoners were taken, observed to be particularly grimy after their long residence on Tumbledown; some had rubbed oil into their clothes to keep out the cold. Seven of the Scots Guards had died, two in the diversion and five in the main battle. Two more men were killed by mortars while helping wounded immediately after the battle. Forty-three Scots Guards were wounded.

The patient Gurkhas had waited all night at their forming-up position to commence their attack on Mount William, which could not start until after the Scots Guards had completed the capture of Tumbledown. When it became obvious that the Tumbledown battle was going to last nearly until dawn, Brigadier Wilson ordered the Gurkhas to move forward out of their exposed position. Lieutenant-Colonel David Morgan led off 'with great relief' and took a path under the northern face of Tumbledown (see the map on page 360), 500 Gurkhas moving in one long file. They were fired upon by shells and mortars and suffered eight men wounded, two seriously. All hope of attacking Mount William that night had faded. A Company, with the battalion's heavy weapons, was able to climb up to a fire-support position at the eastern end of Tumbledown, B Company was moved to an intermediate position; and D Company deployed by platoons ready for an attack on William which would have to be made in daylight. The risks of attacking in daylight were judged to be outweighed by the advantages of pressing on while the Argentinians were still disorganized by their recent loss of Tumbledown. We will leave the Gurkhas preparing for this action.

Two miles to the north-east was being fought the other battle of this night, the attack by 2 Para on the feature known as Wireless Ridge (though there were in fact two separate ridges of high ground there). This would be the second major attack of the war for 2 Para and the battalion had benefited greatly from its previous experience at Goose Green. Wireless Ridge had no steep sides and Lieutenant-Colonel David Chaundler had decided to attack from the north, a left-flanking attack from the British viewpoint. In contrast to Goose Green, the battalion was now able to call upon massive fire support – two batteries of artillery all night and five batteries in an emergency, the gunfire of H.M.S. *Ambuscade*, the mortars of 3 Para as well as their own mortars, and the two Scorpions and two Scimitars of Lieutenant Lord Robin Inness-Ker's troop of the Blues and Royals; the tanks would actually be able to get right into the ridge objectives. 2 Para was also to be given what would prove to be the doubtful benefit of a diversionary raid by the S.A.S. on Stanley Harbour's oil-storage tanks, which were on an extension of Wireless Ridge. The main Argentinian unit facing 2 Para was 7 Regiment, the same unit that had faced 3 Para on Mount Longdon. The main

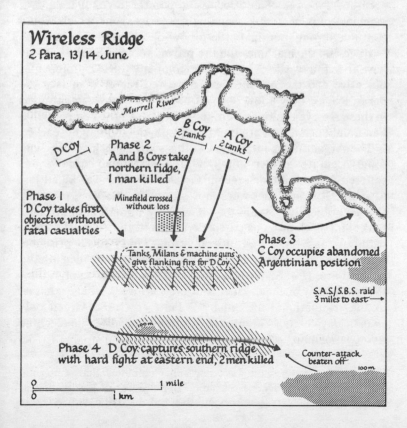

Wireless Ridge
2 Para, 13/14 June

Murrell River

D Coy

B Coy
2 tanks

A Coy
2 tanks

Phase 2
A and B Coys take
northern ridge,
1 man killed

Phase 1
D Coy takes first
objective without
fatal casualties

Minefield crossed
without loss

Phase 3
C Coy occupies abandoned
Argentinian position

Tanks, Milans & machine guns
give flanking fire for D Coy

S.A.S/S.B.S. raid
3 miles to east →

100 m

Phase 4 D Coy captures southern ridge
with hard fight at eastern end, 2 men killed

Counter-attack
beaten off

100 m

0   1 mile
0   1 km

Argentinian positions were facing south, to cover the track along Moody Valley, and this disposition would result in 2 Para's northerly attack initially striking the Argentinian reserve positions.

The Wireless Ridge attack started after the Scots Guards' attack on Tumbledown. Colonel Chaundler, exercising a battalion commander's right, decided upon a 'noisy attack', hoping that the liberal use of fire-power would cause the Argentinians to break before the paras reached the enemy positions. After half an hour of this artillery preparation, Major Philip Neame's D Company crossed its Start Line at 9.45 p.m. local time. They were helped in this move by the Blues and Royals tanks, whose more sophisticated night sights were used to direct the tanks' machine-gun fire on to the Argentinian-held positions. This accurate fire was then used as a pointer for 2 Para's own machine-guns, a useful combination of weapons which would continue throughout the battle for Wireless Ridge. The artillery barrage and the machine-gun fire proved too much for what may have been a weak Argentinian company on D Company's objective and the Argentinians retreated rather than face the paras, leaving only a few dead. It may be that the Argentinians in these reserve positions were the survivors of the company that had fought against 3 Para on Mount Longdon two nights earlier.

The Argentinians responded to this swift British success with some accurate artillery fire and the first fatal casualty was suffered when Colour Sergeant Gordon Findlay was killed just as A and B Companies were moving off, further east, to attack the next objective. Again the effective combination of artillery, tanks and machine-guns persuaded most of the Argentinians to withdraw and there was little fighting. Thirty-seven prisoners were taken; the paras suffered no losses except some men injured by shell fire. Half of Wireless Ridge had now been captured for the loss of only one man killed and a few injured. What a contrast to the tribulations suffered by 2 Para at Goose Green only sixteen days earlier! Lieutenant-Colonel David Chaundler, the new commanding officer, says:

By that stage, everything was going well. Communications were good and I realized how effective the supporting fire was. It is very difficult for peacetime soldiers to appreciate this, but the battalion had Goose

Green behind us by now and it was an essentially all-arms battle and we were able to achieve our objectives with minimum casualties.

It was also a battle of manoeuvre in which we attacked the enemy from different directions and on different axes, so they never did know what was coming at them. That was the difference between doing something for the first and for the second time. Because we had Goose Green behind us, there was a different attitude in the battalion and, while the soldiers were more apprehensive, they were also much more professional because they knew what was going to happen and this was essentially why a quite ambitious plan worked so well.

But the paras now had to tackle the long spine of the southern half of the ridge, probably manned by two Argentinian com panies. The frigate *Ambuscade* had been steadily firing on this objective since the opening of the attack and the fire of the two supporting artillery batteries – 7 and 8 Batteries of 29 Commando Regiment – was also directed there now; 8 Battery was firing so hard that only a constant stream of night-flying helicopters bringing in supplies of ammunition enabled the battery to remain in action. The Blues and Royals too had fired off prodigious amounts of ammunition and some of the tanks had to go back to restock. One had already fired seventeen boxes (of 200 rounds each) of machine-gun ammunition and thirty rounds of 76-mm gun ammunition. Having restocked, the four tanks lined up on the recently captured objectives and, with the paras' own machine guns and Milan anti tank rockets, prepared to 'shoot in' the next phase of the attack. This was to be a right-flanking attack by D Company which would, it was hoped, start at the western end of the southern ridge and sweep all the way along to the eastern end, a distance of nearly three miles, with two half-mile stretches of rocky outcrop in which the main Argentinian defences were situated. This was a clever plan; the tanks and other weapons on the northern ridge were perfectly placed to give ninety-degree flanking support throughout D Company's advance. But it could still be a formidable task for just one company.

Preceded by the supporting fire, Major Neame's company moved up on to the western end of the objective. Yet again, the mass of fire caused the Argentinians to retreat and the first half of the ridge was cleared without encountering any opposition. Half-way along the ridge, in the space between the two rocky outcrops, the company paused while the supporting fire was readjusted. Unfortunately, British shells now fell on part of D Company and one man was killed. The advance resumed. This time the Argentinians stood and fought well and there was a fierce battle for the last part of Wireless Ridge. No personal accounts happen to be available for this action and the brevity of this description should not be taken as an indication that the fighting was any less severe than in other actions described at greater length. D Company nearly ran out of steam at one point but officers and N.C.O.s rallied their men to one last effort and the final part of the company's objective was captured. One more man was fatally wounded – the only man in 2 Para to be killed by small-arms fire that night – and several more men were injured. Most of the Argentinians ran off in the direction of Moody Brook.

D Company were forced to stop at a line short of the true end of Wireless Ridge, this 'unnatural stop line' being imposed because of the possibility that the paras might run into the fire from the nearby S.A.S. diversion, an operation which 2 Para did not hear about until six months later!

Menéndez and his operational commander, Joffre, realized that, if Tumbledown and Wireless Ridge were lost, there was little to stop the British advancing into Stanley and they ordered some major moves. Most of 3 Regiment, whose sector was south of Stanley (see the map on page 327) was ordered to move west and either counter-attack or establish a blocking position on a line in front of Stanley, although there were few natural positions other than Sapper Hill and this was only a mile outside Stanley. Part of 6 Regiment, from the airport peninsula, was ordered to move to the Stanley area and replace 3 Regiment. These were significant decisions, with the Argentinians at last attempting to manoeuvre the large numbers of men held until then on inactive sectors. Some of the moves did take place but it was too late and nothing useful was achieved. Counter-attacks by more immediately available troops were also ordered both on Tumbledown and Wireless Ridge. B Company of 6 Regiment is recorded by Menéndez as having made a 'valiant effort' at Tumbledown, but this is not mentioned by any British participant. The Argentinian effort at Wireless Ridge, however, produced the only counter-attack ever carried out by the Argentinians in the whole war. A company of 3 Regiment, under Major Roberto Berazay, was the main unit designated to the counter-attack and the Recce Troop of the 10th Armoured Cavalry Squadron under Captain Rodrigo Soloaga was to give support with fourteen machine-guns, which had probably been taken from this unit's Panhard armoured cars. But the fire of the British artillery and Milans is blamed for knocking out all of these guns and preventing the cavalrymen's participation in the counter-attack. The most effective officer involved was probably Lieutenant-Colonel Carlos Quevedo, the commander of the 4th Air Portable Battery, the red berets of whose artillerymen led 2 Para to believe that they were counter-attacked by a parachute company. Quevedo had been attempting to rally the Argentinians streaming off Wireless Ridge, but with little success. Fleeing men were claiming that infiltrated British soldiers in captured Argentinian uniforms were calling out in Spanish for the Argentinians to withdraw. Quevedo sent an order out on the local radio net that any man seen giving a verbal order to withdraw which was not confirmed by radio was to be shot at once!

A counter-attack force was eventually prepared, with orders to move up to the location of 7 Regiment's command post, which was still on that part of Wireless Ridge from which 2 Para was held back by its stop line. But, with the British so close and with seven officers at the command post injured, Brigadier-General Joffre authorized the complete abandonment of Wireless Ridge and the cancellation of the counter-attack. But two platoons of Major Berazay's company did not receive the cancellation order and proceeded with their attack. The two platoon commanders were Lieutenant Rodrigo Pérez and Second Lieutenant Diego Arreseigor and it was these officers who led their men up on to Wireless Ridge and made what Major Neame says was 'quite a sporting effort, but one without a sporting chance'. A series of spasmodic attacks in the dark was repulsed by D Company, with only one para being injured. The Argentinian Second Lieutenant Arreseigor was shot in the neck but survived. Heavy British artillery fire was brought down on any likely forming-up places for further counter-attacks and that was the end of the battle for Wireless Ridge, although, as always, Argentinian artillery fire had to be endured on the objectives. The Scots Guards, whose attack on Tumbledown had started earlier than the Wireless Ridge attack, were still fighting hard.

2 Para's capture of Wireless Ridge has been overshadowed by the battalion's earlier battle at Goose Green. It is true that the Argentinian defenders at Wireless Ridge were not all of the same calibre or as numerous as the defenders at Goose Green, but Lieutenant-Colonel Chaundler's imaginative plan, the increased experience of the paras and the cooperation of other arms had brought yet another British success at the modest cost of three men killed and eleven injured. Particular credit is due to the men of D Company who had crossed two Start Lines in full-scale company attacks on this night, had experienced the only close-quarter fighting of the night, and who had beaten off the Argentinian counter-attacks. This company had also fought throughout the Goose Green battle, where it suffered heavy casualties in three separate actions. D Company, 2 Para, thus saw more action than any other sub-unit in the Falklands war and yet its commander, Major Philip Neame (son of General Sir Philip Neame, V.C.), who had not put a foot wrong throughout, went undecorated when so many other Falklands company commanders received the Military Cross.

While the Scots Guards had been attacking the Stanley defences from the south-east and 2 Para from the north-west, a swashbuckling special forces operation had attempted to hit the Argentinians from the north, much closer to Stanley. It was a combined S.A.S., S.B.S. and Royal Marines party, though the S.A.S. provided most of the men involved and Major Cedric Delves, the S.A.S. veteran of the South Georgia glaciers and Pebble Island, was in command. After lying up all day on Cochon Island in Berkeley Sound, seven miles north of Stanley, four Royal Marine 'rigid raiders' came round to Blanco Bay, only two miles from Stanley, the crews following the Argentinian hospital ship *Bahía Paríso* in on their final approach. Here were met S.A.S. and S.B.S. men who had been landed earlier by helicopter and had then moved into this area on foot. Most of these men were left to provide a fire-support base but a small party was embarked in the raiding craft, Major Delves being extremely anxious to proceed with the raid quickly because, if the Argentinians collapsed on this night, he wanted to be the first into Stanley with his men; the S.A.S. had, after all, been amongst the first to land on the Falklands more than six weeks earlier and there was no reason why this honour should not belong to them. The first three craft to be made ready sped away across the water to land on Cortley Ridge, the arm of land which forms the northern side of Stanley Harbour. The intention was to attack any Argentinians encountered and to blow up the fuel-storage tanks located there, this last being a strange intention on the eve of a probable Argentinian collapse and one which would have harmed the Falklanders and the British garrison at Stanley more than the Argentinians.

Unfortunately, the crew of the nearby Argentinian hospital ship spotted the raiders, called out to the land and the British party was fired upon as soon as it landed. Argentinian records show that a marine coastal defence party and an anti-aircraft post were involved. The raiders returned the fire but three of their men were hit and one of the craft lost its propeller blades. Major Delves decided to withdraw. The raiding craft all reached the opposite shore safely, where they were abandoned, and everyone made for the hills from where they were collected by British helicopters next day.

So ended an unsatisfactory operation. The British troops involved were disappointed to be repulsed; the Argentinians believed they had beaten off the first wave of a major landing. The British commanders had been forced to impose damaging restrictions on 2 Para's boundaries at Wireless Ridge and to devote much attention to providing covering artillery fire for the raiders' withdrawal, at some expense to the battles being fought on Tumbledown and Wireless Ridge.

CHAPTER TWENTY-TWO

Surrender

In theory, the British forces were now supposed to prepare for a further night battle, but the remaining Argentinian forces in front of Stanley were fast disintegrating. 2 Para on Wireless Ridge could see 'a whole mass of Argentinians – like a mass of black ants – just walking back to Stanley from all positions, not just Wireless Ridge'. The paras were firing on the Argentinians with every weapon they had and the British artillery was shelling as far as the edge of Stanley. Four Scout helicopters – three from 3 Brigade Air Squadron and one from 656 Squadron – came up and, in a dashing little operation, attacked an Argentinian battery position south of Moody Brook with their wire-guided SS-11 missiles. Several hits were scored on the gun positions and a large explosion was seen in ammunition stocked near by. The helicopters withdrew safely, pursued by mortar fire! Was this the first time that helicopters in the air had been engaged by mortars?

Brigadier Thompson, who was up with 2 Para, saw the Argentinians retiring and he ordered that the two most available units were to move up right to the edge of Stanley. Thus, 2 Para on Wireless Ridge started to move along the road into Stanley on foot and 45 Commando made a forced march from its position on Two Sisters to Sapper Hill. These orders and the Argentinian retirement robbed two other units of the chance to make previously planned attacks. The Gurkhas had been preparing a daylight assault on Mount William but the Argentinians there disappeared. Lieutenant-Colonel David Morgan says:

> If the Argentinians had stood, there would have been a massacre. Our D Company had lost a couple of men wounded in the approach and their blood was up. When a Gurkha sees his friends hurt, he gets really angry and nothing stops him.
>
> That should have been the peak of all our training, the peak of the whole campaign for us. But the boxer was led into the ring for the

championship fight and there was no opponent there. D Company were bloody annoyed for an hour or two but, after a couple of hours or so, the Gurkhas realized that the war was as good as over and they were still alive. My own initial feeling was that of total frustration over the fact that we had come so far to do so little.

The Welsh Guards were denied even the slight glory of occupying Sapper Hill. This battalion had been in reserve to the 5 Brigade attacks on Tumbledown and William but had suffered casualties when it strayed into a minefield. It was because the battalion was still extricating itself from this misadventure that 45 Commando was chosen instead to occupy Sapper Hill.

The British moves went according to plan. 45 Commando reached Sapper Hill just in time to see Welsh Guards being landed on the top by helicopter. Both units remained there. This left 2 Para to claim the honour of being the first British unit actually to reach Stanley when they followed up the Argentinian retreat and reached the outskirts of the town. Corporal Tom Camp, the point section commander of 2 Platoon, was the first man past the first house in Stanley. Lieutenant-Colonel Chaundler had by now ordered that the many Argentinians seen ahead were not to be fired upon.

I wanted to make sure that we did not transgress into the realms of indiscriminate slaughter. Morally, one could not fire on such an utterly routed and demoralized enemy. As we went in, the soldiers started taking off their helmets and putting their red berets on. The red beret was very symbolic. The whole thing was a fairly euphoric moment.

It was something that I really felt the battalion deserved. They had fought the most important battle of the war at Darwin and Goose Green and we had been the only battalion that had been committed twice. I rather felt that if Stanley belonged to anybody it belonged to 2 Para.

Negotiations were, by now, taking place for an Argentinian surrender and the paras were ordered not to go into the town proper so that Brigadier-General Menéndez would not physically be captured by British troops before he had signed the surrender document. The stop-line was at the prominent memorial to the 1914 naval Battle of the Falklands, an appropriate place because the 1982 Falklands war was effectively over.

*

The Argentinian surrender may look inevitable, when viewed in retrospect, but the situation on the Argentinian side was very tense and it was by no means certain, at the time, that they would not choose to fight on in Stanley itself or withdraw and continue the battle from the airfield peninsula, in the hope that the British would run out of steam and settle for some sort of compromise. The exact sequence of events in Stanley that day is not quite clear, because the civilians involved were so exhausted by then that the whole affair has a dream-like quality, and because the Argentinian side is not fully known. Menéndez had discussed the British surrender proposals with Galtieri in preceding days but received only this response, 'You must not speak with the English; the only thing they understand is the language of the gun'.* Now, on the morning of 14 June, Menéndez still had more than 10,000 troops in the Stanley area, outnumbering the British by about two-to-one. Three Argentinian regiments – the 3rd, 6th and 25th – were still largely intact, but most of the other Argentinian troops were either the remnants of the units defeated in the mountains or were administrative troops, although some armoured cars were available. There was plenty of small-arms ammunition and food but artillery ammunition was almost exhausted and the high ground had all been lost. Defensive positions had been prepared in Stanley but it is known that Menéndez had little heart for subjecting the little town to street fighting. According to Menéndez, he decided to contact Galtieri at 8.30 a.m.

I looked at General Joffre and said, 'Oscar, I am going to speak to the Commander-in-Chief.' Oscar said, 'Yes. We cannot go on with this any more.' It was a very bad moment. I went to Government House amid the shelling, wanting to speak privately to Galtieri but could not reach him, only his staff. I told them I wanted to accept Resolution 502. They told me I couldn't do it just like that, but I stressed that the end was in sight and we had to do something fast. I had to wait another hour before I could reach Galtieri himself.†

Resolution 502 was the United Nations call of 2 April for the Argentinians to withdraw from the Falklands.

Menéndez was not only facing the military facts of life but was

* Carlos M. Turolo, *Malvinas – Testimonio de su Gobernador*, p. 258.
† ibid., p. 304.

being subjected to moral pressure from the Task Force and from local civilians, though he does not mention the effects of this in his memoirs. The greatest of credit is due to the young Scots doctor, Alison Bleaney, who was out and about in Stanley that morning, with shells falling just outside the town, with hordes of dazed and unpredictable Argentinian soldiers in every street. Alison Bleaney's main concern was to warn the Task Force that the supposed safe zone for all of Stanley's civilians around the Cathedral, reported as being a fact by the B.B.C., was not in effect and that many civilians might be killed if the British advance continued into Stanley. The Argentinians told her she could attempt to broadcast this news to the Task Force and it eventually did reach Jeremy Moore's headquarters. Meanwhile, Captain Rod Bell's daily radio broadcast from H.M.S. *Fearless* at San Carlos was being listened to in the Stanley radio shack. Present were Eileen Vidal (the regular civilian operator), Alison Bleaney, Bill Etheridge (the Postmaster), and an English-speaking Argentinian officer, Captain Melbourne Hussey, although he forbade the civilians answering Captain Bell to mention his presence. Fortunately Bill Etheridge kept a resumé of the British overtures made that morning.

Fighting must stop on humanitarian grounds – battle in town imminent. No desire to destroy Argentinian forces totally; not in interests of long-term relationship between U.K. and Argentina.

Want to discuss the saving of civilian lives and the International Red Cross proposal for a safe area. You are in a hopeless position, surrounded by sea, totally covered by air. There is little point in continued struggle. Decide to give up now. Our demands are no reflection on the courage of your forces. We have already taken many prisoners and you have lost all the high ground around Stanley.

The world will judge you accordingly if you make no response and there is unnecessary bloodshed in Stanley. Agree a time for a cease-fire now. Do you understand? Cease fire now!

Captain Hussey listened carefully to all this and ordered the civilians to inform the task force that a reply would be made in one hour.

Menéndez managed to speak to Galtieri during that hour's interval. The conversation, according to Menéndez, went something like this:

Galtieri: The British must also be near exhaustion.

Menéndez: I don't know; we are near the end. I want Resolution 502.

Galtieri: It cannot be accepted after the stance we have taken.

Menéndez: I know that, because I have been implementing that stance for two months.

(Admiral Otero came in at that point and reported that the 5th Marine Battalion hardly existed; there were only forty men left in the Sapper Hill area.)

Menéndez: More bad news – Sapper Hill barely has one company to defend it.

Galtieri: Your men should be going forward, not retreating.

Menéndez: I realize that you have no idea of what is going on here.*

Menéndez now spoke with Captain Hussey, who reported the gist of the recent British appeal. Menéndez told him to arrange for a British negotiating team to fly in as soon as possible.

The rest is well known. The British team flew in in a Royal Marine helicopter, with Lieutenant-Colonel Mike Rose of the S.A.S. as chief negotiator, the S.A.S. having had plenty of training and some experience in handling such delicate affairs. Captain Rod Bell's knowledge of Spanish gave him the task of interpreter and a place at a historic meeting. The S.A.S. Mariset communications set allowed London to have a hand in the negotiations – and the British team to have private telephone calls with their families during a lull in the negotiations!

At the end of a strange afternoon, with no firing because of the cease-fire and to the background of some perfect Falklands weather, negotiations were concluded. The British demands that all Argentinian forces on both East and West Falklands be surrendered was agreed. There was a long wait until Major-General Moore flew in by an agreed safe route and the surrender document was signed that evening in the Secretariat building in Stanley. Menéndez and four other Argentinian officers were perfectly dressed, 'like tailor's dummies' according to one observer; Jeremy Moore was in combat kit with a tear in his trousers, his face dirty, tired and strained – a fighting general who had won.

* ibid., pp. 305–6.

The prepared surrender document was timed for 8.59 p.m. local time, 23.59 Zulu (that is, Task Force operational time) but, because of the delay in Moore's arrival, the actual time of signing was 9.30 p.m. local time, 00.30 Zulu. This would actually have given the surrender date as 14 June in Stanley and 15 June for the Task Force, so Menéndez agreed with Moore's request to pretend that the signing had taken place at the originally planned time so that 14 June became the official surrender date for all parties. British photographers were waiting in the corridor, hoping to catch the actual signing, but Jeremy Moore was intent upon securing the surrender of all the Argentinians in the Falklands and would not risk compromising Menéndez's agreement by permitting the indignity of his being photographed in the act of signing this document of defeat.

After the signing, Jeremy Moore was escorted to the brick-built West Store near the Cathedral, where a large number of civilians had gathered for safety. Terry Spruce, the manager of the Falklands Island Company which owned the store, met the general.

We shook hands. He had brought a bottle of whisky for a celebration, although we had plenty. He said, 'I'm Jeremy Moore. I'm sorry it has taken us so long to get here.' It was just one hooley after that. I proposed the first toast 'to Maggie Thatcher and the Task Force' and the second one was to the Queen. I should have done it the other way round. Jeremy Moore then said a few words but I can't remember what he said and I don't think anyone can remember much about it now because of the combination of jubilation and mental exhaustion. It wasn't a physical war for us; it was a mental war.

Next morning, 15 June, the Argentinian troops were moved out of Stanley to the airfield area, which became a huge prisoner-of-war camp, and the first British troops – 2 and 3 Para – moved into Stanley to a subdued reception. There was initial disappointment at the absence of flags and cheering crowds but, as has been stated before, that was not the Falklanders' style and the civilians were, one and all, emotionally exhausted. The paras, and the Royal Marines who followed soon after, were immediately taken into the Falklanders' homes and the real welcome took place quietly there. Many of the British units were exhausted. 3 Para and 45 Commando had trekked the whole way across

East Falkland, each fighting a major battle on the way. The hard men of both units survived well. Brigadier Thompson describes 45 Commando as 'very chirpy' at the end; they were the last unit into Stanley to be allocated covered accommodation. 42 Commando had spent eleven days on the heights of Mounts Challenger and Kent – often short of food, and always wet and cold – before fighting the Mount Harriet battle. There was no room in Stanley for a disappointed 5 Brigade; its battalions went back to Fitzroy and the Gurkhas to Goose Green.

Contact was made with the outlying Argentinian garrisons on that same day. B Company of 40 Commando, who had performed the unglamorous duty of base defence at San Carlos, set out across Falkland Sound for Port Howard by landing craft but had to turn back because of rough seas; they went across later in the day by helicopter and a trawler to accept the surrender of the Argentinian 5th Regiment – a hungry and dispirited unit, one of whose members was Second Lieutenant Menéndez, son of the deposed Argentinian governor at Stanley. The occupation of more distant Fox Bay was entrusted to H.M.S. *Avenger*. Lieutenant-Commander Tony Bolingbroke, with only four other lightly armed men, landed by helicopter and accepted the surrender of the Argentinian 8th Regiment and also liberated the settlement civilians and the exiled 'resistance group' deported by the Argentinians from Stanley. Most of the British warships remained out at sea, at action stations and on guard against a possible Argentinian attack from the mainland where the *Junta* had not accepted that the war was over.

One of the most significant surrenders of Argentinians took place almost unnoticed more than a thousand miles away in the South Sandwich Islands. Captain Chris Nunn of 42 Commando, who had been left with M Company at South Georgia, sailed with part of M Company in *Endurance*, escorted by *Yarmouth*, and occupied the Argentinian scientific and weather station on Cook Island which had been established in 1976. The British Government had been reluctant to evict the Argentinians earlier but the Royal Marines were now able to round up the eleven Argentinians there without a fight and take them away, leaving three British flags flying. The temporary Argentinian occupations of that long chain of islands – Falklands, South Georgia, South Sandwiches – had thus been ended and the Union Jack flew again throughout. But a British ship visiting the area a few months later found that the Argentinians had landed again and left one of their own flags.

The Argentinian soldiers on the Falklands were the first to go home. The administrative problems of feeding more than 10,000 prisoners at Stanley and protecting them from the weather led to the decision to get rid of them as quickly as possible. Most were loaded on to *Canberra* and *Norland* and sailed three days after the fighting finished. Seldom can a defeated army have been returned home so quickly, well before the first of the victors. Five hundred and ninety-three Argentinians, mostly officers and specialists, were retained as a guarantee against a renewed outbreak of hostilities. These men were all returned on the Sealink ferry *St Edmund* a month later.

The British troops were returned home as quickly as possible. The marines of 3 Commando Brigade and the paras made a triumphant return in *Canberra*, entering Southampton on 11 July to a welcome that could not fail to move the thousands of spectators at the port or the millions watching on television. (One result of the war was that for several years *Canberra* had long waiting-lists of passengers – many of them foreigners – waiting to book cruises on the ship which had been right up to the front line at San Carlos.) Other ships and units followed, all to their own welcome. The aircraft-carrier *Invincible* had to stay on station off the Falklands until relieved by her sister ship *Illustrious*, and *Invincible*'s crew sailed home fearful that the British public would have run out of welcomes for Falklands heroes. They need not have worried. The carrier came into Portsmouth on the misty morning of a beautiful September day. I made this note while watching the magnificent welcome on television: 'Looking very smart after 166 days at sea . . . Despite the time that has elapsed since the war, still a huge lump in the throat of this watcher.'

There arose in Britain a controversy over whether the bodies of the soldiers killed in the Falklands should remain there or be returned home, and a long-established policy that the bodies of British servicemen should remain in the country where they had died was challenged. The policy dated back to the closing months of the First World War when a similar argument raged with great intensity among the families of dead soldiers. (It was decided then to seek the opinions of the soldiers still fighting on the Western Front. The response was almost unanimous, that the soldiers would like, if they died, to remain in the battlefield cemeteries with their comrades. The arguments had ceased and the policy was established.) But, for the families of the Falklands dead, 1918 was too long ago and the islands too far away. The families were now given a choice and most asked for the bodies to be brought home. Only sixteen graves were left in the Falklands. The family of Lieutenant-Colonel H. Jones, V.C., who had promised his men that he would see that their bodies would be brought home if they were killed, decided after much heart-searching to leave his body in the Falklands and his is one of fourteen graves in the military cemetery established at San Carlos Settlement. The body of Harrier pilot Lieutenant Nick Taylor is still in his original burial place on Goose Green airfield and the grave of Captain Gavin Hamilton of the S.A.S. remains where he was buried at Port Howard.

No choice was given to the Argentinian families. The *Junta* refused to accept the repatriation of their dead, avoiding the embarrassment of receiving home the dead of a disastrous war and ; .zing an excellent opportunity to remind the world of their claim to the Falklands. A cemetery for the Argentinian dead was established near Darwin and the bodies were given a military funeral with a Catholic service conducted by Monsignor Spraggon from Stanley.

Operation *Corporate* had cost the lives of 255 members of the British forces and of the Falklands civilian community (see table).

Of the deaths, 217 were the result of action with Argentinian forces, twenty-eight of helicopter or aircraft crashes, and ten of the accidental fire of British forces. Comparing these deaths with other conflicts in which Britain has been involved since the Second World War, they are not as high as in Korea (537 deaths), Malaya (525 deaths) and Northern Ireland (352 deaths up to the end of the Falklands war) but the Falklands deaths are more than were suffered in any of the Palestine, Cyprus, Indonesia, Aden or Oman conflicts. The number of men wounded in the Falklands was 777.

British and Falkland Islander fatal casualties

	Officers	Men	Civilians*	Total
Army	7	115	—	122
Royal Navy	13	72	2	87
Royal Marines	2	24	—	26
Merchant Navy	2	7	—	9
Royal Fleet Auxiliary	3	4	—	7
Falkland Islanders	—	—	3	3
Royal Air Force	1	—	—	1
Total	28	222	5	255

*Chinese laundrymen on R.N. ships are classed as civilians but the Chinese on *Atlantic Conveyor* and *Sir Galahad* are counted as ordinary crew members.

The Royal Navy suffered the first casualties and lost four warships sunk with many more damaged; the ships *Atlantic Conveyor*, the *Sir Galahad* and one landing craft were also sunk. The Task Force also lost thirty-four helicopters and fixed-wing aircraft, as shown in the table.

British helicopter and fixed-wing aircraft losses*

Helicopters		Fixed-Wing	
Wessexes	9	Sea Harriers	6
Sea Kings	5	Harrier GR3s (R.A.F.)	4
Chinooks (R.A.F.)	3		
Lynxes	3		
Gazelles (R.M.)	2		
Gazelles (Army)	1		
Scout (R.M.)	1		
Totals	24		10

*Appendix 1 contains details of the lost aircraft and helicopters.

Eight of these losses were due to Argentinian ground fire or missiles, thirteen (all helicopters) were lost when their parent ships were sunk or damaged, and thirteen were destroyed in accidents. I asked Admirals Lewin and Leach (separately) what had been the most pleasant surprises in the war for them. Admiral Lewin's was the performance of the Sea Harriers and their lower casualties than he expected; Admiral Leach's was the smaller number of ship losses than he expected and the low number of men who died in the lost ships reflecting, he said, an improvement in damage-control training since the Second World War when destroyers often lost half or three quarters of their crews when sunk.

The greater losses of the Army and Royal Marines – 148 men dead – occurred in only twenty-seven days (from the S.A.S. helicopter crash in the sea before the San Carlos landings to the end of the war). But sixty-six of the Army's casualties were not on land, being suffered when the S.A.S. helicopter crashed into the sea and when the *Sir Galahad* was bombed. Only fifty-six soldiers and Royal Marines were lost in the six set-piece land attacks of the campaign, the worst casualty list being that of 3 Para on Mount Longdon, with twenty-three men killed in the assault and the subsequent counter shelling. Once the British forces were ashore, they never lost a battle and they completely outclassed the Argentinians. The greatest danger – as had been forecast when the task force sailed – was that of Argentinian air attack on the British ships. All British ship losses and more than half of the helicopter losses were caused by Argentinian aircraft. The table below, showing the cause of the British fatal casualties, further illustrates this point.

Argentinian aircraft: 141 deaths (110 by bombs and cannon fire, 31 by Exocets)
Argentinian ground forces and minefields: 64 deaths
Argentinian land-based naval Exocet: 13 deaths
Other causes: 37 deaths

Unfortunately, one cannot be so precise over the Argentinian casualties. The best figures available are those published in Argentina three weeks after the war, which indicated that 746 Argentinians had been killed. This figure can be broken down as follows, but it must be stressed that these are approximate figures only.

Navy: 393 (368 in *Belgrano*, 22 in smaller ships, 2 naval pilots, one unattributed)
Army: 261
Air Force: 55 (40 aircrew, 15 others)
Marines: 37 (mostly on Tumbledown Mountain)

A further 1,105 soldiers and an unknown number from the other services were wounded or became sick. The number of Argentinians taken prisoner is reliably known. British units carefully documented 12,978 names, a total which includes every Argentinian captured, from the naval personnel and scrap-metal workers taken on South Georgia to the eleven men removed from the South Sandwich Islands after the war. The prisoner-of-war list contains the names of Brigadier-Generals Menéndez, Daher, Joffre and Parada from the army and Castellano of the air force, Admiral Otero, twelve full colonels, twenty-seven lieutenant-colonels and six naval or marine officers of equivalent rank.

The Argentinian material losses were enormous. Almost all of the weapons and equipment taken to the Falklands was lost. The cruiser *Belgrano* and several smaller ships were sunk. More than one hundred fixed-wing aircraft and helicopters were shot down, crashed or were captured in various states of repair on the Falklands. The numbers shot down in action are believed to be seventy. This last subject, Argentinian aircraft losses in combat, has been the subject of intense examination and rivalry between British services and the operators of different weapons systems. The rival claims just cannot be reconciled. I have tried to give a general picture of the progress of the air war in previous chapters but do not feel qualified to comment further, except to say that the combination of Sea Harrier and Sidewinder missile was particularly effective.

What about the justification for the dispatch of the Task Force? A young British sailor writes:

As to me, it seemed totally idiotic for two distant groups of men to be thrust at each other, on a politician's say-so, and then go hell for leather. To start blasting into oblivion and causing death and suffering to all who crossed their paths. All this just didn't make any sense at all – not to me anyway. Perhaps, now it's all over, somebody can tell me why!

For the men of both sides who fought in the Falklands and for the civilians whose home became a battlefield, it was a fantastic experience, all over in just eleven weeks but for many seeming a lifetime. The Anglo-Argentinian War of 1982 might be of little long-term interest to historians compared with the cold-war confrontation of the nuclear superpowers or the seemingly everlasting internal insurgence being suffered by so many unhappy countries. What fascinates so much about the Falklands war was that it was such an unexpected, freakish conflict between two western nations over a colonial dispute.

Much of the empire from which Britain derived so much wealth and power has gone but, in 1982, fragments of that empire still remained Britain's responsibility. The Argentinians had a 150-year-old paper claim to sovereignty, but modern world opinion, expressed through the United Nations, was that self-determination of peoples should now decide sovereignty. The inhabitants of the Falkland Islands had decided to remain British and had entrusted Britain with their protection. Britain had accepted that trust. The Argentinians invaded and would not withdraw. I was travelling in Europe during part of the Falklands war and discussed it with people of several nationalities. The standard view was, 'Why are your two countries fighting for such a tiny, unimportant place?' But when I countered with, 'What else could Britain do once the Argentinians had invaded and then ignored the United Nations call to withdraw?', most agreed that Britain had no option but to send a task force. Who else was going to liberate the inhabitants of Stanley and the settlements? Not every small, weak nation in the world can be brought back to freedom in this way, but how many of them drew a breath of relief when Britain liberated the Falklands? It is far too soon to give a final verdict on the Falklands war – twenty-five years after the event might be a suitable time – but the view might be offered that Britain's show of determination in the Falklands might even, just, have saved a larger conflict. My personal comparison is with the German occupation of the demilitarized Rhineland in 1936 in defiance of the Versailles Treaty, when France, Britain and the other treaty signatories declined to turn out just three battalions of German troops, which is just about all the army Hitler possessed at that time. Hitler took heart at this weakness and moved on step by step, on to a world war which lasted more than five years and cost millions of lives. I realize that the comparison with the Falklands is not entirely appropriate but potential aggressors will surely have taken note of Britain's firmness in 1982.

Any British mistakes in the Falklands pale into insignificance when compared to the blunders of the Argentinian *Junta*, men whose ambitions and decisions caused so much misery and loss of life. But, not for the first time, defeat in war actually led to benefits in the defeated country. Just as Germany's defeat in the Second World War brought the overthrow of Hitler, so the Falklands adventure led directly to the fall of the military dictatorship and the end of internal oppression in Argentina, and to the election a year later of what may prove to be the best civilian government Argentina has ever had. It is unlikely that the Argentinians will give Mrs Thatcher and the British Task Force the credit for this, but their historians may one day admit that a growing stability and improvement in Argentina's affairs came about because of the 1982 war.

Turning to the purely military side of the war, there is little to add to what has already been described in this book. After their initial success in defeating small parties of Royal Marines at Stanley and Grytviken, the Argentinian forces never won a battle, although they did inflict some severe losses on the British ships. After its dispatch from Britain, the Task Force never made a major mistake and won a remarkable victory. It is true that minor errors were made and these did lead to loss of British lives. It was a reminder of the great burden carried by military leaders at all levels. Mistakes on a similar scale in politics or finance would result in votes or money lost and be quickly forgotten; on the battlefield the result is loss of young servicemen's lives. But the British units had fast become battle-wise, improving on hard-won, early experience. The best example of this is to be found in a comparison between 2 Para's traumatic first battle at Goose Green and its smooth victory at Wireless Ridge only sixteen days later.

The soldiers and sailors returned home from the Falklands, some to medals and promotion, most to further service, some to retirement – all to look back and reflect on that Falklands experience. Some men had been badly shaken by it all, having joined for a technical career and a steady wage; the Recruiting Officer never says that the ultimate duty of all servicemen is to be ready to fight and maybe to die. One small unit composed of older, technical men came back from the Falklands with three of its number under warning of court-martial for refusal to obey orders under the stress of the bombing period at San Carlos; the charges were quietly dropped on return to England. One senior officer, also at San Carlos, suffered an emotional breakdown on his return and was liable to break into floods of tears for several months. The regular fighting units, however, took the war in their stride and emerged the stronger. One officer in the paras says that 'the Falklands was really like one long exercise, nothing compared to Northern Ireland', but a sixteen-year-old marine, on a short-term junior engagement from which he had been considering leaving because 'there was no challenge', signed up for a full nine-year period of service as soon as he returned from the Falklands 'because that was what I had joined for and I would hate to be left back in England if it happened again'.

I asked many men for their views 'looking back on it later'. There was a general resentment at what they saw as the political failure of both sides which led to the war but an acceptance that, after the Argentinians refused to withdraw, the Task Force had to be sent, and most men were proud at having been involved. There was a natural sorrow over lost and mutilated friends but no personal resentment against the Argentinian servicemen. A minority thought it a total waste. It would be interesting to ask their views again in ten years' time. Let this chapter be concluded with the view of a Falkland Islander, Alan Miller from Port San Carlos.

All of us here feel extremely humble at the incredible cost of sending the Task Force, so much being accomplished by so many for just so few. All through history, Britain has waited until she has been kicked, made a fool of, and almost too late before doing something about it. But once the British bulldog gets his teeth into something, look out! Three-to-one against at Goose Green and similar odds in the Battle for Stanley. Obviously British is Still Best.

We have a golden opportunity now that our future is secure to really build and develop our little country, so that those 252 men who gave their lives so that we could be free did not do so in vain.

The future looks good and we will be ever grateful to you wonderful British people.

A Visit to the Battlefields

Studying battlefields is for me both a hobby and one of the most interesting parts of my work. In September 1983 I had the opportunity to visit the Falklands and, with the help of the local army authorities, particularly the Army Air Corps, I was able to visit the San Carlos area, Goose Green, Mount Kent and Wireless Ridge. A visit to Mount Longdon was cancelled at the last minute, because a new minefield was discovered that morning and the area closed – but I was able to view Mount Longdon, Two Sisters, Mount Harriet, Tumbledown Mountain and the *Sir Galahad* disaster site at Fitzroy from a hovering helicopter. The flight over the peaks defending Stanley emphasized two points – the first how important Tumbledown was in the Argentinian defensive system and the second that Stanley was always in sight, yet it took the British forces fifteen days of preparation and fighting, during which 124 British lives were lost, to cover a distance which can be crossed in a few minutes by helicopter. Most of these peaks have their memorials to the soldiers who died there, just like the memorials I have seen so often on the Somme, around Arras and Ypres, and in Normandy.

I had plenty of time to study San Carlos Settlement where 2 Para and 40 Commando came ashore on D-Day and which became an important base for the remainder of the war. The site of 3 Brigade's headquarter bunkers, gouged out of a bed of gorse, could be clearly identified. The settlement farm buildings showed many signs of the military presence. The floor of the shearing shed had hundreds of small black circles where troops had used their hexi-cookers and a sheep pen in the shed was marked 'Sgt's Mess 81 Coy RAOC'. A shed near by had a roughly painted sign, '2 PARA GP, FIRST HERE, FIRST INTO BATTLE, FIRST INTO STANLEY', to which later units have added irreverent amendments. Another sign near a stream from which troops took water for purifying reads: 'WATER, OPEN FROM 1100–2200. JUNTA SUCKS'. Pat Short, the settlement manager, states that he will never paint over these signs nor will he fill in the large pits near by, once occupied by 79 Battery's guns. These will be left, partly for the interest of visitors and partly because the pits are the right size and shape to provide shelter for sheep at lambing time. Smaller slit trenches have been filled in, however; sheep falling in these cannot get out again. The Short household still has a strong interest in the war and its aftermath. Four kittens were called 'Harrier', 'Lynx', 'Wessex' and 'Tibbitabs', though 'Sea King' and 'Vulcan' were under consideration for 'Tibbitabs'.

I took a long walk round the shore overlooking the San Carlos anchorage. I reproduce here the notes I made at the time.

My first impression – the great distances between locations, over the water to Ajax Bay and over a much larger hill than I expected to Port San Carlos – all a surprise after using maps of the area for so long. Fanning Head – much further away.

Superb anchorage still being used as R.N. servicing area. *Stena Inspector* with *Fort Austin* (?) and a Type 21 frigate, the rest of the huge anchorage empty and superbly peaceful where Mr Short counted twenty-three ships at one time. Two buoys off Ajax Bay marking the wreck of *Antelope*. Great ranges of hills, little dots of Rapier posts on the crests. A solitary Lynx flies across to Ajax. The air as clear as crystal – no pollution. Fine for me all afternoon – cloud to the north-west over West Falkland and a rain shower on Sussex Mountain, leaving a brilliant rainbow. Kelp beds – not dense. Grey geese taking off, surprised by me.

Walk along the shore line, parts of it littered with flotsam and debris; make a note of these:

A naval 'woolly-pully'. Very charred respirator case, 'Made by Remploy 1977'. Soggy paperback – Black Camelot. *Dozens of shoes and 'trainers' sometimes inexplicably in pairs. Many ration sachets, 'Beef Stock Drink 5gs' obviously not popular. Jerricans and water cans. Field dressing, plastic medical drip of some sort. Peculiar packets sealed round and round with black masking tape; opened one – just a piece of rubber inside. Hundreds of beer cans, a Blue Nun wine bottle. Long strip of fancy red carpeting. Returnable Courage keg 'Sparkling Beer'. Small field mattress with Spanish-language label. Blue water-proof trousers – charred, a petty officer's shirt, a naval beret – no badge. Very long, rubber fire-fighting glove, then not far away, its pair. Plastic ammunition containers of many kinds. Small fire-extinguisher cylinders. Smashed-up furniture. Many green (but one red) zip-up, foam-filled seat covers or small mattresses, one marked '*Argonaut*'. Then, as from another world, an old kelper's cart. Finally, the remains of a wreath of imitation lilies and daffodils, possibly laid on the water on the*

*recent pilgrimage by relatives and now cast up on Blue Beach where
40 Commando made its landing on D-Day.*

Mr Short later told me that much of the flotsam came from
Argonaut, which was moored near by while her crew cut away
the decks above an unexploded bomb, but some of the material
probably came from *Antelope*, which blew up between San
Carlos and Ajax Bay.

But the most unforgettable part of San Carlos is the permanent
cemetery now established on the smooth green between Mr
Short's house and what was once Blue Beach. Here are buried
the bodies of fourteen men whose families wished them to be
left in the Falklands. On one side are seven soldiers: Lieutenant-
Colonel Jones, V.C.; three other men of 2 Para; and three men
from the Army Air Corps helicopter which crashed while flying
5 Brigade's Signals Officer forward to Fitzroy. On the other side
of the cemetery are seven Royal Marines: five from 45 Commando; one from the Commando Logistic Regiment; and
Lieutenant Nunn, who was killed when his helicopter was shot
down at Goose Green. (At least one further grave, from a post-
war accident, has since been added.) The wall of the cemetery
is made from stone brought from Fox Bay, the best-quality stone
in the Falklands. Pat Short was offered the normal gardener's
fee for looking after the cemetery but firmly refused any
payment, 'It is the least I can do for the boys who chucked the
Argies out for us.' Mr Short has planted daffodils and crocuses
around the cemetery. The grave of 'H' Jones has on it some
heather, brought from his native Devonshire by Mrs Jones; it is
growing well. Captain Chris Dent's grave has a photograph of
his small son on it. The headstone of Corporal Ken Evans, of 45
Commando, has this message, '*Saya chinta awak*, Ken' – 'I love
you, Ken' in Malay.

My visit to Goose Green was a more hurried affair, a planned
leisurely walking of 2 Para's battlefield being curtailed by a last-
minute revision to the helicopter schedule. But I was able to
have a good look at Darwin Hill, where A Company had such a
hard fight and where Lieutenant-Colonel Jones was killed. A
group of rocks marked the place where he fell. The trench he
was attacking – now filled in – is the usual tumble of stones
and turf, no different from hundreds of other old Argentinian

trenches on the Falklands. Scattered medical equipment was lying in the nearby gorse bushes where the wounded were treated. Like so many 'hill' positions which now give their name to battles fought, Darwin Hill is no more than a small fold in the ground which would not be looked at twice in the English countryside.

Nearer to Goose Green Settlement, on the edge of the grass airfield, is the lone grave of Lieutenant Nick Taylor, the Sea Harrier pilot shot down there on 4 May 1982. When his family heard that the Goose Green people were anxious to look after the grave, the authorities made the unusual concession of allowing Nick Taylor's body to remain in the original grave where the Argentinians had buried him. In a little fenced enclosure on a windswept pasture, alongside a corrugated-iron lambing windbreak, the wooden cross erected by the Argentinians stands now alongside a British war graves headstone. The family inscription on his headstone reads: *In proud memory of a dearly loved husband, son and brother shot down and killed whilst flying for the country he loved*. The note I made as I stood there was, 'How apt is Rupert Brooke's "There is some corner of a foreign field . . ." ' Just a hundred yards away is the large hole made by the Argentinian Aeromacchi shot down during the Battle of Goose Green, whose pilot, Lieutenant Miguel, also died.

The shortness of my visit to Goose Green prevented me walking to the Argentinian cemetery near Darwin but I was able to view it from the air and also to talk to the army officer in charge of battlefield affairs and the burial of the Argentinian bodies which were still being found more than a year after the war. Eleven bodies were found in 1983 up to the time of my visit in September. Most were not identifiable but the last one found before my visit was that of a Skyhawk pilot, Captain Jorge García, shot down near San Carlos on 25 May 1982 and only recently found in his dinghy on a beach in West Falkland. Each newly discovered Argentinian body is given a military funeral with a blessing by Father Monaghan, the assistant Catholic priest at Stanley.

My final battlefield visit was to Wireless Ridge, driven out there from Stanley in a friendly civilian's Land-Rover. Again, I reproduce the actual notes I made that afternoon.

As before, the area far more extensive than I had expected. Not a continuous series of defended positions, but posts on crests dominating the open spaces in front. The smaller positions in crevices with slabs of stone built into sangars. In one, an empty belt and a pile of spent cartridges. All around, shell and mortar holes deep in the peat, full of black water, very large chunks of turf on the lips showing how deep the shell has burst in the soft soil. But little sign of where shells had hit the rock outcrops. A piece of ploughed-up ground, thirty-feet long, where a cluster bomb has burst. Saline drips, warning trip-wires with soft drink cans hung in batches of three or four. 0·5 m-g tripods, giving a good impression of the size of this much-feared Argentinian weapon. A mortar base-plate with dozens of cardboard mortar-bomb containers. Many empty SLR mags. A recoil-less cannon with a long cleaning rod near by.

Behind each of the defensive positions, larger shelters where Argentinians had lived, still full of personal belongings – boots, socks, toilet articles, good-quality shaving brushes, ponchos, blankets, a few officers' mattresses, a fancy toilet bag, disposable razors, water-bottles, often floating in the black water of small ponds, a fine set of leather harness which I will take home. But no steel helmet which I had set my heart on, I am told that American helmets are on sale in Stanley at £15, the same design as used in the war but not the genuine battlefield souvenir I was looking for. Documents: 'Marcha de las Malvinas' – a patriotic song; a letter from 'Néstor' written on 9 June 1982 but never posted; comics; a range-table for a mortar. Rats, dead in a pond or scuttling into the tussock grass – never here, I am told, before the debris of war came.

On a high part of the ridge, a metal cross with a brass plate containing the names of three men of 2 Para who died. Stacked against the memorial, rusty weapons and a short belt of ammo. Not far away a British soldier has painted on the side of a large rock: 'THAT'L TEACH YA.'

Walked on as far as a still uncleared minefield where we had to stop and then down to the track near Moody Brook stream and through an Argentinian artillery position. Hundreds of rusty cartridge cases, no good for cutting down for ashtrays like the British brass ones, too rusty; by contrast the green-plastic shell covers, like new. A heap of stout ammunition boxes – some Spanish, some American – angry thoughts at the International arms trade.

Back to the old Marine Barracks at Moody Brook, now a vast dumping-ground for Argentinian and British vehicles and a few helicopters, all wrecked in the war. Much grafitti; one catches my eye: 'WARLORD DEC '82, EXPLOSIVE DUSTMAN.'

In Retrospect

The Falklands War was fought in 1982. The original version of my book was written in 1984, and published in 1985. I have had two further years of other work since then, in the summers taking parties of visitors out to the 1914–18 battlefields, and in the winters preparing a book on the R.A.F. Bomber Command offensive against Berlin in 1943–4. Now I come back to re-examine that war in the South Atlantic with fresh eyes. With this greater perspective and being more free now from close contact with the events of 1982, I find that some of my views have developed and sharpened. (I stress again two things: that this book is mainly a military study, and that I cannot write these comments without the benefit of a hindsight not enjoyed by the commanders at the time.)

These are my latest thoughts on the major features of the war:

The sadness of the setbacks and of the casualties to both sides. The general excellence of the British forces compared to their Argentinian opponents.

The incredible speed with which the Task Force was deployed, and the retaking of South Georgia less than a month after the Argentinians had captured it – the recapture being so nearly jeopardized by the over-optimism of the S.A.S.

The fortuitous location of Ascension Island and the value of the United States' supplies and fuel delivered there. The careful preparation work by Admiral Woodward's ships while awaiting the arrival of 3 Commando Brigade from Ascension.

The high quality of the airmanship of the R.A.F.'s Black Buck flights. The persistent refusal of part of the naval lobby to accept that the first bomb dropped by Flight Lieutenant Withers and his crew did hit the runway and did deny the runway's use to

Argentinian high performance jets. Let those narrow minds study the maps and the radii of action of the Argentinian jets and of the Sea Harriers, and imagine the effect upon the outcome of the war if Woodward's ships had been forced back 200 miles by the Argentinian use of the Stanley runway.

The decisive effect upon the Argentinian Navy of the *Belgrano* sinking. The loss of *Sheffield*, and the leaping into prominence of the Exocet factor.

The misleading of the Argentinians over the British landing plans and the eventual success of the landings.

The self-sacrifice of the Navy during the 'Bomb Alley' period. The fortunate availability of the Sea Harrier and the effectiveness of its Sidewinder missile.

The success of 2 Para at Goose Green, marred by the costly error at Darwin Hill.

The unnecessary sinking of the *Coventry* and the crippling loss of helicopters on *Atlantic Conveyor*.

The brilliant handling, by Brigadier Thompson, of his brigade in the move forward to the Stanley defences. The hardships of the trek across East Falkland and of the freezing conditions for the men on Mounts Kent and Challenger.

The frustration and difficulties of Brigadier Wilson, trying too hard to wrest some glory for the Army in this 'Navy–Marine party' (an Army officer's words). The unsoundness of the southern thrust which culminated in the *Galahad* horror. (More about this later.)

The true brilliance of those underpublicized victories in the final hours of the war – Mount Longdon, Two Sisters, Mount Harriet, Wireless Ridge, Tumbledown Mountain.

And, after the war, the obsession of the Ministry of Defence in concealing the mistakes made, a policy which led to so much anguish for the relatives and headline opportunities for the press. Surely it would be better to admit these at the time, or immediately afterwards, so that the comparatively few mistakes can be judged against the greater picture of success in the war. There is a golden rule of history: 'It all comes out in the end'.

The wastefulness of inter-service rivalry.

THE SOUTHERN THRUST

There is one aspect of the war on which I would like to comment in more detail.

Many of one's moves in warfare are dictated by the actions of the enemy or by other outside factors. If the Falklands campaign had to be fought again, there was one British instigated move which, with hindsight, would best not have been made.

A critical moment can be identified. On the afternoon of 2 June, part of 2 Para leapt forward from Goose Green to Fitzroy and Bluff Cove, thirty-four miles nearer to Stanley. (The full description is on pages 296–8.) It was a fine performance by 2 Para and the helicopter crews concerned, and the operation received much publicity. The move could not have been made without a Chinook helicopter that had been virtually 'hi-jacked' by Brigadier Wilson.

The background to this turning point was that Brigadier Thompson had already moved most of 3 Commando Brigade, on foot and by helicopter, up to the Stanley defences by a northern route. (See map on page 280.) But Major-General Moore, the newly arrived overall commander, had promised the also newly arrived Brigadier Wilson that 5 Brigade could come up to the front line *separately in the south* and have an equal chance with 3 Commando Brigade in the final attack on Stanley. Moore needed the main strength of both brigades up at the front, because of the known Argentinian strength. There were two reasons why a southern move up by 5 Brigade was desirable. One was to keep the Argentinian attention focused on the south, inducing them to believe that the main British attack on Stanley would eventually come along the track which came in from the south-west, when, in fact, the main strength of the final attack would come from the west and north-west. The second reason was that 5 Brigade was the only Army brigade in what, until now, had been almost exclusively a Navy–Marine operation. Wilson was desperate for his brigade not to follow in the foot-steps of the marines, and perhaps be still in the rear if the Argentinians collapsed quickly, but to move up as rapidly as possible to be alongside the marines at the moment of victory. Moore was almost as anxious not to be seen to be favouring his brother Royal Marines. For these two reasons, he did not order the recall of 2 Para that afternoon but decided to send the rest of 5 Brigade after them in the south by whatever means possible.

The main problem now was that those means forward were limited. The two Guards battalions, who were next to move up, and later the Gurkhas, could travel either by air (helicopter), by sea, or on foot across open country; there were no troop-carrying lorries and no proper roads. The swift helicopter lift which had taken 2 Para forward was not again available. The limited number of helicopters left after the loss of *Atlantic Conveyor* were all hard at work on existing commitments at the San Carlos base or supporting 3 Commando Brigade, which was already well forward.

Brigadier Wilson said that he would march his units forward; after all, 3 Para and 45 Commando had earlier marched a similar distance in the north. It was here that the differences in geography and terrain between the northern and southern routes became crucial. Overland, the two routes were of similar lengths, but the paras and marines who had marched the northern route had received the benefit of two civilian settlements – Douglas and Teal – on their gruelling trek across the atrocious Falklands terrain. Furthermore, the guardsmen were not as suited to such marches as the paras and marines. An attempt by the Welsh Guards to cover the short distance from San Carlos to Goose Green on foot failed dismally.

The only remaining way to move up 5 Brigade and then support it was by sea, but, here again, geography was not on the British side. 3 Commando Brigade in the north had already opened up a forward base at Teal Inlet, where sea-borne supplies could be received, the small 'Sir' landing ships being able to sail comfortably between San Carlos and Teal during the hours of darkness. The southern sea route to Fitzroy and Bluff Cove was two and half times as long because of the huge bulge of the sub-island of Lafonia. (The situation is best seen on the map on page 197.) As has been described earlier, the larger and faster *Intrepid* and *Fearless* could not be risked so far forward in daylight and these two ships could not unload and return to San Carlos before dawn. There was a further difference between Teal and the southern harbours. Ships lying in Teal Inlet had the duel advantage of a Rapier anti-aircraft missile defence and freedom from direct observation by the Argentinians. Fitzroy and Bluff Cove were only scantily defended, and Fitzroy was clearly visible to the Argentinians on their nearby hill positions. When *Sir Galahad* could not reach Bluff Cove on the morning of 8 June, it was forced into Fitzroy, seen by the Argentinians, and bombed.

So the bold thrust forward a week earlier, when Brigadier Wilson sent 2 Para off on its afternoon dash by helicopter, led directly to that terrible disaster. Poor Tony Wilson, who had been seen off from England with exhortations by his Army superiors to obtain some glory for his service, returned in disfavour. His name was noticeably absent from the long list of postwar awards. Despite having been the only Army commander to have seen action in the war, he was allowed to retire a few months later.

It is making full use of hindsight to say that Jeremy Moore should have recalled 2 Para that first afternoon. He would certainly have risked the fury of the Army and would have had to abandon his plan of keeping Argentinian attention focused to the south. But again, with hindsight, 5 Brigade could easily have been transported by the shorter northern sea route to Teal Inlet and then deployed by short local marches or helicopter lifts to their allocated sectors of the front facing the Stanley defences. That deployment could probably have been achieved just as quickly as the laborious and dangerous shuttling of various ships which eventually brought 5 Brigade up into the line in the south. It is ironic that the unit which started off the southern move, 2 Para, was later transferred, by just such a short helicopter flight and local march, to the extreme northern end of the battlefront in the final attack on the Stanley defences.

Appendix 1

Task Force Order of Battle and Roll of Honour

The purpose of this appendix is to list the British ships and units which served in the Operation *Corporate* Task Force, to record the ships and aircraft lost and to honour the names of all the British servicemen who died. The dead are listed in alphabetical order in each incident and their home locations are given. Decorations shown were mostly those awarded posthumously after the war.

PART ONE
R.N. Ships of the Original Task Force

(The following abbreviations are used: A.B. – Able Seaman, Lt – Lieutenant, Lt-Cdr – Lieutenant-Commander, P.O. – Petty Officer.)

Alacrity   Type 21 Frigate

Antelope   Type 21 Frigate   Bombed in San Carlos Water on 23 May, blew up and sank.
Fatal Casualty: Steward M. R. Stephens, Mansfield, Notts. The Royal Engineers sergeant who died while attempting to defuse a bomb is listed under 'Army Units'.

Antrim   County Class Destroyer   Damaged by air attack in Falkland Sound on 21 May: there were no fatal casualties. Wessex helicopter XP 142 was damaged and never flew again. It is now in the Fleet Air Arm Museum at Yeovilton.

Ardent   Type 21 Frigate   Bombed in Falkland Sound on 21 May and sunk. Lynx helicopter XZ 251 was lost.
Fatal Casualties: A.B.(Sonar) D. K. Armstrong, Prudhoe, Northumberland; Lt-Cdr R. W. Banfield, Liskeard, Cornwall; A.B.(Sonar) A. R. Barr, Bridgwater, Somerset; P. O. Engineering Mechanician P. Brouard, Crewkerne, Somerset; Cook R. J. S. Dunkerley, Windsor, Berks; Acting Leading Cook M. P. Foote, Havant, Hants; Marine Engineering Mechanic (Mechanical) 1 S. H. Ford, Poole, Dorset; Acting Steward S. Hanson, Ecclesfield, Sheffield, S. Yorkshire; A.B.(Sonar) S. K. Hayward, Barrow-in-Furness, Cumbria; A.B.(Electronic Warfare) S. Heyes, St Budeaux, Devon; Weapons Engineering Mechanic (Radio) 1 S. J. Lawson, Whitley Bay, Northumberland; Marine Engineering

Mechanic (Mechanical) 2 A. R. Leighton, Margate, Kent; Air Engineering Mechanician 1 A. McAuley, Yeovil, Somerset; Acting Leading Seaman (Radar) M. S. Mullen, Roby, Liverpool, Merseyside; Lt B. Murphy, Yeovil, Somerset; Leading Physical Training Instructor G. T. Nelson, Saltash, Cornwall; Acting P.O. Weapons Engineering Mechanic (Radio) A. K. Palmer, Truro, Cornwall; Cook J. R. Roberts, Llanberis, Gwynedd; Lt-Cdr J. M. Sephton, D.S.C., Preston, Dorset; Acting Leading Marine Engineering Mechanic (Mechanical) S. J. White, Washington, Tyne and Wear; Acting Leading Marine Engineering Mechanic (Electrical) G. Whitford, Blackburn, Lancs; Marine Engineering Mechanic (Mechanical) G. S. Williams, Kidlington, Oxon.

Argonaut   Leander Class Frigate   Bombed and damaged in Falkland Sound on 21 May.
Fatal Casualties: A.B.(Radar) I. M. Boldy, Derby; Seaman (Missile) M. J. Stuart, Bredon, Hereford and Worcester (killed on his eighteenth birthday).

Arrow   Type 21 Frigate

Brilliant   Type 22 Frigate   Damaged by air attack in Falkland Sound on 21 May.

Broadsword   Type 22 Frigate   Hit by bomb, which did not explode, while operating off northern coast of West Falkland on 25 May.

Conqueror   Valiant Class Submarine

Coventry   Type 42 Destroyer   Bombed and sunk while operating off northern coast of West Falkland on 25 May. Lynx helicopter XZ 242 was lost.
Fatal Casualties: Marine Engineering Mechanic (Mechanical) 1 F. O. Armes, Norwich, Norfolk; Acting Chief Weapons Engineering Artificer J. D. L. Caddy, Eastleigh, Hants; Marine Engineering Artificer (Mechanical) 1 P. B. Callus, Emsworth, Hants; Acting P.O. Catering Accountant S. Dawson, Scunthorpe, Humberside; Acting Weapons Engineering Mechanic (Radio) 1 J. R. Dobson, Exeter, Devon; P.O.(Sonar) M. G. Fowler, Southsea, Hants; Weapons Engineering Mechanic (Ordnance) 1 I. P. Hall, Cowley, Oxon; Lt R. R. Heath, Gosport, Hants; Acting Weapons Engineering Mechanician 1 D. J. A. Ozbirn, Bishop's Waltham, Hants; Lt-Cdr G. S. Robinson-Moltke, Petersfield, Hants; Leading Radio Operator (Warfare) B. J. Still, Co. Laoise, Eire; Marine Engineering Artificer 2 G. L. J. Stockwell, Herne Bay, Kent; Acting Weapons Engineering Artificer 1 D. A. Strickland, Harrow, Greater London; A.B.(Electronic Warfare) A.D. Sunderland, Sherborne, Dorset; Marine Engineering Mechanic (Mechanical) 2 S. Tonkin, Sheffield, S. Yorkshire; Acting Cook I. Turnbull, Hartlepool, Cleveland; Acting Weapons Engineering Artificer 2 P. P. White, Pangbourne, Berks; Weapons Engineering Artificer (Apprentice) I. R. Williams, Neston, Cheshire; Laundryman Kyo Ben Kwo, Shaukiwan, Hong Kong.

Endurance   Ice Patrol Ship

Fearless   Amphibious Assault Ship   Acted as Command Ship of the Amphibious Group. Her Landing Craft Utility 'Foxtrot Four' was bombed and sunk in Choiseul Sound on 8 June with the following fatal casualties:

Royal Marines: Marine R. D. Griffin, Sheffield, S. Yorkshire; Colour-Sgt B. R. Johnston, Q.G.M., Exmouth, Devon; Sgt R. J. Rotherham, Cerne Abbas, Dorset; Marine A. J. Rundle, Cheadle, Cheshire. Royal Navy: Marine Engineering Artificer (Propulsion) A. S. James, Bishop's Waltham, Hants; Acting Leading Marine Engineering Mechanic (Mechanical) D. Miller, Pagham, W. Sussex.

Glamorgan   County Class Destroyer   Damaged by land-launched Exocet missile on the Stanley gunline on 12 June. Wessex helicopter XM 837 was destroyed.

Fatal Casualties: P.O. Air Engineering Mechanic (Electrical) M. J. Adcock, Fortuneswell, Dorset; Cook B. Easton, Portsmouth, Hants; Air Engineering Mechanic (Mechanical) 1 M. Henderson, Cumbuslang, Glasgow, Strathclyde; Air Engineering Mechanic (Radio) 1 B. P. Hinge. Bristol, Avon; Local Acting Chief Air Engineering Mechanician D. Lee, Leeds, W. Yorkshire; Air Engineering Artificer (Mechanical) 2 K. I. McCallum, Portland, Dorset; Cook B. G. Malcolm, Gosport, Hants; Marine Engineering Mechanic (Mechanical) 2 T. W. Perkins, Cardiff, S. Glamorgan; Leading Cook M. Sambles, Portsmouth, Hants; Leading Cook A. E. Sillence, Doncaster, S. Yorkshire; Steward J. D. Stroud, Gosport, Hants; Lt D. H. R. Tinker, Rochester, Kent; P.O. Aircrewman C. P. Vickers, Wyke Regis, Dorset.

Glasgow   Type 42 Destroyer   Damaged by bomb off Stanley on 12 May: no casualties.

Hermes   Aircraft-Carrier   Flagship of Task Group 317.8 operating off the Falklands from 1 May until the end of the war. (Casualties to air components of *Hermes* and *Invincible* are listed under 'Fleet Air Arm'.)

Intrepid   Amphibious Assault Ship

Invincible   Aircraft-Carrier   Acted as Air Defence Commander's ship for Task Group 317.8.

Plymouth   Rothesay Class Frigate   Bombed and damaged in Falkland Sound on 8 June.

Sheffield   Type 42 Destroyer   Hit by Exocet missile on 4 May; sank on 10 May.

Fatal Casualties: Lt-Cdr D. I. Balfour, Grayshott, Hants; P.O. Marine Engineering Mechanic (Mechanical) D. R. Briggs, D.S.M., Lee-on-Solent, Hants; Catering Assistant D. Cope, Stourport-on-Severn, Hereford and Worcester; Weapons Engineering Artificer 1 A. C. Eggington, Purbrook, Hants; Sub-Lt R. C. Emly, Havant, Hants; P.O. Cook R. Fagan, Stubbington, Hants; Cook N. A. Goodall, Enfield, Greater London; Leading Marine Engineering Mechanic

(Mechanical) A. J. Knowles, Gosport, Hants; Leading Cook T. Marshall, Gosport, Hants; Petty Officer Weapons Engineering Mechanic A. R. Norman, Gosport, Hants; Cook D. E. Osborne, Portsmouth, Hants; Weapons Electrical Artificer 1 K. R. F. Sullivan, Porchester, Hants; Cook A. C. Swallow, Bembridge, Isle of Wight; Acting Chief Weapons Mechanician M. Till, Stubbington, Hants; Weapons Engineering Mechanician 2 B. J. Wallis, Porchester, Hants; Leading Cook A. K. Wellstead, Portsmouth, Hants; Master-at-Arms B. Welsh, Gateshead, Tyne and Wear; Cook K. J. Williams, Gosport, Hants; Lt-Cdr J. S. Woodhead, D.S.C., Stubbington, Hants; Lai Chi Keung, Hong Kong.

Spartan   Swiftsure Class Submarine

Splendid   Swiftsure Class Submarine

Yarmouth   Rothesay Class Frigate

PART TWO
Later Naval Additions to Task Force

Active, *Ambuscade* and *Avenger*   Type 21 Frigates

Andromeda, *Minerva* and *Penelope*   Leander Class Frigates

Bristol   Type 82 Destroyer

Cardiff and *Exeter*   Type 42 Destroyers

Courageous and *Valiant*, Valiant Class Submarines, and *Onyx*, Oberon Class Submarine

Dumbarton Castle and *Leeds Castle*   Castle Class Patrol Vessels

Hecla, *Herald* and *Hydra*   Hecla Class Survey Ships acting as ambulance ships, ferrying wounded to Montevideo.

The deep-sea trawlers *Cordella*, *Farnella*, *Junella*, *Northella* and *Pict* were taken up from trade and operated as R.N. mine-countermeasures vessels.

PART THREE
Fleet Air Arm Squadrons

800 Squadron   Operated Sea Harriers from *Hermes* in both the ground-support and air-defence roles. The squadron can probably be credited with the destruction of thirteen Argentinian aircraft and one helicopter in the air and with two aircraft and two helicopters on the ground.
Own Aircraft Casualties:
Sea Harrier XZ 450   Shot down by ground fire while attacking Goose Green airfield on 4 May. The pilot, Lt N. Taylor, from Closworth, Dorset, was killed.
Sea Harrier ZA 192   Aircraft lost in unknown circumstances, possibly destroyed

by own bomb detonation, after taking off for operations on 23 May. Lt-Cdr G. W. J. Batt, D.S.C., from Yeovil, Somerset, was killed.

801 Squadron   Operated Sea Harriers from *Invincible*, mainly in the air-defence role. Destroyed seven Argentinian aircraft in the air and one helicopter on the ground.
Own Aircraft Casualties:
Sea Harriers XZ 452 and 453   Lost at sea while on patrol on 6 May in unknown circumstances, possibly a collision. The pilots, Lt-Cdr J. E. Eyton-Jones, from Nottingham, and Lt W. A. Curtis, from East Chinnock, Somerset, were killed.
Sea Harrier ZA 174   Lost over the side of *Invincible*; Lt-Cdr M. Broadwater was rescued.
Sea Harrier XZ 456   Shot down by Roland missile on 1 June and crashed in the sea off Stanley; F/Lt I. Mortimer was rescued by a task-force helicopter.

(Nos. 809 and 899 Squadrons provided Sea Harriers which operated under the control of 800 and 801 Squadrons. No. 1 Squadron R.A.F., provided Harrier GR3s which operated under 800 Squadron control; details of the GR3 operations are included under the 'R.A.F.' heading on page 405.)

820 Squadron   Operated Sea King 5s from *Invincible*. Flew more than 1,400 sorties in providing near-continuous anti-submarine cover for the task force over a sixty-four-day period without suffering any casualties.

824 Squadron   Operated Sea King 2s. Taken to the Falklands in *Fort Grange* and *Olmeda* and used in the transport role. No casualties.

825 Squadron   Formed from 706 Training Squadron at Culdrose. Embarked Sea King 2s in *Atlantic Causeway* and *QE2* and used ashore in later stages of the war, including the *Sir Galahad* rescue. No casualties.

826 Squadron   Operated Sea King 5s from *Hermes* and *Fort Austin* in the anti-submarine role and supporting land operations.
Sea King ZA 132   Lost at sea while on anti-submarine patrol on 12 May. Crew rescued.
Sea King XZ 573   Lost at sea on night of 18 May. Crew rescued.

845 Squadron   Operated Wessex 5s from various ships in transport and land-support roles.
Wessexes XT 464 and 473 crashed on the Fortuna Glacier, South Georgia, in a storm on 22 April and were abandoned. Crews and passengers were rescued.

846 Squadron   Operated Sea King 4s and 5s from *Hermes* and other ships in Special Forces insertion and land forces support roles.
Sea King 4 ZA 311   Crashed in sea while moving stores around task force on the evening of 23 April. The pilot was rescued but P.O. Aircrewman K. S. Casey, of Portland, Dorset, died.
Sea King 4 ZA 290   Landed near Punta Arenas, Chile, after secret operation

on 18 May and destroyed by crew. Crew unhurt and repatriated by Chileans.
Sea King 5 Z A 294   Crashed into sea while transferring troops on 19 May. The
aircrewman, Cpl M. D. Love, D.S.M., of Walton-Le-Dale, Lancs, died, together
with eighteen other men whose names are recorded elsewhere.

847 Squadron   Formed at Yeovilton, mainly from 771 and 772 Training Squadrons. Equipped with Wessex 5s, which were embarked in *Atlantic Causeway* and
Engadine. Supported final phases of land operations without suffering casualties.

848 Squadron   Formed at Yeovilton, mainly from 707 Training Squadron.
Equipped with Wessex 5s, which were embarked in *Atlantic Conveyor* and other
ships as reinforcements for operations with land forces. Six Wessex 5s – X S 480,
495, 499 and 512, and XT 476 and 483 – were lost in *Atlantic Conveyor*.
(Squadron personnel lost in *Atlantic Conveyor* are listed in that ship's entry.)

The following squadrons had helicopters flying from task-force ships as single-
helicopter detachments: 702 and 815 (Lynxes), 737 (Wessex 3s), 829 (Wasps).
Helicopters lost in *Antrim*, *Ardent*, *Coventry* and *Glamorgan* have already been
noted. Lynx X Z 700 was lost in *Atlantic Conveyor*.

PART FOUR
The Royal Fleet Auxiliary

Logistic Landing Ships: *Sir Bedivere, Sir Galahad, Sir Geraint, Sir Lancelot, Sir
Percivale, Sir Tristram.*
   Sir Galahad and *Sir Tristram* were bombed at Fitzroy on 8 June.
Fatal Casualties: *Sir Galahad* – 3rd Engineering Officer C. F. Hailwood,
Farnborough, Hants; 2nd Engineering Officer P. A. Henry, Berwick-upon-
Tweed, Northumberland; 3rd Engineering Officer A. J. Morris, Poole, Dorset;
Electrical Fitter Leung Chau, Kowloon, Hong Kong; Butcher Sung Yuk Fai,
Kowloon, Hong Kong. Forty military casualties are recorded elsewhere. *Sir
Tristram* – Bosun Yu Sik Chee, Kowloon, Hong Kong; Seaman Yeung Swi Kami,
New Territories, Hong Kong.
Fleet and Support Tankers: *Appleleaf, Bayleaf, Blue Rover, Brambleleaf,
Olmeda, Olna, Pearleaf, Plumleaf, Tidepool, Tidespring.*
Supply Ships: *Fort Austin, Fort Grange, Regent, Resource, Stromness.*
Helicopter Support Ship: *Engadine.*
   The Royal Maritime Auxiliary Service ships *Goosander* (heavy-lift salvage
ship) and *Typhoon* (long-range salvage tug) also served with the Task Force.

PART FIVE
Ships Taken Up from Trade

Liners: *Canberra, Queen Elizabeth II, Uganda.*
Passenger/General Cargo Ships: *Norland, Rangatira, St Edmund, St Helena.*
Container and Cargo Ships: *Astronomer, Atlantic Causeway, Atlantic Conveyor*
(see below), *Avelona Star, Baltic Ferry, Contender Bezant, Elk, Europic Ferry,
Geestport, Laertes, Lycaon, Nordic Ferry, Saxonia, Strathewe, Tor Caledonia.*

Tankers: *Alvega, Anco Charger, Balder London, British Avon, British Dart, British Esk, British Tamar, British Tay, British Test, British Trent, British Wye* (bombed and damaged on 29 May: no casualties), *Fort Toronto, G. A. Walker, Scottish Eagle, Shell Eburna.*

Offshore Support Vessels: *British Enterprise III, Stena Inspector, Stena Seaspread, Wimpey Seahorse.*

Tugs: *Irishman, Salvageman, Yorkshireman.*

Cable Ship: *Iris.*

The following fatal casualties were suffered when *Atlantic Conveyor* was hit by an Exocet missile on 25 May: Merchant Navy – Bosun J. B. Dobson, Exmouth, Devon; Mechanic F. Foulkes, Great Plumpton, near Kirkham, Lancs; Steward D. R. S. Hawkins, Newquay, Cornwall; Mechanic J. Hughes, Gosport, Hants; Captain (Ship's Master) I. H. North, D.S.C., Doncaster, S. Yorkshire; Mechanic E. N. Vickers, Middlesbrough, Cleveland; Royal Navy – Air Engineering Mechanic (Radio) 1 A. U. Anslow (845 Squadron), Wolverhampton, W. Midlands; Chief P.O. Writer E. Flanagan, Gillingham, Kent; Leading Air Engineering Mechanic (Electrical) 1 D. Pryce (845 Squadron), Salisbury, Wilts; Royal Fleet Auxiliary – Radio Officer R. Hoole, Sutton-in-Ashfield, Notts; Seamen Ng Por and Chan Chi Shing, Hong Kong.

Ten helicopters were lost in *Atlantic Conveyor*. Six Wessexes and three Chinooks are recorded elsewhere; the tenth was Lynx XZ 700.

PART SIX
Royal Marine Units

(The following common abbreviations are used here and in the 'Army Units' section: Capt. – Captain, Lt – Lieutenant, Sgt – Sergeant, L/Sgt – Lance-Sergeant, Cpl – Corporal, L/Cpl – Lance-Corporal, Mne – Marine, Pte – Private, Gdsm. – Guardsman)

INFANTRY UNITS

40 Commando   Landed at San Carlos on D-Day and used mainly for defence of San Carlos base area.

Fatal Casuality: Mne S. G. McAndrews, Wythenshawe, Greater Manchester, killed in an air raid at San Carlos on 27 May.

42 Commando   Detached M Company for the capture of South Georgia. Main unit landed at Port San Carlos on D-Day. Moved to Mount Kent area by helicopter and fought the Battle of Mount Harriet.

Fatal Casualties: Cpl J. Smith, Torquay, Devon, killed by shell fire on 11 June; Cpl L. G. Watts, Watford, Herts, killed in the Mount Harriet action.

45 Commando   Landed at Ajax Bay on D-Day. 'Yomped' most of the way across East Falkland and fought the Battle of Two Sisters.

Fatal Casualties: In air raid at Ajax Bay on 27 May – Mne P. D. Callan, Great Sutton, Cheshire; Sgt R. Enefer, Plympton, Devon; Cpl K. Evans, Waterlooville, Hants; Mne P. B. McKay, Macduff, Grampian; Mne D. Wilson, Glenrothes, Fife. In patrol incident near Mount Kent on 11 June – Cpl P. R. Fitton, Bilston,

W. Midlands; Sgt R. A. Leeming, Arbroath, Tayside; Mne K. Phillips, Dartford, Kent; Cpl A. B. Uren, Lapford, Devon. In action on Two Sisters – Mne C. G. Macpherson, Oban, Strathclyde; Mne M. J. Nowak, Derby; Cpl I. F. Spencer, Arbroath, Tayside.

3 Commando Brigade Air Squadron   Equipped with nine Gazelle and six Scout helicopters and operated in support of the land campaign throughout.
Helicopter and Aircrew Casualties:
Gazelle XX 411   Shot down by ground fire near Port San Carlos on 21 May. The pilot, Sgt A. P. Evans from Landrake, Cornwall, was killed.
Gazelle XX 402   Shot down by ground fire while attacking Argentinian troops near Port San Carlos on 21 May. The crew, Lt K. D. Francis from Fyfield, Essex, and L/Cpl B. P. Giffin from Christchurch, Hants, were killed.
Scout XT 629   Shot down by Pucará near Camilla Creek while supporting 2 Para on 28 May. The pilot, Lt R. J. Nunn, D.F.C., from Trenewan, Cornwall, was killed.

Special Boat Squadron   Carried out many special operations in the recapture of both South Georgia and the Falklands.
Fatal Casualty: Sgt I. N. ('Kiwi') Hunt, of New Zealand and Poole, Dorset, was killed in an accidental patrol clash with the S.A.S. near Teal on 2 June.

Commando Logistic Regiment   Established supply base at Ajax Bay.
Fatal Casualty: Mne C. Davison, Killingworth, Newcastle-upon-Tyne, Tyne and Wear, was killed in the 27 May air raid at Ajax Bay.

OTHER ROYAL MARINE UNITS

3 Commando Brigade Headquarters and Signals Squadron, 1st Raiding Squadron, Mountain and Arctic Warfare Cadre, Air Defence Troop, Y Troop (electronic warfare), Commando Forces Band (stretcher-bearers) and minor detachments. There were no fatal casualties in these units.

PART SEVEN
Army Units

2nd Battalion The Parachute Regiment   Landed at San Carlos on D-Day and fought the Battles of Goose Green and Wireless Ridge.
Fatal Casualties: At Goose Green – Lt J. A. Barry, St Albans, Herts; L/Cpl G. D. Bingley, M.M., Beverley, Humberside; L/Cpl A. Cork, Canterbury, Kent; Capt. C. Dent, Whitley Bay, Tyne and Wear; Pte S. J. Dixon, Basildon, Essex; Pte M. W. Fletcher, Stockport, Cheshire; Cpl D. Hardman, Hamilton, Strathclyde; Pte M. Holman-Smith, Bodmin, Cornwall; Pte S. Illingsworth, D.C.M., Doncaster, S. Yorkshire; Lt-Col. H. Jones, V.C., O.B.E., Kingswear, Devon; Pte T. Mechan, Glasgow, Strathclyde; Cpl S. N. Prior, Brighton, E. Sussex; L/Cpl N. R. Smith, Cheltenham, Glos; Cpl P. S. Sullivan, Ravensthorpe, Northants; Capt. D. A. Wood, Gillingham, Kent. On Wireless Ridge – Colour/Sgt G. P. M. Findlay, Grimsby, Humberside; Pte D. A. Parr, Lowestoft, Suffolk; Pte F. Slough, Reading, Berks.

3rd Battalion The Parachute Regiment   Landed at Port San Carlos on D-Day. Marched across East Falkland and fought the Battle of Mount Longdon.
Fatal Casualties on Mount Longdon: Pte R. J. Absolon, M.M., Weybridge, Surrey; Pte G. Bull, Brixworth, Northants; Pte J. S. Burt, Walthamstow, Greater London; Pte J. D. Crow, Tonbridge, Kent; Pte M. S. Dodsworth, Walsall, Staffs; Pte A. D. Greenwood, Withington, Greater Manchester; Pte N. Grose, Gosport, Hants; Pte P. J. Hedicker, Aldershot, Hants; L/Cpl P. D. Higgs, Ashbury, Oxon; Cpl S. Hope, Heywood, Greater Manchester; Pte T. R. Jenkins, Ross-on-Wye, Hereford and Worcester; Pte C. E. Jones, Greens Norton, Northants; Pte S. I. Laing, Lanchester, Durham; L/Cpl C. K. Lovett, Worthing, W. Sussex; Cpl K. J. McCarthy, Frome, Somerset; Sgt I. J. McKay, V.C., Rotherham, S. Yorks; Cpl S. P. F. McLaughlin, Wallasey, Merseyside; L/Cpl J. H. Murdoch, Renfrew, Strathclyde; L/Cpl D. E. Scott, Winslow, Bucks; Pte I. P. Scrivens, Yeovil, Somerset; Pte P. A. West, Newcastle-upon-Tyne, Tyne and Wear; Craftsman A. Shaw, R.E.M.E., from Corby, Northants, attached to 3 Para.

1st Battalion Welsh Guards   Landed at San Carlos, ferried by ship to Bluff Cove and Fitzroy, part of battalion bombed in the *Sir Galahad*.
Fatal Casualties on *Sir Galahad*: L/Cpl A. Burke, Wrexham, Clwyd; L/Sgt J. R. Carlyle, Ruthin, Clwyd; Gdsm. I. A. Dale, Pontypridd, Mid Glamorgan; Gdsm. M. J. Dunphy, Llechryd, Dyfed; Gdsm. P. Edwards, Denbigh, Clwyd; Sgt C. Elley, Pontypridd, Mid Glamorgan; Gdsm. M. Gibby, Waitstown, Rhondda, Mid Glamorgan; Gdsm. G. C. Grace, Newport, Gwent; Gdsm. P. Green, Rhyl, Clwyd; Gdsm. G. M. Griffiths, Llandeilo, Dyfed; Gdsm. D. N. Hughes, Wrexham, Clwyd; Gdsm. G. Hughes, Llanfairfechan, Gwynedd; Gdsm. B. Jasper, Pontypridd, Mid Glamorgan; Gdsm. A. Keeble, Pontyclun, Mid Glamorgan; L/Sgt K. Keoghane, Newport, Gwent; Gdsm. M. J. Marks, Stanford-le-Hope, Essex; Gdsm. C. Mordecai, Maesteg, Mid Glamorgan; L/Cpl S. J. Newbury, Pentwyn, Cardiff, S. Glamorgan; Gdsm. G. D. Nicholson, Bridgend, Mid Glamorgan; Gdsm. C. C. Parsons, Cardiff, S. Glamorgan; Gdsm. E. J. Phillips, Carmarthen, Dyfed; Gdsm. G. W. Poole, Pontypridd, Mid Glamorgan; Gdsm. N. A. Rowberry, Cardiff, S. Glamorgan; L/Cpl P. A. Sweet, Cwmbach, Mid Glamorgan; Gdsm. G. K. Thomas, Cardiff, S. Glamorgan; L/Cpl N. D. M. Thomas, Llanelli, Dyfed; Gdsm. R. G. Thomas, Barry, S. Glamorgan; Gdsm. A. Walker, York, N. Yorkshire; L/Cpl C. F. Ward, Feltham, Greater London; Gdsm. J. F. Weaver, Port Talbot, W. Glamorgan; Sgt M. Wigley, Connah's Quay, Clwyd; Gdsm. D. R. Williams, Holyhead, Anglesey, Gwynedd. Attached troops who died on *Sir Galahad*: Army Catering Corps – L/Cpl B. C. Bullers, Walsall, Staffs; Pte A. M. Connett, Manchester; Pte M. A. Jones, Carmarthen, Dyfed; Pte R. W. Middlewick, Brighton, E. Sussex. R.E.M.E. – Craftsman M. W. Rollins, Birmingham, W. Midlands; L/Cpl A. R. Streatfield, Woking, Surrey. L/Cpl C. C. Thomas died of wounds inflicted by shell fire on 13 June.

2nd Battalion Scots Guards   Landed at San Carlos, ferried by ship to Bluff Cove and fought the Battle of Tumbledown Mountain.
Fatal Casualties on Tumbledown: Gdsm. D. J. Denholm, Edinburgh; Gdsm. D. Malcolmson, Irvine, Strathclyde; L/Sgt C. Mitchell, Edinburgh; Gdsm. J. B. C. Reynolds, D.C.M., Renfrew, Strathclyde; Sgt J. Simeon, Knightswood, Glasgow;

Gdsm. A. G. Stirling, Pollock, Glasgow; Gdsm. R. Tanbini, Dundee, Tayside; Warrant Officer II D. Wight, Edinburgh.

1/7th Gurkha Rifles Landed at San Carlos, carried out minor operations around Goose Green, ferried by sea to Bluff Cove and were preparing to assault Mount William when cease-fire was ordered. The battalion suffered no fatal casualties in the war but one Gurkha was killed later while carrying out battlefield clearance at Goose Green.

22nd Special Air Service Regiment D and G Squadrons were deployed and took part in many special operations including the recapture of South Georgia, the raid on Pebble Island, a raid on Stanley Harbour and much reconnaissance work.
Fatal Casualties: In Sea King Z A 294 which crashed in the sea while transferring troops on 19 May – Cpl R. E. Armstrong, Hereford; Sgt J. L. Arthy, Hereford; W.O. I M. Atkinson, Hereford; Cpl W. J. Begley, Hereford; Sgt P. A. Bunker, Hereford; Cpl R. A. Burns, Dundee; Sgt P. P. Currass, Hereford; Sgt S. A. I. Davidson, Hereford; W.O. II L. Gallagher, Hereford; Sgt W. C. Hatton, Hereford; Sgt W. J. Hughes, Hereford; Sgt P. Jones, Hereford; L/Cpl P. N. Lightfoot, Hereford; Cpl M. V. McHugh, Stockport, Manchester; Cpl J. Newton, Hereford; S/Sgt P. O'Connor, Hereford; Cpl S. J. G. Sykes, Huddersfield, Yorks; Cpl E. T. Walpole, Hereford.
Capt. G. J. Hamilton, M.C., Hereford, was killed near Port Howard on 10 June.

Royal Engineers 59 Independent Commando Squadron and parts of 33, 36 and 38 Engineer Regiments were deployed.
Fatal Casualties: 9 Parachute Squadron – Cpl A. G. McIlvenny, Rainham, Kent, and Sapper W. D. Tarbard, Derby, on the *Sir Galahad*; Cpl S. Wilson, Aldershot, Hants, with 3 Para on Mount Longdon; L/Cpl J. B. Pashley, Eckington, Derby, with 2nd Scots Guards in diversion action near Tumbledown. 49 Squadron – Sgt J. Prescott, Hindley, Lancs, in H.M.S. *Antelope*. 59 Independent Commando Squadron – Sapper P. K. Ghandi, Wembley, London, in the air raid at San Carlos; Cpl M. Melia, Plymouth, Devon, with 2 Para at Goose Green; Sapper C. A. Jones, Cinderford, Glos, with 45 Commando on Two Sisters.

Royal Signals The 5th Infantry Brigade Headquarters and Signals Squadron and parts of 14th and 30th Signal Regiments and of 602 Signal Troop were deployed.
Fatal Casualties: With 603 Tactical Air Control Party – Cpl D. F. McCormack, Glasgow, in S.A.S. helicopter crash into sea on 19 May. 5th Infantry Brigade Headquarters and Signals Squadron – S/Sgt J. I. Baker, Rothwell, Northants, and Maj. M. L. Forge, Rochester, Kent, in 656 Squadron helicopter which was accidentally shot down by H.M.S. *Cardiff* on the night of 5/6 June.

Royal Army Medical Corps 16 Field Ambulance and parts of 19 Field Ambulance were deployed.
Fatal Casualties: L/Cpl I. R. Farrell, Everton, Liverpool, and Maj. R. Nutbeem, Aldershot, Hants, both of 16 Field Ambulance, in the *Sir Galahad*.

656 Squadron, Army Air Corps Deployed three Scout helicopters with 3 Commando Brigade and then six Gazelles and three Scouts with 5 Brigade. Gazelle XX 377. Accidentally shot down by Sea Dart of H.M.S. *Cardiff* on the night of 5/6 June. The crew – S/Sgt C. A. Griffin, Netheravon, Wilts, and L/Cpl S. J. Cockton, Camberley, Surrey – were killed, together with the two Royal Signals passengers.

Royal Artillery 29th Commando Regiment (H.Q., 7, 8, 79 and 148 Batteries), 4th Field Regiment (H.Q., 29 and 97 Batteries), 'T' Battery of 12th Air Defence Regiment and parts of 32nd Guided Weapons and 49th Field Regiments were deployed. No fatal casualties.

Blues and Royals 3 and 4 Troops, with four Scorpions and four Scimitars and one recovery vehicle, were deployed. No fatal casualties.

Other Army Units Royal Corps of Transport: parts of 17 Port and 29 Movements Regiments, part of 47 Air Despatch Squadron, 407 Troop. Royal Army Ordnance Corps: 81 Ordnance Company and part of 9 Ordnance Battalion. Royal Electrical and Mechanical Engineers: 10 Field Workshop and part of 70 Aircraft Workshop. Part of 160 Provost Company, Royal Military Police. 6 Field Cash Office, Royal Army Pay Corps. 601, 602 and 603 Tactical Air Control Parties. The only fatal casualties in these units were two men of 603 Tactical Air Control Party whose names are recorded under 'Royal Signals' and 'Royal Air Force'.

PART EIGHT
Royal Air Force

F/Lt G. W. Hawkins, of Wokingham, Berks, and R.A.F. Upavon, was the only R.A.F. fatal casualty in Operation *Corporate*. He died in the Sea King which crashed into the sea on 19 May.

1 Squadron Operated Harrier GR3s in ground-attack role from *Hermes* and from ground strip at Port San Carlos. Flew approximately 120–130 operational sorties.
Aircraft Casualties:
Harrier XZ 972 Shot down in the sea off Port Howard on 21 May. The pilot, F/Lt G. Glover, was injured and taken prisoner.
Harrier XZ 988 Hit by ground fire over Goose Green on 27 May and crashed seven miles away. S/Ldr R. Iveson was later picked up by a task-force helicopter.
Harrier XZ 989 Hit by ground fire near Mount Harriet on 30 May, lost fuel and crashed in the sea. S/Ldr J. J. Pook ejected and was picked up by a task-force helicopter.

18 Squadron Deployed five Chinook heavy lift helicopters on *Atlantic Conveyor*. One was left to work at Ascension and three – ZA 706, ZA 716 and ZA 719 – were lost on *Atlantic Conveyor*. The only Chinook to reach the Falklands – ZA 718 – supported the land forces in the final weeks of the war.

Vulcan Detachment Aircraft and crews of 44, 50 and 101 Squadrons were deployed to Ascension Island. The detachment carried out three bombing and two missile raids in the Stanley area; a further missile raid had to be aborted just north of the Falklands.

Victor Detachment   Aircraft and crews of 55 and 57 Squadrons and of the Victor Operational Conversion Unit carried out a very large number of refuelling missions between the United Kingdom and Ascension and between Ascension and the Falklands, as well as carrying out very long-range reconnaissance flights during the operation to recapture South Georgia.

Nimrod Detachment   Aircraft and crews of 42, 120, 201 and 206 Squadrons flew a total of 402 training and operational missions in support of Operation *Corporate*. An aircraft of 42 Squadron flew the first of many maritime reconnaissance missions from Ascension and other squadrons flew very long-range missions in support of the task force and of the Black Buck operations.

Air Transport Force   VC10s of 10 Squadron and Hercules aircraft of 24, 30, 47 and 70 Squadrons flew a large number of sorties between the United Kingdom and Ascension. In addition, 47 Squadron flew twenty-nine very long-range sorties south of Ascension, dropping urgent supplies and key personnel into the sea near task-force ships.

Ascension Island Defence and Support   29 Squadron provided Phantom FGR2s for local defence and 202 Squadron provided Sea Kings for search and rescue work.

Other R.A.F. Units   Elements of Nos. 3, 15 and 63 R.A.F. Regiment Squadrons, the R.A.F. Explosive Ordnance Disposal Team and other R.A.F. detachments took part in Operation *Corporate*.

PART NINE
Falkland Islands

The British Roll of Honour is concluded with the names of the three Falkland Islanders who were accidentally killed by shell fire on the night of 11/12 June: Mrs Doreen Bonner, Mrs Mary Goodwin, Mrs Susan Whitley.

Appendix 2

Argentinian Forces in the Falklands

A full Argentinian Order of Battle has not yet been published but the following details are believed to be reliable.

III BRIGADE
(Commander: Brigadier-General Omar Edgardo Parada)

It is possible that the whole of this brigade was recruited in the province of Corrientes; the 4th and 12th Regiments were certainly recruited there.

The brigade was airlifted to the Falklands in the third week of April and was primarily intended to provide the main garrison on West Falkland and at Goose Green. Brigade Headquarters was intended to move to Goose Green but never left Stanley.

4th Regiment (Lieutenant-Colonel Diego Alejandro Soria)
Allocated the most westerly sector of the defences outside Stanley and therefore occupied the most ill-prepared defences and took the first shock of the British advance, suffering casualties in patrol actions, by artillery and Harrier attacks, and in the fighting on Two Sisters and Mount Harriet.

5th Regiment (Colonel Juan Ramón Mabragana)
Stationed in and around Port Howard. Possibly suffered minor casualties through air action and naval shelling, but not involved in ground action. Probably suffered from shortage of rations and sickness.

8th Regiment (Colonel Ernesto Alyendro Repossi)
Stationed at Fox Bay and had similar experience to the 5th Regiment.

12th Regiment (Lieutenant-Colonel Ítalo Angel Piaggi)
Based mainly at Goose Green/Darwin and maintained an outpost on Fanning Head. One company remained in the Mount Challenger area until helicoptered to Goose Green during the British attack there.

The regiment suffered heavy casualties in the fighting at Goose Green and most of those men who were not killed surrendered to 2 Para.

X BRIGADE
(Brigadier-General Oscar Luiz Joffre)

Brigade Headquarters was flown to Stanley on 11 April. The main part of the brigade quickly followed and was concentrated around Stanley. Brigade Headquarters was housed at the Town Hall for four days before moving to the intended permanent location at Moody Brook. British naval shelling forced the headquarters to leave on 30 April and it moved to Stanley House in Stanley from where Brigadier-General Joffre and his staff conducted the defensive operations around Stanley.

3rd Regiment   (Lieutenant-Colonel Davíd Ubaldo Comini)
Recruited in the La Tablada area of Buenos Aires province. The regiment started its move to the Falklands on 11 April and may have been the first reinforcement sent after Britain announced the dispatch of the Task Force. The regiment's main sector was immediately south of Stanley with the task of repelling any British landing on the south coast. This sector was never attacked by British ground forces, though casualties were suffered through naval shelling and by the dispatch of companies to bolster the sectors farther west when these were threatened by the British advance.

6th Regiment   (Lieutenant-Colonel Jorge Halperín)
Allocated to the defence of the Stanley airfield peninsula, it was thus not in direct action with British forces although regularly harassed by air attack and shelling.

7th Regiment   (Lieutenant-Colonel Omar Giménez)
Recruited in the La Plata province. Allocated to the sector which included Mount Longdon, Wireless Ridge and the narrow peninsula north of Stanley Harbour. Involved in heavy fighting with 3 Para on Mount Longdon and with 2 Para on Wireless Ridge.

INDEPENDENT INFANTRY UNITS

2nd Marine Infantry Battalion   (Commander not known)
An amphibious specialist unit which provided the main landing force in the Argentinian invasion on 2 April. The battalion was withdrawn to the mainland immediately afterwards and took no part in the later fighting.

5th Marine Infantry Battalion
(*Capitán de fragata* Carlos Hugo Robacio)
This unit, trained for operations on land, was flown in to replace the 5th Marines. It was allocated the sector which included Tumbledown Mountain, Mount William and Sapper Hill and had a tough battle with the Scots Guards on Tumbledown as well as suffering artillery and air attack.

25th Regiment   (Colonel Mohamed Ali Seinerdin)
This was a symbolic national unit, being made up with troops from every province in Argentina. It was flown into Stanley on 2 April but was not involved

in any fighting on that day. It was thereafter intended to provide the main garrison in the Falklands while an Argentinian civil administration was being established, but that role was overtaken by the arrival of the British Task Force and the subsequent fighting. C Company was sent to Goose Green and was engulfed in the battle there but most of the rest of the regiment remained in the area of Stanley airfield where its members were regularly bombed and shelled.

OTHER UNITS

Los Búzos Tácticos, sometimes known as the *Buzo Táctico*
(*Capitán de fragata* Pedro Edgardo Giachino)
This specialist marine unit is described in Argentinian documents as being 'The Marine Amphibious Reconnaissance Company and Marine Tactical Divers'. It landed at Mullet Creek on the night of 1/2 April and assaulted Moody Brook Barracks, which were empty, and Government House, which was defended by Royal Marines. It was at Government House that Captain Giachino and some of his men were killed. The unit was then withdrawn to its main base at Mar del Plata on the mainland.

Artillery

The Argentinian artillery units involved in the main fighting were the 3rd Artillery Battalion (Lieutenant-Colonel Martín Antonio Balza), the 4th Air Portable Group (Lieutenant-Colonel Carlos Alberto Quevedo) and the 601st Anti-Aircraft Battalion (Lieutenant-Colonel Héctor Lubín Arias).

Minor Units

The 601st and 602nd Commando Companies provided detached parties for reconnaissance and 'special forces' work and were used as 'stiffeners' – particularly as snipers – in the main Argentinian defensive positions. It is believed that the 601st operated mainly with III Brigade and the 602nd with X Brigade.

The 9th (at Port Howard and Fox Bay), 10th and 601st Engineer Companies, the 10th Armoured Cavalry Squadron (almost immobile at Stanley) and the 181st Military Police and Intelligence Company also served in the Falklands.

Acknowledgements

I am very grateful for the help given by the following men and women who were involved either in the Argentinian invasion and occupation of the Falklands and South Georgia or in the subsequent campaign to reoccupy those places. The different groups are included roughly in the order of their involvement. Ranks shown were those of April–June 1982; titles and decorations were those held at that time or awarded immediately after the war.

BRITISH FORCES IN THE SOUTH ATLANTIC, 2 APRIL 1982

H.M.S. *Endurance:* Capt. N. J. Barker, C.B.E.; *Royal Marines, Stanley:* Cpl C. F. Bryan, Maj. M. J. Norman; *Royal Marines, South Georgia:* Mne S. P. Chubb. (The Royal Marines subsequently served in J Coy, 42 Commando in Operation *Corporate.*)

FALKLAND ISLANDERS

Stanley: Basil and Betty Biggs, Myriam Booth, Rex Browning, Nidge Buckett, B.E.M., and Ron Buckett, Annie and Tony Chater, Gerald Cheek, Bill Etheridge, John and Veronica Fowler, Bob and Judy Gilbert, Jill Harris and Les Harris, B.E.M., Paul Howe, Sir Rex Hunt, C.M.G., Alison, Barbara and Nanette King, Anton Livermore, David McLeod, George and Velma Malcolm, Sydney Miller, Father Augustine Monaghan, Harold Rowlands, Rob and Val Rutterford, Monsignor Daniel Spraggon, O.B.E., Joan and Terry Spruce, Brian Summers, Owen Summers, Eileen Vidal, B.E.M., and Leona Vidal, Stuart Wallace, Patrick Watts, M.B.E.
Burntside House: Gerald and Kay Morrison. *Estancia House:* Richard Stevens. *Goose Green:* Gavin Browning, Ken Clitheroe, Finlay Ferguson, Eric Goss, M.B.E., and Shirley Goss. *Keppel Island:* Sam Miller. *Port San Carlos:* Alan Miller (died 1984). *Salvador Settlement:* Robin and Saul Pitaluga. *San Carlos:* Pat Short and family.

COMMANDS AND STAFF

Chief of the Defence Staff: Admiral of the Fleet Sir Terence Lewin, G.C.B., M.V.O., D.S.C. *First Sea Lord:* Admiral Sir Henry Leach, K.C.B. *Task Force Commander:* Admiral Sir John Fieldhouse, G.C.B., G.B.E. *Staff at Northwood:* Cdr F. B. Goodson, O.B.E., Capt. R. J. Husk, Lt-Cdr N. R. Messinger, R.D., Cdr T. G. Maltby. *Commander Land Forces:* Major-General Sir Jeremy Moore, K.C.B., O.B.E., M.C. *General Moore's Staff:* Col. I. S. Baxter, C.B.E., Col. R. F. Preston. *Air Commander:* Air Marshal Sir John Curtiss, K.C.B., K.B.E. *Commander Task Group 317.8:* Rear-Admiral Sir John Woodward, K.C.B. *Admiral Woodward's Staff:* Cdr C. R. Hunneyball. *Commodore Amphibious Warfare:* Commodore M. C. Clapp, C.B., A.D.C. *Commodore Clapp's Staff:* Lt-Cdr T. J. Stanning, Maj. A. Todd.

ROYAL NAVAL SHIPS

Alacrity: Cdr C. J. S. Craig, D.S.C. *Antelope:* C.P.O. A. J. Baker, C.P.O. R. A. Shadbolt, Cdr N. J. Tobin, D.S.C., Lt-Cdr H. N. Watson. *Antrim:* Lt-Cdr D. Ford, Detachment Sgt/Maj. G. R. Rowe, R.M., A.B.(M) K. L. Rowe, Cdr R. J. Sandford, O.B.E. *Ardent:* C.M.E.A.(P) M. J. Cox, W.E.M.1 D. P. Lee, C.P.O. N. D. Lee, Cdr A. W. J. West, D.S.C. *Argonaut:* Capt. C. H. Layman, D.S.O. *Arrow:* Cdr P. J. Bootherstone, D.S.C. *Avenger:* Master-at-Arms W. Jarvis, Capt. H. M. White. *Brilliant:* Capt. J. F. Coward, D.S.O. *Bristol:* M.E.M.1(M) M. J. Fordham, Capt. A. Grose, L/A C.M.E.M.N.(L) L. W. Mayor. *Broadsword:* Capt. W. R. Canning, D.S.O. *Cardiff:* Capt. M. G. T. Harris. *Conqueror:* Cdr C. L. Wreford-Brown, D.S.O. *Coventry:* Capt. D. Hart-Dyke, M.V.O., A.B. S. Ingleby, Sub-Lt A. G. Moll, P.O.(M) B. C. Savage, R.Op1stCl. P. D. Smith, Ldg Smn(M) K. Stuart. *Fearless:* L.W.E.M.(R) J. H. Jones, Lt-Cdr J. D. D. Whitehead. *Glamorgan:* Capt. M. E. Barrow, D.S.O., P.O.W.E.M.(O) C. M. H. James, P.O.W.E.A.2 R. J. Vere. *Glasgow:* C.P.O. T. W. Thackrah. *Hermes:* C.P.O. M. K. Creek, C.P.O. A. L. Taylor, Naval Airman 1 A. J. Wroot. *Intrepid.* Sgt R. C. J. Bucksey, R.M., Ldg Smn(R) A. J. Lee. *Invincible:* Lt-Cdr G. W. Frazer, Lt-Cdr P. J. Ross, Yeoman P. B. Tattum. *Plymouth:* L.M.E.M.(L) A. Millar, Capt. D. Pentreath, D.S.O. *Sheffield:* Ldg Smn(EW) A. D. Gilchrist, Lt I. K. Goddard, C.P.O. A. H. Hirji, A.B. C. P. Megson, A.B. M. Morton, Capt. J. F. T. G. Salt, P.O.(R) P. Sheppard. *Spartan:* Cdr J. B. Taylor. *Yarmouth:* Cook A. S. Kay, Ldg Smn G. Lote, Lt-Cdr M. J. C. Page.

FLEET AIR ARM SQUADRONS

800 Squadron: F/Lt D. H. S. Morgan, D.S.C. *801 Squadron:* Lt-Cdr D. D. Braithwaite, Lt B. D. Haigh, Lt M. W. Watson. *845 Squadron:* P.O.A.C.M.N. J. A. Balls, B.E.M., Lt M. J. Crabtree. *846 Squadron:* P.O.A.C.M.N. A. B. T. Ashdown, P.O.A.C.M.N. R. Burnett, C.P.O.A.C.M.N. M. J. Tupper, D.S.M. *848 Squadron* (on *Atlantic Conveyor*): L.A.E.M.(R) M. R. Green, A.E.M.N.1 N. D. Stronach.

416   Acknowledgements

ROYAL FLEET AUXILIARY AND MERCHANT SHIPS

Atlantic Conveyor: P.O. L. A. F. Cobbett (see also 848 Squadron above). *Sir Galahad:* L.W.E.M.(O) M. Lough, Capt. P. J. G. Roberts, D.S.O. (see also L/Cpl Gilbert under 'The Royal Marines' and L/Cpl Skinner under 'The Army' below). *Sir Lancelot:* C.P.O. G. P. Nicklin. *Uganda:* First Officer J. P. Harris, Lt-Cdr R. A. Pollard.

THE ROYAL AIR FORCE

1 Squadron: F/Lt J. W. Glover, F/Lt T. A. Harper, S/Ldr R. D. Iveson, S/Ldr J. J. Pook, D.F.C., W/Cdr P. T. Squire, D.F.C., A.F.C. *18 Squadron:* F/O N. T. Bausor, F/Lt A. A. Lawless. *44 Squadron:* S/Ldr A. C. Montgomery. *47 Squadron:* F/Lt A. K. Desai, F/Lt C. P. A. Harris. *50 Squadron:* F/Lt D. A. Castle, F/O C. C. M. Lackman, S/Ldr C. N. McDougall, D.F.C., S/Ldr R. J. Reeve, F/Lt R. Trevaskus. *55 Squadron:* F/Lt A. M. Skelton. *57 Squadron:* W/Cdr A. M. Bowman, M.B.E., F/Lt P. G. Heath, F/Lt P. A. Standing, F/Lt H. M. Williams. *101 Squadron:* F/Lt G. C. Graham, F/O P. L. Taylor, F/Lt W. F. M. Withers, D.F.C., F/Lt R. D. Wright. *206 Squadron:* W/Cdr D. Emmerson, A.F.C. *No. 1 Explosive Ordnance Unit:* F/Lt A. J. Swan, Q.G.M. *Other R.A.F.:* F/Lt D. S. Davenall, tanker planning staff at Ascension; S/Ldr A. L. Gordon, Air Liaison Officer at 5 Brigade H.Q.; S/Ldr B. S. Morris, O.B.E., A.F.C., Harrier GR3 Air Liaison Officer on *Hermes* and later commander of the forward Harrier strip at Port San Carlos.

THE ROYAL MARINES

42 Commando Detachment in South Georgia Reoccupation: Maj. J. M. G. Sheriden, L/Cpl N. R. Young. *3 Commando Brigade H.Q. and attached units:* Capt. R. D. Bell, Capt. R. J. Boswell, Lt-Cdr M. J. Callaghan, Maj. J. S. Chester, O.B.E., Capt. D. M. Constance, P.O. P. E. Holdgate, Sgt D. Munnelly, Lt R. F. Playford, Capt. M. J. Samuelson, Maj. S. E. Southby-Tailyour, Brig. J. H. A. Thompson, C.B., O.B.E., A.D.C. *40 Commando:* Capt. S. J. D. Bush, Mne M. P. Spence, Capt. R. J. Williams. *42 Commando:* Capt. P. M. Babbington, M.C., Lt-Cdr N. T. Brown, Mne M. A. Curtis, Cpl M. Eccles, M.M., Mne M. A. Hagyard, Capt. I. McNeill, C.S.M. C. J. March, Mne N. P. Rees, Capt. D. G. Wheen, Mne M. A. Wright. (See also *Royal Marines, Stanley* and *South Georgia* under 'British Forces in the South Atlantic, 2 April 1982' above.) *45 Commando:* Lt C. I. Dytor, M.C., Capt. I. R. Gardiner, C.S.M. T. M. Gibson, Mne B. T. Hughes, Sgt D. J. Malone, Mne S. G. Oyitch. *1st Raiding Squadron:* Capt. C. I. J. Baxter, L/Cpl C. B. Gilbert, Cpl N. P. Smith. *3 Commando Brigade Air Squadron:* Capt. P. L. Bancroft, Lt-Cdr G. R. A. Coryton, Capt. A. B. Newcombe, Capt. N. E. Pounds. *Commando Logistic Regiment:* Maj. T. P. P. Knott, M.C.

THE ARMY

2 Para: Lt-Col D. R. Chaundler, 2/Lt M. A. Coe, Lt C. S. Connor, M.C., Capt. A. P. Coulson, L/Cpl G. M. Johnston, Maj. C. P. B. Keeble, D.S.O., Maj. P. Neame, L/Cpl M. Robbins, Pte D. L. Rogerson, Pte M. Sheepwash. *3 Para:* Cpl I. P. Bailey, M.M., Maj. D. A. Collett, M.C., Pte K. M. Eaton, C/Sgt B. Faulkner, D.C.M., Capt. A. R. Freer, L/Cpl A. J. Goring, Cpl G. Heaton, 2/Lt I. C. Moore. *2nd Scots Guards:* L/Sgt I. Davidson, Lt A. H. J. Fraser, Sgt R. F. Jackson, M.M., L/Sgt T. McGuinness, C.S.M. E. M. McKay, L/Sgt I. Miller, Sgt A. C. Naismith, C.S.M. W. Nicol, D.C.M., 2/Lt C. S. T. Page, Lt-Col M. I. E. Scott, D.S.O., 2/Lt J. Stuart. *1/7th Gurkhas:* Lt-Col D. P. de C. Morgan, O.B.E. *4th Field Regt:* Maj. A. J. Rice. *12th Air Defence Regt:* Maj. D. R. I. Berry, Sgt W. C. Boyd, Sgt G. J. Morgan, Sgt R. W. Pearson, Capt. T. D. R. Turner, Maj. J. H. Wilkinson. *29th Commando Regt:* Maj. B. Armitage, Bombardier G. J. Powell, Maj. J. N. G. Starmer-Smith, Cpl N. J. Wells. *Blues and Royals:* Trooper P. R. Fugatt, Cpl of Horse P. Stretton, Cpl of Horse S. P. Thomson. *656 Squadron Army Air Corps:* L/Cpl S. B. Cholerton, Cpl M. Lord, W.O.II M. J. Sharp, Maj. C. S. Sibun, Sgt J. R. A. Sutherland, Sgt R. J. Walker. *9 Parachute Squadron R.E.:* Maj. C. M. Davies, M.B.E., L/Cpl W. A. Skinner. *11 Field Squadron:* Sapper T. W. Hall. *16th Field Ambulance:* Maj. C. G. Batty, M.B.E. *602 Tactical Air Control Party:* Maj. A. S. Hughes.

PERSONAL ACKNOWLEDGEMENTS

I am particularly grateful to my wife Mary, for help with preparation of maps, checking typescripts and compiling the index; to Jean Thomas, for translation of Spanish books and articles; to Janet Mountain, for her diligent typing of two drafts of the book's typescript (her ninth book for me); to my daughter Anne Bell, for research at the Fleet Air Arm Museum, where she is a member of the staff; and to A. Duncan of Falkland, Scotland, for research into the origin of Falkland's name.

I am pleased to acknowledge the cooperation of the Public Relations staffs of the Ministry of Defence and of individual branches of the services; this help was essential for my work and was, in most cases, willingly and efficiently given. I would particularly like to thank Harry Backler, Lt-Col. David Dunn (who also served in the Task Force on P.R. duties), Richard Gardner, Sue Hamlin, Graeme Hammond (also with the Task Force), Capt. Barry Hawgood and Sgt Rick Haynes of the Royal Marines, Michael Pentreath, Brig. David Ramsbotham, Chris Shepherd, Mike Stuart, Tony Talbot and Tony Warner. I would also like to thank the following service personnel who, although they did not serve with the Task Force, have been very helpful: Col. Derek Brownson, R.E., Maj. G. McGregor Dallas, R.H.F., Warrant Officer I Dave Gibson, Coldstream Guards, Lt-Cdr Mike Wignall, R.N., and Lt David Humphrey, R.N.

I would like to record my thanks to the following bodies, which have provided valuable help with research requests: B P Shipping Ltd, British Antarctic Survey, Cunard Shipping Services Ltd, Lloyd's Register of Shipping, Naval Historical Library, P & O, Royal Geographical Society, Royal Marines Secretariat, United Nations Information Centre, London.

Bibliography

The following works have been consulted:

OFFICIAL PUBLICATIONS

Basic Facts about the United Nations, United Nations, 1980
The Despatch by Admiral Sir John Fieldhouse, G.C.B.. G.B.E., Commander of the Task Force Operations in the South Atlantic, April to June 1982, *London Gazette*, December 1982
The Disputed Islands, HMSO, 1982
The Falkland Islands and Dependencies, HMSO, March 1982 (pamphlet)
The Falkland Islands. The Facts, HMSO, May 1982
Falkland Islands Review (The Franks Report), HMSO, January 1983

OTHER WORKS

Ethell, Jeffrey, and Price, Alfred, *Air War South Atlantic*, Sidgwick & Jackson, 1983
Frost, Major-General John, *2 Para Falklands*, Buchan & Enright, 1983
Hastings, Max, and Jenkins, Simon, *The Battle for the Falklands*, Michael Joseph, 1983
Jolly, Surgeon-Commander Rick, *The Red and Green Life Machine*, Century, 1983
Thompson, Major-General Julian, *No Picnic*, Leo Cooper/Secker & Warburg, 1985
The *Sunday Times* Insight Team, *The Falklands War*, Sphere, 1982

Destefani, Rear-Admiral Laurio H., *The Malvinas, the South Georgias and the South Sandwich Islands. The Conflict with Britain*, Edipress, Buenos Aires, 1982
Kon, Daniel, *Los Chicos de la Guerra*, New English Library, 1983
Turolo, Carlos M., *Malvinas – Testimonio de su Gobernador*, Sudamericana 1983

Index